Lanthanide and Actinide Chemistry and Spectroscopy

Norman M. Edelstein, EDITOR

Lawrence Berkeley Laboratory

Based on a symposium

sponsored by the Division

of Inorganic Chemistry at

the 178th Meeting of the

American Chemical Society,

Washington, D.C.,

September 10–13, 1979.

ACS SYMPOSIUM SERIES **131**

AMERICAN CHEMICAL SOCIETY
WASHINGTON, D. C.   1980

Library of Congress CIP Data

Lanthanide and actinide chemistry and spectroscopy.
  (ACS symposium series; 131 ISSN 0097–6156)

  Includes bibliographies and index.

  1. Rare earth metals—Congresses. 2. Actinide ele-
ments—Congresses. 3. Spectrum analysis—Congresses.
  I. Edelstein, Norman M., 1936–   . II. American
Chemical Society. Division of Inorganic Chemistry. III.
Series: American Chemical Society. ACS symposium
series; 131.

QD172.R2L27   546'.41   80–17468
ISBN 0–8412–0568–X   ACSMC8   131 1–472 1980

ACS Symposium Series

M. Joan Comstock, *Series Editor*

FOREWORD

The ACS SYMPOSIUM SERIES was founded in 1974 to provide
a medium for publishing symposia quickly in book form. The
format of the Series parallels that of the continuing ADVANCES
IN CHEMISTRY SERIES except that in order to save time the
papers are not typeset but are reproduced as they are submitted by the authors in camera-ready form. Papers are reviewed under the supervision of the Editors with the assistance
of the Series Advisory Board and are selected to maintain the
integrity of the symposia; however, verbatim reproductions of
previously published papers are not accepted. Both reviews
and reports of research are acceptable since symposia may
embrace both types of presentation.

CONTENTS

v

PREFACE

The last published symposium on lanthanide and actinide chemistry, sponsored by the Division of Inorganic Chemistry and the Division of Nuclear Chemistry and Technology of the American Chemical Society, was held in 1966. The purpose of this earlier symposium was "to summarize the significant areas of current chemical research. . . ." The same statement may be made about the present symposium; however, the topics covered differ considerably. For example, there was not one chapter on organolanthanide or organoactinide chemistry in the earlier symposium, while in the present volume a goodly fraction of the chapters are on this topic. Further, the availability of significant amounts of the transcurium elements have led to the elucidation of the properties of the elements and their compounds with atomic numbers greater than 96. Also, as in other areas of science, new, sophisticated instrumentation is in the process of revolutionizing the quality and type of data obtained on the f-block elements and compounds.

This volume is intended to introduce the nonspecialist chemist to recent trends in lanthanide and actinide chemistry and spectroscopy, to summarize this work, and to identify directions for future studies. Inevitably, the chapters in this collection reflect (to some extent) the tastes of the organizer.

I would like to thank the participants in the symposium for their contributions, Dr. William T. Carnall for his help in organizing the spectroscopy part of the symposium, and Drs. John Fackler, Gary Long, and Leonard Interrante of the Division of Inorganic Chemistry for their efforts on behalf of the symposium and the publication of the proceedings. Acknowledgment is made to the Donors of the Petroleum Research Fund, administered by the American Chemical Society, for partial support of this symposium.

Lawrence Berkeley Laboratory
Berkeley, CA 94720

December 21, 1979.

NORMAN M. EDELSTEIN

ORGANOACTINIDE AND
ORGANOLANTHANIDE CHEMISTRY

Nonclassical Activation of Carbon Monoxide by Organoactinides

TOBIN J. MARKS, JUAN M. MANRIQUEZ, and PAUL J. FAGAN
Department of Chemistry, Northwestern University, Evanston, IL 60201

VICTOR W. DAY, CYNTHIA S. DAY, and SARAH H. VOLLMER
Department of Chemistry, University of Nebraska, Lincoln, NE 68588

Abstract

This article reviews recent results on the carbonylation
chemistry of bis(pentamethylcyclopentadienyl) thorium and uranium
hydrocarbyl and dialkylamide complexes. Facile migratory insertion of carbon monoxide into metal-carbon and metal-nitrogen bonds
is observed. In several cases bihaptoacyl and bihaptocarbamoyl
complexes were isolated and characterized by single crystal X-ray
diffraction. The great strength of the metal-oxygen bonding in
these species is evident in metrical and spectral data, as well
as in the reaction chemistry, which is decidedly alkoxycarbene-
like. In the case of the bis(pentamethylcyclopentadienyl) actinide dialkyls, the final carbonylation products are C-C coupled
cis-1,2-enediolate complexes, while for the corresponding bis(dialkylamides), the products are bis(carbamoyl) species. Both
types of compound have been characterized by X-ray diffraction.
The carbon monoxide chemistry observed here may be of relevance
to mechanistic discussions of catalytic CO reduction, especially
that involving actinide oxide or actinide oxide supported catalysts.

Introduction

Our recent research in actinide organometallic chemistry
(1-5) has sought to exploit those features of f-element ions
which differ from transition metal ions. The goal of our effort
has been to discover and to understand to what degree the large
ionic radii and f valence orbitals might foster a unique new
organometallic chemistry. Exploration has been at both the chemical and physicochemical levels with the central issues concerning
the properties of actinide-to-carbon sigma bonds and related
functionalities. We have learned that the thermal stability and
chemical reactivity of these linkages can be modulated to a considerable degree (and often in opposite directions) by changes in
the supporting ligands within the actinide ion coordination sphere.

0–8412–0568–X/80/47–131–003$06.25/0

Thus, while the coordinative saturation of the triscyclopentadienyl alkyls, alkenyls, alkynyls, and aryls (hydrocarbyls) of thorium and uranium, $M(\eta^5-C_5H_5)_3R$ (6,7,8,9), affords considerably enhanced thermal stability over that of the simple homoleptic derivatives (10,11,12), it is at the expense of chemical reactivity.

In an effort to more finely tune the coordinative saturation of actinide hydrocarbyls and to provide greater than one metal-carbon bond for reaction, we have initiated an investigation of biscyclopentadienyl thorium and uranium chemistry (6,13,14). Systems based upon the pentamethylcyclopentadienyl ligand have proved to be some of the most interesting and form the basis of this article. The advantages of the $\eta^5-(CH_3)_5C_5$ ligand are that it makes far greater steric demands than $\eta^5-C_5H_5$ (thus reducing the number of large, bulky ligands which can be accommodated at the metal center) while imparting far greater solubility and crystallizability. It also appears that the methyl $C(sp^3)$-H bonds of this ligand are more inert with respect to scission than cyclopentadienyl $C(sp^2)$-H bonds; this has the effect of hindering a common thermal decomposition process, intramolecular hydrogen atom abstraction (7,8,15,16,17), hence of preserving the metal-to-carbon sigma bond for other chemical transformations. The net result is that pentamethylcyclopentadienyl actinide hydrocarbyls form the basis for an elaborate and extremely reactive new class of organometallic compounds.

The purpose of this article is to review the chemical, physicochemical, and structural properties of bis(pentamethylcyclopentadienyl) actinide compounds with respect to one reagent: carbon monoxide. The interaction of organometallic complexes with carbon monoxide is a subject of enormous technological importance. Vast quantities of acetic acid, alcohols, esters, and other important chemicals are presently produced using organic feedstocks, carbon monoxide, and homogeneous catalysts of the Group VIII transition metals (18,19,20). Much of this chemistry is now well-understood from model studies and is based upon the "classical" migratory insertion reaction of carbon monoxide into a metal-to-carbon sigma bond to form an acyl derivative (A) (equation (1)) (21,22,23). An industrially important example of this type of chemistry is the rhodium catalyzed hydroformylation cycle illustrated in Figure 1 (18). It is not clear, however, that the

$$\underset{M}{\overset{\displaystyle CH_3}{\underset{|}{}}} + CO \rightleftharpoons \underset{M \leftarrow CO}{\overset{\displaystyle CH_3}{\underset{|}{}}} \rightleftharpoons \underset{M-C=O}{\overset{\displaystyle CH_3}{\underset{|}{}}} \tag{1}$$

$$\underline{A}$$

classic picture of CO activation established for low-valent, "soft", mononuclear, Group VIII metal complexes is complete or accurate in describing the mechanisms of Fischer-Tropsch (24-28), methanation (24-28), ethylene glycol synthesis (29), and other reactions in which drastic changes in the CO molecule such as facile deoxygenation and homologation are occurring. Clearly there is a necessity to develop new carbon monoxide chemistry and to elucidate new reaction patterns. Such research is essential to understanding the fundamental aspects of processes which will be of ever-increasing importance in an economy shifting to coal-based feedstocks. It will be seen that the carbonylation reactions of bis(pentamethylcyclopentadienyl) actinide hydrocarbyls and related compounds differ dramatically from the "classical" pattern and afford a better insight into the reactivity of carbon monoxide at metal centers which exhibit both high oxygen affinity and high coordinative unsaturation. In the sections which follow we consider first the chemical and then the structural aspects of this problem.

Synthesis and Chemistry

The sequence shown in equations (2) and (3) offers an effective route to monomeric, highly crystalline, thermally stable thorium and uranium organometallics with either one or two metal-

$$2(CH_3)_5C_5^- + MCl_4 \xrightarrow[100^\circ C]{\text{toluene}} M[(CH_3)_5C_5]_2Cl_2 + 2Cl^- \qquad (2)$$

$$M = Th, U$$

$$M[(CH_3)_5C_5]_2Cl_2 + 2RLi \xrightarrow[THF]{\text{ether or}} M[(CH_3)_5C_5]_2R_2 + 2LiCl \qquad (3)$$

$$M = Th, \quad R = CH_3, \ CH_2Si(CH_3)_3, \ CH_2C(CH_3)_3, \ CH_2C_6H_5, \ C_6H_5$$
$$M = U, \quad R = CH_3, \ CH_2Si(CH_3)_3, \ CH_2C_6H_5$$

carbon sigma bonds (6,30,31). All compounds shown in these and subsequent reactions were thoroughly characterized by elemental analysis, cryoscopic molecular weight in benzene (solubility permitting), infrared and NMR spectroscopy, and, in several cases, by single crystal X-ray diffraction. Structures B and C are proposed for these new compounds in solution.

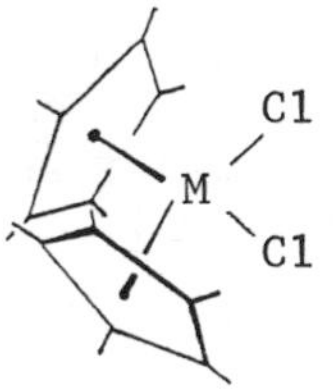

B

M = U

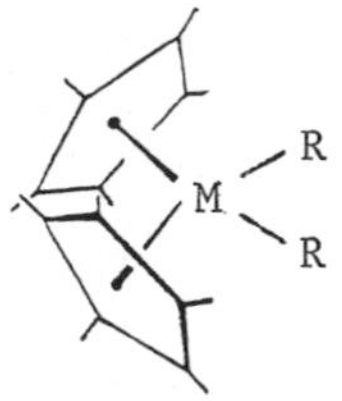

C

M = Th,U

The reaction of $Th[(CH_3)_5C_5]_2(CH_3)_2$ and $U[(CH_3)_5C_5]_2(CH_3)_2$ with carbon monoxide is quite rapid ($\underline{6},\underline{32},\underline{33}$). At -80^0C in toluene solution, these compounds absorb 2.0 equivalents of carbon monoxide (at less than one atmosphere pressure) within 1 hour. Upon warming to room temperature, the dimeric products ($\underline{1}$) are isolated in essentially quantitative yield (equation (4)). The infrared spectra ($\nu_{C=C}$ = 1655 cm^{-1}; ν_{C-O} = 1252, 1220 cm^{-1}) as well as the single methyl resonance in the ^{1}H NMR spectrum strongly suggests that C-C coupling of the inserted carbon monoxide

$$2M[(CH_3)_5C_5]_2(CH_3)_2 + 4CO \xrightarrow[-80^0]{\text{toluene}} \{M[(CH_3)_5C_5]_2(OC(CH_3)=C(CH_3)O)\}_2$$

(4)

$\underline{1a}$ M = Th (**colorless crystals**)
$\underline{1b}$ M = U (brown crystals)

molecules has occurred to form enediolate moieties ($\underline{D}$). Confirmation of this hypothesis was achieved by single crystal X-ray dif-

$$\begin{array}{ccc} CH_3 & & CH_3 \\ \diagdown C & = & C \diagdown \\ -O \diagup & & \diagup O- \end{array}$$

$\underline{D}$

fraction studies on $\underline{1a}$ ($\underline{6},\underline{32},\underline{33}$). As can be seen in Figure 2, four carbon monoxide molecules have been coupled to form four thorium-oxygen bonds and, stereospecifically, two cis-substituted carbon-carbon double bonds. Two $Th[(CH_3)_5C_5]_2$ units in the common "bent sandwich" configuration ($\underline{34}$) are components of a ten-atom metallocycle. The enediolate ligands are essentially planar with genuine C-C double bonds ($C_1-C_2 = C_1'-C_2' = 1.33(2)$Å). Further structural remarks are reserved for the following section.

As a prelude to discussing additional f-element chemistry, it is at this point worth noting the results of carbonylation experiments with the biscyclopentadienyls of early transtion metals. As is the case for the actinides, these elements in the higher oxidation states exhibit a great affinity for oxygen-donating ligands ($\underline{35},\underline{36}$), and their chemistry will place further actinide results in a more meaningful perspective. Floriani and coworkers have carried out an extensive investigation of the reaction of biscyclopentadienyl titanium, zirconium, and hafnium bishydrocarbyls and halohydrocarbyls with carbon monoxide (equations (5) and (6)) ($\underline{37},\underline{38},\underline{39}$). Only monocarbonylation is observed. Similar

$$M(C_5H_5)_2R_2 + CO \rightleftharpoons M(C_5H_5)_2(COR)R \qquad (5)$$

M = Zr, Hf
R = CH_3, $CH_2C_6H_5$, (C_6H_5 not reversible)

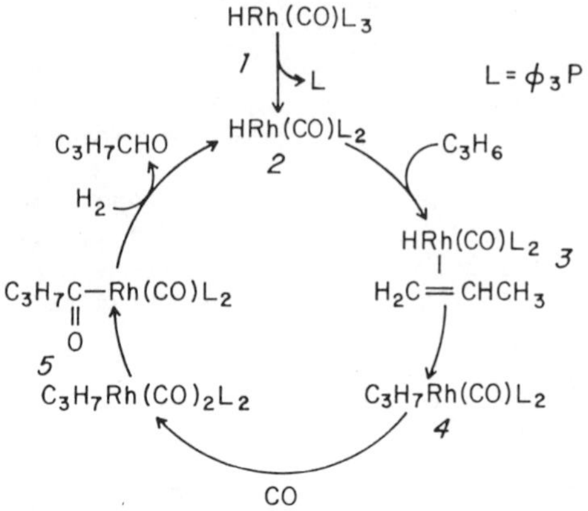

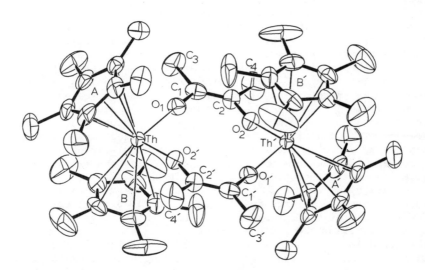

Journal of Molecular Catalysis

Figure 1. A proposed mechanism for the rhodium-catalyzed hydroformylation of propylene (18)

Journal of the American Chemical Society

Figure 2. ORTEP drawing of the nonhydrogen atoms for the [Th(η^5-(CH$_3$)$_5$-C$_5$)$_2$(μ-O$_2$C$_2$(CH$_3$)$_2$)]$_2$ molecule, 1a. All atoms are represented by thermal vibrational ellipsoids drawn to encompass 50% of the electron density. Atoms of a given type labelled with a prime (′) are related to those labelled without by the crystallographic inversion center midway between the two thorium atoms. The crystallographically independent pentamethylcyclopentadienyl ligands are labelled A and B, respectively (32).

$$M(C_5H_5)_2(R)X + CO \longrightarrow M(C_5H_5)_2(COR)X \qquad (6)$$

$$M = Ti$$
$$X = Cl, \ R = CH_3, \ C_2H_5, (CH_2C_6H_5 \text{ reversible})$$
$$X = I, \ R = C_2H_5$$

results have recently been reported by Lappert, et. al. (40). The insertion products are not simple acyls as A, but are bihaptoacyls (E, F) in which both carbon and oxygen atoms are bound to the metal ion. Such bonding is evidenced in the vibrational spectra of

metal acyls by a lowering of the C-O stretching frequency. Thus, typical frequencies for nonconjugated transition metal monohapto-acyls fall in the range 1630-1680 cm^{-1} (41,42) while nonconjugated transition metal bihaptoacyls are generally in the range 1530-1620 cm^{-1} (37,38,39). This decrease in C-O force constant and presumably bond order can be rationalized in terms of the contribution of valence bond resonance hybrid F . The molecular structures of Ti(C$_5$H$_5$)$_2$(COCH$_3$)Cl (38) and Zr(C$_5$H$_5$)$_2$(COCH$_3$)CH$_3$ (37) have been studied by X-ray diffraction and the results are presented in Figures 3 and 4, respectively. Although the acyl coordination is clearly bihapto, in both cases the metal-carbon distance is ca. 0.1Å shorter than the metal-oxygen distance. It should be noted that oxygen coordination allows the metal ions to achieve 18 electron valence shells. This saturation may be a crucial reason for the reluctance of the monoacyls to insert a second molecule of CO. Trends in the position of the equilibria in equations (5) and (6) can be explained in terms of the relative metal-carbon bond strengths in reactants and products as well as the extent of conjugation in the insertion product (37,38).

Lauher and Hoffmann (43) have studied the M(C$_5$H$_5$)$_2$R$_2$ + CO insertion reaction by extended Hückel molecular orbital calculations. The approach of the carbon monoxide lone pair is expected to be most favorable in the direction where there is best overlap with the M(C$_5$H$_5$)$_2$R$_2$ lowest unoccupied molecular orbital. This direction is along the perpendicular to the ring centroid-metal-ring centroid plane (34,43,44,45) and is expected, after R migration, to yield a product of structure G, i.e., with the C-O vector pointing away from the unreacted R ligand. An unsolved problem concerning this insertion process is why only products of structure H have so far been identified (cf. Figures 3 and 4). A fleeting intermediate, very possibly of structure G, has been noted in the reaction of Zr(C$_5$H$_5$)$_2$(p-CH$_3$C$_6$H$_4$)$_2$ with carbon monoxide (46).

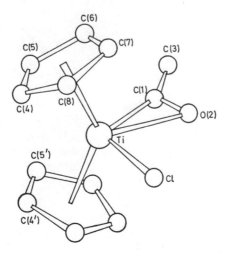

Journal of the Chemical Society, Dalton Translation

Figure 3.   Molecular structure of Ti(η^5-C_5H_5)$_2$(η^2-COCH$_3$)Cl (38)

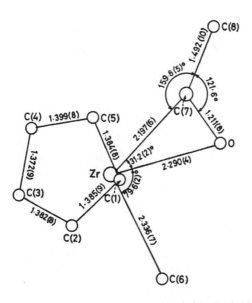

Journal of the Chemical Society, Dalton Translation

Figure 4.   Molecular structure of Zr(η^5-C_5H_5)$_2$(η^2-COCH$_3$)CH$_3$ viewed perpendicular to the Zr(COCH$_3$)(CH$_3$) plane (37)

G   H

Bercaw and coworkers have studied the reaction of bis(penta-methylcyclopentadienyl)zirconium dialkyls with carbon monoxide (47,48). As in the case of the aforementioned biscyclopentadi-enyls, reversible monoinsertion is observed (equation (7)) yield-ing a bihaptoacyl (ν_{CO} = 1537 cm^{-1}). Interestingly, when car-

$$Zr[(CH_3)_5C_5]_2(CH_3)_2 + CO \rightleftharpoons Zr[(CH_3)_5C_5]_2(COCH_3)(CH_3) \qquad (7)$$

bonylation is carried out at 75° C and 1 atm. of CO pressure for 24 hr., a product is isolated, which on the basis of chemical and spectroscopic data was assigned the monomeric enediolate structure I.

I

At this juncture there appear to be some distinct similari-ties between the early transition metal and actinide carbonylation results. The enediolate products are, with the exception of mole-cularity, the same (cf. 1 and I), although the insertion/coupling reaction appears to be much more rapid for thorium and uranium. It was decided to explore these similarities and differences in greater depth, and to shed light on crucial questions such as how endiolate species arise. Were organoactinide bihaptoacyls in-volved? Were bis(bihaptoacyl) species involved? Why were the actinide coupling products dimers and the zirconium coupling prod-uct a monomer? Structure sensitivity experiments were next carried out with variously substituted organoactinides. The coordination of bulky alkyl groups affords monomeric coupling products (equa-tion (8)) (32,33) as found in the zirconium system (I). Since the bulky substitutents inhibit intermolecular C-C fusion, it was next attempted to inhibit intramolecular fusion by deleting an alkyl functionality.

Chloromonoalkyl derivatives can be synthesized by the procedure shown in equation (9) (6,49).

$$\text{(8)}$$

$$
\begin{aligned}
&\underset{\sim}{2} \quad M=Th, \ R=CH_2C(CH_3)_3\\
&\underset{\sim}{3a} \ M=Th, \ R=CH_2Si(CH_3)_3\\
&\underset{\sim}{3b} \ M=U, R=CH_2Si(CH_3)_3
\end{aligned}
$$

$$M[(CH_3)_5C_5]_2Cl_2 + RLi \xrightarrow{\text{ether}} M[(CH_3)_5C_5]_2(Cl)R + LiCl \qquad (9)$$

$$
\begin{aligned}
&\underset{\sim}{4} \quad M=Th, \ R=CH_2C(CH_3)_3\\
&\underset{\sim}{5a} \ M=Th, \ R=CH_2Si(CH_3)_3\\
&\underset{\sim}{5b} \ M=U, \ R=CH_2Si(CH_3)_3
\end{aligned}
$$

The reaction of $\underset{\sim}{4}$ with one equivalent of carbon monoxide proceeds rapidly and irreversibly at room temperature to yield a product which, on the basis of the infrared spectrum, can be assigned a bihaptoacyl coordination geometry (equation (10)) (33). An inter-

$$\text{(10)}$$

esting spectral property of this complex is that the acyl C-O stretching frequency of 1469 cm^{-1} for $\underset{\sim}{6}$ (confirmed by ^{13}C substitution) is lower than in the analogous transition metal compounds; this suggests a significantly greater contribution from the carbenoid resonance hybrid $\underline{K}$. The molecular structure of $\underset{\sim}{6}$, elucidated by X-ray diffraction, is presented in Figure 5 (33). It reveals several unique features. First, the orientation of the C-O vector in the bihaptoacyl ligand is for the first time away from the un-reacted ligand, i.e. in orientation G, which is that predicted for transition metal ions in the calculations of Lauher and Hoffmann (43). Second, the Th-O(2.37(2)Å) and Th-C(2.44(3))Å metrical para-meters reveal that, unlike the previously discussed titanium and zirconium bihaptoacyls, the metal-oxygen distance is not apprecia-bly longer than the metal-carbon distance and is probably slightly shorter. This observation is in accord with the previously men-tioned vibrational spectroscopic data which also suggest a greater importance for the carbenoid structure K, i.e., greater metal-oxygen interaction. At 100°C in toluene solution, 6 slowly rear-ranges to a product 7, which on the basis of infrared and ^{1}H NMR studies (including those using 6 prepared from ^{13}CO) is assigned a cis-enolate structure (equation (11)) (33). The nature of this

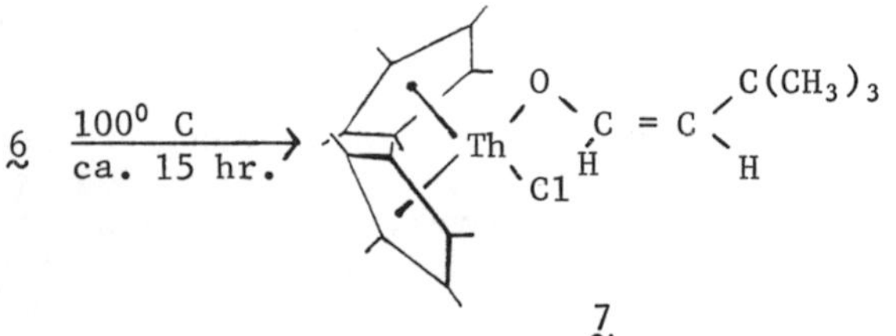

$$6 \xrightarrow[\text{ca. 15 hr.}]{100° \text{ C}} \qquad (11)$$

7

product can be understood in terms of the importance of carbenoid species K. Such hydrogen atom migration reactions are typical of alkoxycarbene chemistry (50,51,52,53).

Carbonylation of the trimethylsilylmethyl derivatives 5a and 5b provides further information on the reactivity of organoactin-ide bihaptoacyls (6,33). The reaction of these compounds with carbon monoxide takes place at -78° C (equation (12)).

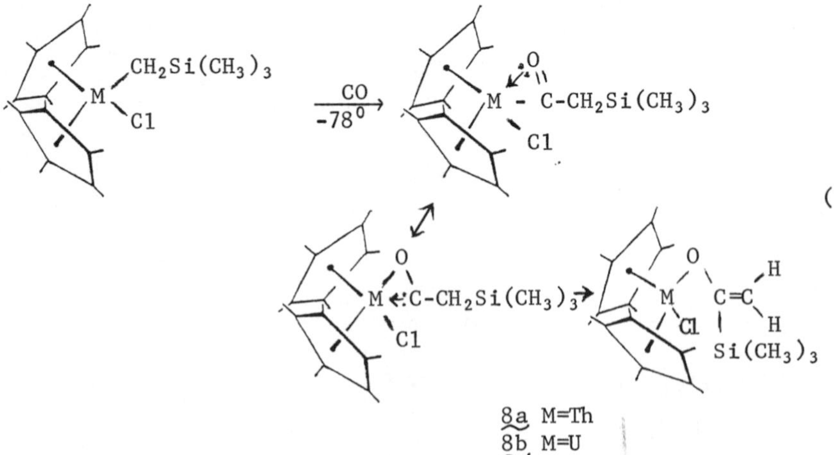

$$\xrightarrow[-78°]{CO} \qquad (12)$$

8a M=Th
8b M=U

An unstable intermediate can be detected at low temperature by ^{1}H NMR which is attributed to a bihaptoacyl compound. On standing in solution at room temperature, the intermediate rearranges to another compound. From infrared and proton NMR data (including measurements on products derived from $\underset{\sim}{5}$ prepared with ^{13}CO) the trimethysilylenolate species $\underset{\sim}{8}$, produced by $Si(CH_3)_3$ migration, is assigned. As expected for rearrangements involving a carbene-like center, a third row element will have a greater migratory aptitude (54,55,56,57).

The carbonylation studies on the bis(pentamethylcyclopentadienyl) actinide dialkyls and chloroalkyls provide strong support for the involvement of highly activated, carbene-like bihaptoacyls in the CO migratory insertion process and in the ultimate formation of enediolate products. Still, on the basis of this evidence, several points remain unclear. The relationship of the two different bihaptoacyl configurations, G and H, in the reactivity patterns and in the contrasts between d- and f-metal centers is still not well-defined. The nature of the monoinsertion products has been clarified, but the pathway to the enediolate has not. Although the coupling of two bihaptoacyl units represents the most plausible route to the enediolate, no evidence for a bis(bihaptoacyl) species could be obtained. In an effort to elucidate these points and, simultaneously, to explore what other metal-to-element bonds might suffer migratory insertion within the highly unsaturated bis(pentamethylcyclopentadienyl) actinide environment, carbonylation of dialkyl amide (-NR$_2$) derivatives (58,59,60,61) was undertaken. One goal was to employ the bulk of the -NR$_2$ moiety and the probable stability of a planar, conjugated, amide-like insertion product to thwart C-C fusion, and to thus allow isolation of a bis(bihapto) insertion product.

Bis(pentamethylcyclopentadienyl) thorium and uranium chlorodialkylamides and bis(dialkylamides) were prepared from the corresponding dichlorides (6,31) by the route of equation (13). The highly air-sensitive new compounds

$$M[(CH_3)_5C_5]_2Cl_2 + x\ LiNR_2 \xrightarrow[25^0]{\text{diethyl ether}}$$

$$M[(CH_3)_5C_5]_2(NR_2)_xCl_{2-x} + x\ LiCl \tag{13}$$

$$
\begin{array}{llll}
\underset{\sim}{9a} & x = 1, & M = Th, & R = C_2H_5 \\
\underset{\sim}{9b} & x = 1, & M = U, & R = C_2H_5 \\
\underset{\sim}{10a} & x = 1, & M = Th, & R = CH_3 \\
\underset{\sim}{10b} & x = 1, & M = U, & R = CH_3 \\
\underset{\sim}{11a} & x = 2, & M = Th, & R = CH_3 \\
\underset{\sim}{11b} & x = 2, & M = U, & R = CH_3 \\
\end{array}
$$

were isolated and characterized in the usual manner (62,63). The chlorodialkylamides react with CO(1 atm.) at 95^{0}C to yield carbamoyl (CONR$_2$) (64,65) insertion products $\underset{\sim}{12}$ and $\underset{\sim}{13}$ (equation (14)). The C-O stretching frequencies of these compounds, verified by ^{13}CO

substitution, are low ($1516-1559$ cm^{-1}) for carbamoyl complexes
(more typical values are in the range $1565-1615$ cm^{-1} ($\underline{64},\underline{65}$)) and
suggest, by analogy to the metal acyls, bihapto coordination ($\underline{66}$)

$$\qquad (14)$$

$\underline{12}$a M = Th, R = C$_2$H$_5$ $\underline{12}$b M = U, R = C$_2$H$_5$
$\underline{13}$a M = Th, R = CH$_3$ $\underline{13}$b M = U, R = CH$_3$

of the inserted CO functionality. It will be seen that in all
cases, the perturbation in ν_{CO} on going from the monohapto to bihapto structure is far less for the carbamoyls than for the acyls.
The magnetic non-equivalence of the N-alkyl groups in $\underline{12}$ and $\underline{13}$
indicates that rotation about the C-N bond, a process known for
transition metal carbamoyl complexes ($\underline{65}$), is slow on the NMR
timescale at room temperature. The molecular structure of $\underline{12a}$ has
been studied by X-ray diffraction and the remarkable result is illustrated in Figure 6. Two bihaptocarbamoyl conformers, corre-
sponding to structures $\underline{G}$ and $\underline{H}$, are both present in the crystal of
$\underline{12a}$; there is approximately equal population of either configuration. This case is the first unambiguous indication that a biscyclopentadienyl metal complex can exist in both structures. Further structural details are deferred to the following section. The
dynamics of $\underline{G} \rightleftharpoons \underline{H}$ interconversion were also investigated for $\underline{12b}$
and $\underline{13b}$ in CF$_2$Cl$_2$/toluene-d$_8$ solution by ^{1}H NMR spectroscopy (equa-
tion (15)) (62,63). The large U(IV)-induced isotropic shifts al-
low observation of both conformers at temperatures below ca.-90^0 C

$$\qquad (15)$$

(at 90 MHz). Raising the temperature produces broadening and
coalescence of the respective $\underline{G}$ and $\underline{H}$ signals as process (15) becomes rapid on the NMR timescale. For $\underline{13b}$, application of standard line-shape analysis relationships ($\underline{67}$) yields $\tau = 9.6 \times 10^{-3}$
sec at -65^0C, and Δ G‡ = 10 ± 1 kcal/mole.
    Although there is greater steric bulk about the metal ion, the
bis(dimethylamido) complexes, $\underline{11}$, are more reactive toward CO
than the chloro-substituted analogues, $\underline{10}$, and monocarbonylation
is complete within 2 hr. at 0^0 C and 1 atm CO (equation (16)).
The new complexes were characterized by the standard techniques
($\underline{62},\underline{63}$); from the vibrational spectral data (ν_{CO} = 1521 cm^{-1}) the
bihaptocarbamoyl structure is again assigned. These complexes re-
act reversibly with a second equivalent of CO in toluene solution

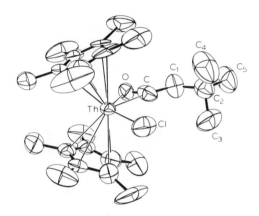

Figure 5. *ORTEP drawing of the nonhydrogen atoms of* $Th[\eta^5-(CH_3)_5C_5]_2$-$[\eta^2-COCH_2C(CH_3)_3]Cl$, **6a.** *All atoms are represented by thermal vibrational ellipsoids drawn to encompass 50% of the electron density.*

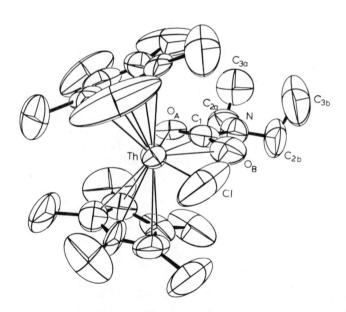

Figure 6. *Perspective ORTEP plot of the nonhydrogen atoms for the* $Th[\eta^5-(CH_3)_5C_5]_2[\eta^2-CON(C_2H_5)_2]Cl$ *molecule,* **12a.** *All atoms are represented by thermal vibrational ellipsoids drawn to encompass 50% of the electron density. The two half-weighted positional possibilities for the oxygen atom of the disordered diethylcarbamoyl ligand are designated by O_A and O_B; these correspond to the G and H isomeric forms of* **12a** *respectively (62).*

$$\text{(16)}$$

$$\underset{\sim}{14a} \quad M = Th, \quad R = CH_3$$
$$\underset{\sim}{14b} \quad M = U, \quad R = CH_3$$

at 65^0 C to yield bis(carbamoyl) complexes (equation (17)). Bihapto ligation is assigned on the basis of infrared spectral data

$$\text{(17)}$$

$$\underset{\sim}{15a} \quad M = Th, \quad R = CH_3$$
$$\underset{\sim}{15b} \quad M = U, \quad R = CH_3$$

($\nu_{CO} = 1523$ cm^{-1}). In addition, compound $\underline{15b}$ has been studied by
single X-ray diffraction $(\underline{68})$. As can be seen in Figure 7, the
double insertion product is indeed a bis(bihaptocarbamoyl). Furthermore, the orientation of the inserted CO units is ideal for an
intramolecular coupling reaction to produce an enediolate. As in
$\underline{6a}$, there is evidence that the Th-O distance (2.363(9) Å) is shorter
than the Th-C distance (2.418(15)Å). The η^2-CONR$_2$ ligands in $\underline{15b}$
are essentially planar as found in $\underline{12a}$ (Figure 6). All attempts
to detect rotation about the C-NR$_2$ bonds in any of the actinide
bihaptocarbamoyl complexes by high temperature NMR studies have
so far been unsuccessful. We estimate that $\Delta G^{\ddagger} \gtrsim 23$ kcal/mole for
this process, hence there is a substantial barrier to major ex-
cursions from η^2-CONR$_2$ planarity. In view of these observations
it may not be surprising that all attempts to couple the η^2 CON
(CH$_3$)$_2$ units by C-C fusion in $\underline{15a}$ and $\underline{15b}$, which may require rotation about the C-N bond, have so far been unsuccessful. Heating
$\underline{15}$ at temperatures as high as 100^0 C under high vacuum causes de-
carbonylation to form $\underline{14a}$ and $\underline{14b}$, respectively. Prolonged heating under carbon monoxide produces decomposition and the precip-
itation of ill-defined, insoluble products. The observation of
sequential mono and bis(bihaptocarbamoyl) formation in the actinide dialkylamide chemistry is the strongest evidence to date that
analogous processes in the actinide dialkyl systems lead to enediolate products. In terms of the contrast between d- and f-element

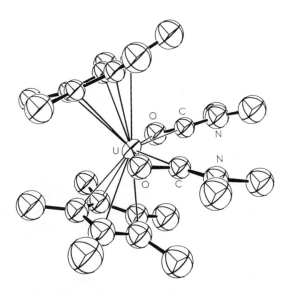

Figure 7. ORTEP drawing of the nonhydrogen atoms in $U[\eta^5-(CH_3)_5C_5]_2-[\eta^2-CON(CH_3)_2]_2$, **15b**. *All atoms are represented by thermal vibrational ellipsoids drawn to encompass 50% of the electron density.*

reactivity patterns, the greater tendency of the actinide ions to expand the coordination sphere is particularly evident in the structure of $\underline{15}$: electron counting reveals it to be formally a 20 electron valence system for thorium and a 22 electron valence system for uranium (counting the two 5f electrons). In addition, the formal coordination number of 10 is very rare for biscyclopentadienyl transition metal compounds, the only well-documented example to our knowledge being $Hf(CH_3C_5H_4)_2(BH_4)_2$ ($\underline{69}$).

Structural Aspects

Sufficient data are now available to discuss two important structural aspects involved in the activation of CO by bis(pentamethylcyclopentadienyl) actinide derivatives. First, the structural features accompanying various degrees of coordinative unsaturation in the organoactinide reactants are discussed. Second, the nature of the nonclassically bound CO in organoactinide migratory insertion products is considered and compared with d-element analogues.

The marked chemical reactivity of the bis(pentamethylcyclopentadienyl) actinide derivatives reflects, among other factors, a ligation pattern with sufficient coordinative unsaturation to promote chemistry, but without such unsaturation that instability of the molecule results. Thus, as an illustration of this problem, we have already shown ($\underline{14}$) that the highly unsaturated compound $U(C_5H_5)_2Cl_2$ is unstable with respect to $U(C_5H_5)_3Cl$ and $U(C_5H_5)Cl_3 \cdot$ 2L (L = a Lewis base such as THF). In contrast, biscyclopentadienyl uranium complexes with higher coordination numbers such as $U(C_5H_5)_2(acetylacetonate)_2$ ($\underline{70}$) and $U(C_5H_5)_2[Re(COCH_3)_2(CO)_4]_2$ ($\underline{71}$) or with bulky monodentate ligands such as $U(C_5H_5)_2[N(C_2H_5)_2]_2$ ($\underline{72}$) do not suffer ligand redistribution.

An important parameter in controlling the number and relative orientations of other ligands in bis(cyclopentadienyl) actinide complexes, hence in controlling to a degree the coordinative unsaturation, reactivity and stability, is the ring center of gravity-metal-ring center of gravity angle, C_g-M-C_g. Metrical data for a number of biscyclopentadienyl and ring-substituted biscyclopentadienyl actinide complexes are compiled in Table I. The compounds with the smallest C_g-M-C_g values contain cyclopentadienyl rings joined by a methylene bridge. Here high coordination numbers are possible even with relatively bulky ligands (Cl$^-$). As will be discussed more quantitatively elsewhere ($\underline{73}$), considerable displacement of the ligands above and below the plane which bisects C_g-M-C_g can take place. When the cyclopentadienyl rings are not joined, C_g-M-C_g increases and high coordination numbers are only possible with more compact bidentate ligands such as acetylacetonate (Table I). Also in these cases the non-cyclopentadienyl ligands are more tightly compressed toward the plane bisecting C_g-M-C_g. Finally, the six pentamethylcyclopentadienyl structure determinations (Table I) reveal a further increase in C_g-M-C_g and

relatively high coordination numbers only with non-bulky ligands
(H^-) or with multidentate ligands having a very small "bite"
(η^2-COX). The non-cyclopentadienyl ligands are closely constrained
to an equatorial girdle in the C_g-M-C_g bisecting plane. Chemical
ramifications of these structural results are that the M[(CH_3)$_5C_5$]$_2R_2$
and M[(CH_3)$_5C_5$]$_2$[NR'$_2$]$_2$ carbonylation chemistry is likely to be
occurring in the narrow equatorial girdle and that this girdle
will not accommodate sterically bulky reagents. On the other hand,
molecules with smaller C_g-M-C_g angles will be susceptible to approach from a greater range of directions and by more sterically
demanding ligands.

As noted earlier, the classical migratory insertion product
for transition metal alkyl carbonylation is a monohaptoacyl (A).
There is now considerable structural information available for
such species, and representative data are set out in Table II. It
can be seen that the configuration about the acyl carbon atom is
that expected for sp^2 hydribridization with, in some cases, slight
expansion of the M-C-O angle and slight contraction of the M-C-X
angle (X=C). The corresponding valence angles in acetone are
$\not< $ C-C-C=116.2° (74). As the interaction between the acyl oxygen
atom and the metal ion becomes significant (the M-O distance approaches the M-C distance in magnitude), the M-C-X angle increases
and the M-C-O angle decreases. Since the O-C-X angle remains approximately constant, the net effect is a rotation of the acyl
group about an axis perpendicular to the OCX plane. As the bihaptoacyl interaction increases in magnitude, the C-O stretching
frequency decreases. Unfortunately, any accompanying changes in
the C-O bond distance are beyond the accuracy of most of the
structure determinations. Reference to Table II indicates that
ν_{CO} is also sensitive to the identity of the metal and to the
other ligands in the complex; this frequency has also been shown
to be sensitive to the electronic properties of the hydrocarbyl
moiety (37,38). The most extreme instance of interaction with the
acyl oxygen atom is in the organoactinide Th[(CH_3)$_5C_5$]$_2$(C1)COCH$_2$C-
(CH_3)$_3$, 6. Here the metal-oxygen distance (2.37(2)Å) approaches
the Th-O single bond distance in 1a (2.154(8)Å) and the M-C-C
angle (169(2)°!) approaches linearity.

Data for monohapto and bihapto carbamoyl complexes are also
compiled in Table II. As in the case of the acyls, the O-C-X
angle remains relatively constant, and the effect of the metal-
oxygen interaction is to rotate the carbamoyl moiety about an axis
perpendicular to the OCN plane. The valence angles about the carbon atom of the inserted CO in the actinide biscarbamoyl, 15b, are
nearly identical to those in the actinide acyl 6. Again, the
magnitude of the metal-oxygen interaction is evidenced by a severe
contraction in the M-C-O angle and an expansion of the M-C-N
angle toward linearity. Any alteration of the carbamoyl C-O distances upon bihapto ligation cannot be discerned in the diffraction data. The decrease in ν_{CO} upon oxygen coordination does not
appear to be as great as in the acyl derivatives.

TABLE I. COMPARISON OF COORDINATION GEOMETRIES IN BISCYCLO-
PENTADIENYL ACTINIDE COMPLEXES [a]

Parameter Compound	C_g-M-C_g	M-C(ring)(Å)	M-X(Å)	Coordination Number
$Li^+(THF)_2U_2[CH_2(C_5H_4)_2]_2Cl_5^{-b}$	105°	2.72(5)	2.68(1)(Cl_T), 2.83(1)(Cl_B)	10
$U[CH_2(C_5H_4)_2](2,2'\text{-bipy})Cl_2^c$	106°	2.72(3)	2.68(2)(N), 2.71(1)(Cl)	10
$U(C_5H_5)_2[Re(COCH_3)_2(CO)_4]_2$	120°	2.77(6)	2.37(3)(0)	10
$\{Th[(CH_3)_5C_5]_2H_2\}_2^e$	130°	2.83(1)	2.03(1)(H_T), 2.29(3)(H_B)	9
$\{U[(CH_3)_5C_5]_2Cl\}_3^f$	128°	2.76(3)	2.901(5)(Cl)	8
$\{Th[(CH_3)_5C_5]_2[O_2C_2(CH_3)_2]\}_2^g$	129°	2.83(6)	2.150(4)(O)	8
$Th[(CH_3)_5C_5]_2(Cl)[COCH_2C(CH_3)_3]^h$	138°	2.80(3)	2.37(2)(O), 2.44(3)(C) 2.672(5)(Cl)	9
$Th[(CH_3)_5C_5]_2(Cl)[CON(C_2H_5)_2]^i$	138°	2.78(4)	2.42(4)(O), 2.32(5)(C) 2.59(2)(Cl)	9
$U[(CH_3)_5C_5]_2[CON(CH_3)_2]_2^j$	138°	2.788(13)	2.363(9)(O), 2.418(15)(C)	10

a T=terminally bonded ligand; B=bridge bonded ligand.

b Reference 13.

c Day, C.S.; Day, V.W.; Ernst, R.D.; Marks, T.J., manuscript in preparation.

d Reference 71.

e Broach, R.W.; Schultz, A.J.; Williams, J.M.; Brown, G.M.; Manriquez, J.M.; Fagan, P.J.; Marks, T.J., *Science*, 1979, 203, 172-174.

f Manriquez, J.M:; Fagan, P.J.; Marks, T.J.; Vollmer, S.H.; Day, C.S.; Day, V.W., J. Am. Chem. Soc., 1979, 101, 5075-5078.

g Reference 32.

h Reference 33.

i Reference 62. Disordered structure.

j Reference 68.

TABLE II. COMPARISON OF METRICAL AND SPECTRAL PARAMETERS
IN ACYL AND CARBAMOYL COMPLEXES

$$M-C\overset{O}{\underset{X}{|}}$$

Parameter Compound and Type	M-C(Å)	M-O(Å)	C-O(Å)	∠M-C-X	∠M-C-O	∠O-C-X	ν_{CO}(cm^{-1})
η^1-acyl							
Fe(HBpz$_3$)(CO)$_2$COCH$_3$ [a]	1.968(5)	>2.8	1.193(6)	119.0(4)°	124.3(4)°	116.7(5)°	1683,1648
Fe$_2$(C$_5$H$_4$C$_6$H$_8$CO)(CO)$_5$ [b]	1.9596(30)	>2.8	1.206(4)	115.68(20)°	126.94(23)°	117.37(26)°	1630
Ni[(CH$_3$)$_2$C$_2$O]$_2$(C$_5$H$_5$)$_2$ [c]	1.895(2)	>2.6	1.191(2)	112.03(13)°	125.29(15)°	122.67(18)°	1616,1591
Mo(C$_5$H$_5$)(CO)$_2$P(C$_6$H$_5$)$_3$COCH$_3$ [d]	2.284(14)	>3.1	1.211(16)	121.2(9)°	120.9(10)°	117.7(12)°	
η^2-acyl							
{Mo(Cl)(CO)$_2$[P(CH$_3$)$_3$][COCH$_2$Si(CH$_3$)$_3$]}$_2$ [e]	2.023(3)	2.292(2)	1.19(1)	146.7(3)°	86.1(2)°	127.2(3)°	1585
Ti(C$_5$H$_5$)$_2$(Cl)COCH$_3$ [f]	2.07(2)	2.194(14)	1.18(2)	154.0(16)°	79.7(6)°	126.3(13)°	1620
Zr(C$_5$H$_5$)$_2$(CH$_3$)COCH$_3$ [g]	2.197(6)	2.290(4)	1.211(8)	159.8(5)°	78.6(4)°	121.6(6)°	1545
Th[(CH$_3$)$_5$C$_5$]$_2$(Cl)COCH$_2$C(CH$_3$)$_3$ [h]	2.44(3)	2.37(2)	1.18(3)	169 (2)°	73 (1)°	118 (2)°	1469
η^1-carbamoyl							
PtCl$_2$(CO)CON(i-Pr)$_2$$^-$ [i]	1.96(2)	>2.6	1.36(3)	126 (3)°	115 (3)°	118 (4)°	1550
cis-Mn(CO)$_4$(NH$_2$CH$_3$)CONHCH$_3$ [j,k]	2.072(11)	>2.9	1.251(10)	121.4(7)°	121.2(7)°	117.4(8)°	1610 or 1540
m	2.072(11)	>2.9	1.228(15)	123.1(12)°	120.1(15)°	116.7(20)°	1610 or 1540
η^2-carbamoyl							
Mo(NCS)$_4$(NO)CON(CH$_3$)$_2$$^{-2}$ [l]	2.029(6)	2.078(6)	1.322(7)	162.7°	73.3°	123.96(60)°	1676
U[(CH$_3$)$_3$C$_5$]$_2$[CON(CH$_3$)$_2$]$_2$ [m]	2.418(15)	2.363(9)	1.286(16)	168 (1)°	72.1(8)°	120.0(1)°	1523
Th[(CH$_3$)$_5$C$_5$]$_2$(Cl)[CON(C$_2$H$_5$)$_2$] [n]	2.32(5)	2.42(4)	1.37(5)				1516

[a] Cotton, F. A.; Frenz, B. A.; Shaver, A., Inorg. Chim. Acta., 1973, 7, 161-169.

[b] Churchill, M. R.; Chang, S. W. Y., Inorg. Chem., 1975, 14, 1680-1685.

[c] Churchill, M. R.; DeBoer, B. G.; Hackbarth, J. J., Inorg. Chem., 1974, 13, 2098-2105.

[d] Churchill, M. R.; Fennessey, J. P., Inorg. Chem., 1968, 7, 953-959.

[e] Carmona-Guzman, E.; Wilkinson, G.; Atwood, J. L.; Rogers, R. D.; Hunter, W. E.; Zaworotko, M. J., J. Chem. Soc. Chem. Comm., 1978, 465-466, and Atwood, J. L., private communication.

[f] Reference 38.

[g] Reference 37.

[h] Reference 33.

[i] Reference 64.

[j] Chipman, D. M.; Jacobson, R. A., Inorg. Chim. Acta., 1967, 1, 393-396 (tetragonal modification).

[k] Breneman, G. L.; Chipman, D. M.; Gailes, C. J.; Jacobson, R. A., Inorg. Chim. Acta., 1969, 3, 447-450 (monoclinic modification).

[l] Reference 66.

[m] Reference 68.

[n] Reference 62. Disordered structure.

Conclusions

This work underscores the very high chemical reactivity that derives from actinide hydrocarbyl and related complexes containing the appropriate supporting ligands. In the case of carbon monoxide chemistry, facile migratory insertion reactions are ubiquitous. This CO activation process does not adhere to the classical transition metal pattern, but rather the high oxygen affinity and coordinative unsaturation of the thorium and uranium centers gives rise to bihaptoacyl and bihaptocarbamoyl complexes. The tendency of the bihaptoacyls to react as alkoxycarbenes is a striking facet of the chemistry and one that is qualitatively reminiscent of early transition metal organometallics. Given the same ligand array as a d-element ion it is not surprising that the larger actinide ions would be more unsaturated and more reactive. In support of this notion, the spectral, structural, and chemical data strongly argue that the perturbation of the bihaptoacyls and carbamoyls toward a carbene-like species i.e., an increased contribution from resonance hybrid F, is greater for the f-element systems. It is possible that differences in metal-ligand orbital overlap as well as in the tendency to undergo redox processes also contribute to reactivity contrasts between the d and f systems.

The present carbonylation results with f-element organometallics are relevant to mechanistic discussions of catalytic CO reduction at two levels. In terms of general mechanistic schemes (of which a large number exist) (24-29, 75) the organoactinide reactions suggest modes for CO reactivity in situations in which the catalyst exhibits high oxygen affinity and high coordinative unsaturation. Considering the evidence in heterogeneous systems for dissociative CO adsorption (25,76), labelled alcohol and ketone deoxygenation (26,77), labelled ketene deoxygenation (26, 78, 79), as well as surface alkoxide and carboxylate (or possibly bihaptoacyl) formation (80,81), the high oxygen affinity of many or most CO reduction catalyst surfaces is an entirely reasonable assumption. The necessity of high coordinative unsaturation is supported by the above observations and by kinetic data which indicate that CO inhibits many of the reduction catalysts (i.e., it competes with sites needed for CO dissociation and/or hydrogen adsorption) (24-28).

The present results suggest a ready means for catalytic alcohol formation via carbene-like bihapto formyl and acyl species (e.g., equations (18) and (19)). Precedent exists for the alkoxide formation step of equation (19) (47,48,82,83). Chain growth could occur via the insertion of an unsaturated surface site into a H_3C-OM(or R-OM) bond (an oxidative addition) to yield a metalcarbon bond, followed by further carbonylation, as illustrated in equations (20) and (21). There is good precedent for the insertion of metal ions into carbon-oxygen bonds (84,85,86). Hydro-

$$M-H \xrightarrow{CO} M \overset{O}{\underset{\longleftarrow:C-H}{\diagup}} \xrightarrow{H_2} M^{\diagdown O \diagdown} CH_3 \xrightarrow{M-H} CH_3OH \tag{18}$$

$$\xrightarrow[\text{or } H_2]{MH} M^{\diagdown O \diagdown} CH_3 \xrightarrow{M-H} CH_3OH \tag{19}$$

$$\overset{O-CH_3}{\underset{M-M}{\diagup}} \longrightarrow \overset{O \quad CH_3}{\underset{M \quad M}{\diagup \diagdown \diagup}} \tag{20}$$

$$\overset{O \quad CH_3}{\underset{M \quad M}{\diagup \diagdown \diagup}} \xrightarrow{CO} \overset{O \quad O \diagdown}{\underset{M \quad M}{\diagup \diagdown \diagup \cdot C}} {}^{\diagdown CH_3} \longrightarrow \text{etc.} \tag{21}$$

genolysis of the metal-carbon bond formed in equation (20) would produce saturated hydrocarbons, β-hydride elimination within an ethyl or larger group would produce olefins; both products are observed in the Fischer-Tropsch reaction (24-28). The enediolate formation reaction reported here suggests ways by which glycol (29) or hydrocarbon formation might occur (equations (22)-(25)).

$$\overset{H \quad H}{\underset{M \quad M}{\underset{\displaystyle |}{\overset{\displaystyle |}{}}}} \xrightarrow{2CO} \overset{\overset{\displaystyle H \quad H}{\underset{\displaystyle C \quad C}{| \quad |}}}{\underset{M \quad M}{\underset{\displaystyle O \quad O}{\diagdown \diagup \diagdown}}} \longrightarrow \overset{\overset{\displaystyle H \diagdown \quad \diagup H}{C = C}}{\underset{M \qquad M}{\underset{\displaystyle O \qquad O}{\diagdown \quad \diagup}}} \tag{22}$$

$$\overset{\overset{\displaystyle H \diagdown \quad \diagup H}{C = C}}{\underset{M \qquad M}{\underset{\displaystyle O \qquad O}{\diagdown \quad \diagup}}} \xrightarrow[H_2]{MH \text{ or}} \overset{CH_2 - CH_2}{\underset{OH \quad OH}{| \quad\quad |}} \tag{23}$$

$$\xrightarrow{\Delta} M^{\diagup O} \; {}^{O \diagdown}M + HC \equiv CH \xrightarrow[MH]{H_2 \text{ or}} \text{etc.} \tag{24}$$

$$\xrightarrow{H_2} \overset{OH \quad OH}{\underset{M \quad M}{| \quad |}} + H_2C{=}CH_2 \longrightarrow \text{etc.} \tag{25}$$

Reactions analogous to the reverse of equation (24) are well documented (87,88). Unsaturated hydrocarbons such as ethylene are readily incorporated into products under catalytic conditions (24-28).

In regard to specific catalytic systems for CO reduction and homologation, the organoactinide carbonylation results described here are particularly relevant to the "isosynthesis" reaction (24-28, 89-93). In this catalytic process, synthesis gas (CO + H$_2$) is converted over thoria, ThO$_2$, (alone or promoted with K$_2$CO$_3$ or Al$_2$O$_3$) into branched paraffins, olefins, alcohols, and aromatics. Until recently, the lack of precedent for thorium-hydrogen and thorium-carbon bonds as well as any carbonylation chemistry thereof, has rendered mechanistic discussion of isosynthesis impossible. The role of thoria as a support in transition metal catalyzed CO reduction (24-28) may also involve some of the chemistry discussed here.

Acknowledgments

We thank the National Science Foundation (T.J.M., CHE 76-84494 A01) and the University of Nebraska Computer Center (V.W.D.) for generous support of this research. T.J.M. and V.W.D. are Camille and Henry Dreyfus Teacher-Scholars.

Literature Cited

1. Marks, T.J.; Fischer, R.D., Eds. "Organometallics of the f-Elements," Reidel Publishing Co., Dordrecht, Holland, 1979.
2. Marks, T.J. Prog. Inorg. Chem., 1978, 24, 51-107; ibid., 1979, 25, 224-333.
3. Marks, T.J. Acc. Chem. Res., 1976, 9, 223-230.
4. Tsutsui, M.; Ely, N.; Dubois, R. Acc. Chem. Res., 1976, 9, 217-222.
5. Baker, E.C.; Halstead, G.W.; Raymond, K.N. Struct. Bonding, (Berlin), 1976, 25, 23-68.
6. Fagan, P.J.; Manriquez, J.M.; Marks, T.J. in reference 1, Chapt. 4.
7. Marks, T.J.; Seyam, A.M.; Kolb, J.R. J. Am. Chem. Soc., 1973, 95, 5529-5539.
8. Marks, T.J.; Wachter, W.A. J. Am. Chem. Soc., 1976, 98, 703-710.
9. Kalina, D.G.; Marks, T.J.; Wachter, W.A. J. Am. Chem. Soc., 1977, 99, 3877-3879.
10. Marks, T.J.; Seyam, A.M. J. Organometal. Chem., 1974, 67, 61-66.
11. Kohler, E.; Bruser, W.; Thiele, K.H. J. Organometal. Chem., 1974, 76, 235-240.
12. Sigurdson, E.R.; Wilkinson, G. J. Chem. Soc. Dalton Trans., 1977, 812-818.
13. Secaur, C.A.; Day, V.W.; Ernst, R.D.; Kennelly, W.J.; Marks, T.J. J. Am. Chem. Soc., 1976, 98, 3713-3715.
14. Ernst, R.D.; Kennelly, W.J.; Day, C.S.; Day, V.W.; Marks, T.J. J. Am. Chem. Soc., 1979, 101, 2656-2664.
15. Davidson, P.J.; Lappert, M.F.; Pearce, R. Chem. Rev., 1976, 76, 219-242, and references therein.

16. Erskine, G.J.; Hartgerink, J.; Weinberg, E.L.; McCowan, J.D. J. Organometal. Chem., 1979, 170, 51-61, and references therein.
17. Kohler, F.H.; Prossdorf, W.; Schubert, U.; Neugebauer, D. Angew. Chem. Int. Ed. Engl., 1978, 17, 850-851.
18. Parshall, G.W. J. Mol. Catal., 1978, 4, 243-270.
19. Eisenberg, R.; Hendriksen, D.E. Advan. Catal., 1979, 28, in press.
20. Falbe, J. "Carbon Monoxide in Organic Synthesis," Springer Verlag, Berlin, 1970.
21. Calderazzo, F. Angew. Chem. Int. Ed., 1977, 16, 299-311.
22. Heck, R.F., "Organotransition Metal Chemistry," Academic Press, N.Y., 1974, Chapt. IX.
23. Wojcicki, A. Advan. Organometal. Chem., 1973, 11, 87-145.
24. Masters, C. Advan. Organometal. Chem., 1979, 17, 61-103.
25. Ponec, V. Catal. Rev.-Sci. Eng., 1978, 18, 151-171.
26. Schulz, H. Erdol, Kohle, Erdgas, Petrochem., 1977, 30, 123-131.
27. Vannice, M.A. Catal. Rev.-Sci. Eng., 1976, 14, 153-191, and references therein.
28. Henrici-Olive, G.; Olive, S. Angew. Chem. Int. Ed. Engl., 1976, 15, 136.
29. Pruett, R.L. Ann. N.Y. Acad. Sci., 1977, 295, 239-248.
30. Manriquez, J.M.; Fagan, P.J.; Marks, T.J. J. Am. Chem. Soc., 1978, 100, 3939-3941.
31. Manriquez, J.M.; Fagan, P.J.; Marks, T.J. manuscript in preparation.
32. Manriquez, J.M.: Fagan, P.J.; Marks, T.J.; Day, C.S.; Day, V.W. J. Amer. Chem. Soc., 1978, 100, 7112-7114.
33. Manriquez, J.M.; Fagan, P.J.; Marks, T.J.; Day, C.S.; Vollmer, S.H.; Day, V.W., manuscript in preparation.
34. Petersen, J.L.; Lichtenberger, D.L.; Fenske, R.F.; Dahl, L.F. J. Am. Chem. Soc., 1975, 97, 6433-6441.
35. Keppert, D.L., "The Early Transition Metals," Academic Press, N.Y., 1972, Chapt. 1.
36. Pearson, R.G.,Ed., "Hard and Soft Acids and Bases," Dowden, Hutchinson, and Ross, Stroudsberg, PA, 1973.
37. Fachinetti, G.; Floriani, C.; Stoeckli-Evans, H. J. Chem. Soc., Dalton Trans., 1977, 2297-2302.
38. Fachinetti, G.; Fochi, G.; Floriani, C. J. Chem. Soc., Dalton Trans., 1977, 1946-1950.
39. Fachinetti, G.; Floriana, C. J. Organometal. Chem., 1974, 71, C5-C7.
40. Lappert, M.F.; Luong-Thi, N.T.; MiLne,C.R.C., J. Organometal. Chem., 1979, 174, C35-C37.
41. Maslowsky, E.,Jr. "Vibrational Spectra of Organometallic Compounds," Wiley-Interscience, N.Y., 1977, p. 155.
42. Green, M.L.H. "Organometallic Compounds," Vol. 2, Methuen, London, 1968, p. 257-261.

43. Lauher, J.W.; Hoffmann, R. J. Am. Chem. Soc., 1976, 98, 1729-1742.
44. Petersen, J.L.; Dahl, L.F. J. Am. Chem. Soc., 1975, 97, 6422-6433.
45. Green, M.L.H.; Douglas, W.E., J. Chem. Soc., Dalton Trans., 1972, 1796-1800.
46. Erker, G.; Rosenfeldt, F. Angew. Chem. Int. Ed. Engl., 1978, 17, 605-606.
47. Manriquez, J.M.; McAlister, D.R.; Sanner, R.D.; Bercaw, J.E. J. Am. Chem. Soc., 1978, 100, 2716-2724.
48. Manriquez, J.M.; McAlister, D.R.; Sanner, R.D.; Bercaw, J.E. J. Am. Chem. Soc., 1976, 98, 6733-6735.
49. Manriquez, J.M.; Fagan, P.J.; Marks, T.J. manuscript in preparation.
50. Baron, W.J.; DeCamp, M.R.; Hendrick, M.E.; Jones, M.,Jr.; Levin, R.H.; Sohn, M.B.; in "Carbenes," Jones, J., Jr.; Moss, R.A.; eds., Wiley-Interscience, N.Y., 1973, Vol. I, p. 128.
51. Moss, R.A.; in reference 50, p. 280.
52. Kirmse, W. "Carbene Chemistry," Academic Press, N.Y., 1971, Chapt. 3, Section E.
53. Hoffmann, R.W. Angew. Chem. Int. Ed., 1971, 10, 529-540.
54. Wentrup, C., Topics Curr. Chem., 1976, 62, 173-251.
55. Reference 50, p. 72.
56. Reference 52, Chapt. 12.
57. Robson, J.H.; Schechter, H., J. Am. Chem. Soc., 1967, 89, 7112-7113.
58. Eller, P.G.; Bradley, D.C.; Hursthouse, M.B.; Meek, D.W., Coord. Chem. Rev., 1977, 24, 1095.
59. Bradley, D.C., Advan. Chem. Ser., 1976, 150, 266-272.
60. Bradley, D.C. Chem. Brit., 1975, 11, 393-397.
61. Bradley, D.C. Advan. Inorg. Chem. Radiochem., 1972, 15, 259-322.
62. Manriquez, J.M.; Fagan, P.J.; Marks, T.J.; Vollmer, S.H.; Day, C.S.; Day, V.W. submitted for publication.
63. Manriquez, J.M.; Fagan, P.J.; Marks, T.J. unpublished results.
64. Dell'Amico, D.B.; Calderazzo, F.; Pelizzi, G., Inorg. Chem., 1979, 18, 1165-1168, and references therein.
65. Angelici, R.J., Acc. Chem. Res., 1972, 5, 335-341.
66. Müller, A.; Ulrich, S.; Werner, E. Inorg. Chim. Acta., 1979, 32, L65-L66.
67. Anet, F.A.L.; Basus, V.J. J. Mag. Resonan., 1978, 32, 339-343, and references therein.
68. Manriquez, J.M.; Fagan, P.J.; Marks, T.J.; Vollmer, S.H.; Day, C.S.; Day, V.W. unpublished results.
69. Johnson, P.L.; Cohen, S.A.; Marks, T.J.; Williams, J.M. J. Am. Chem. Soc., 1978, 100, 2709-2716.
70. Day, C.S.; Stults, B.R.; Marianelli, R.S.; Day, V.W., manuscript in preparation.
71. Arduini, A.; Takats, J.; Lukehart, C.M.; Vollmer, S.H.; Day, V.W., submitted for publication.

72. Jamerson, J.D.; Takats, J., <u>J. Organometal. Chem.</u>, 1974, 78, C23-C25.
73. Manriquez, J.M.; Fagan, P.J.; Marks, T.J.; Day, C.S.; Vollmer, S.H.; Day V.W., manuscript in preparation.
74. "Tables of Interatomic Distances and Configurations in Molecules and Ions," <u>Chem. Soc., Spec. Publ.</u>, 1965, 18, m99s.
75. Casey, C.P.; Andrews, M.A.p McAlister, D.R., <u>J. Am. Chem. Soc.</u>, 1979, 101, 3373, and references therein.
76. Bioloen, P.; Helle, J.N.; Sachtler, W.H.M. <u>J. Catal.</u>, in press.
77. Kummer, J.T.; Emmett, P.H. <u>J. Amer. Chem. Soc.</u>, 1953, 75, 5177-5183.
78. Blyholder, G.; Emmett, P.H. <u>J. Phys. Chem.</u>, 1959, 63, 962-965.
79. Blyholder, G.; Emmett, P.H. <u>J. Phys. Chem.</u>, 1960, 64, 470-472.
80. Blyholder, G.; Goodsel, A.J. <u>J. Catal.</u>, 1971, 23, 374-378.
81. Blyholder, G.; Shihabi, D.; Wyatt, W.V.; Bartlett, R. <u>J. Catal.</u>, 1976, 43, 122-130.
82. Fachinetti, G.; Floriani, C.; Roselli, A.; Pucci, S. <u>J. Chem. Soc., Chem. Comm.</u>, 1978, 269-270.
83. Wolczanski, P.T.; Threlkel, R.S.; Bercaw, J.E. <u>J. Amer. Chem. Soc.</u>, 1979, 101, 218-220.
84. Schlodder, R.; Ibers, J.A.; Lenorda, M.; Graziani, M. <u>J. Amer. Chem. Soc.</u>, 1974, 96, 6893-6900.
85. Heck, R.F., "Organotransition Metal Chemistry," Academic Press, N.Y., 1974, p. 255-260.
86. Noyori, R. in "Transition Metal Organometallics in Organic Synthesis," Alper, H., ed., Academic Press, 1976, Vol. 1, p. 145-146.
87. Fieser, L.F.; Fieser, M. "Reagents for Organic Synthesis," Wiley, N.Y., 1967, Vol. 1, 759-764.
88. Sharpless, K.B.; Williams, D.R. <u>Tetrahedron Lett</u>, 1975, 3045-3046.
89. Natta, G.; Colombo, U.; Pasquon, I. in "Catalysis," Emmett, P.H., ed., Reinhold, N.Y., 1957, Vol. 5, Chapter 2.
90. Cohn, E.M. in "Catalysis," Emmett, P.H., ed., Reinhold, N.Y., 1956, Vol. 4, Chapt. 3.
91. Pichler, H.; Ziesecke, H-H.; Traeger, B. <u>Brennstoff-Chem.</u>, 1950, 31, 361-374.
92. Pichler, H.; Ziesecke, K.-H.; Fitzenthaler, E. <u>Brennstoff-Chem.</u>, 1949, 30, 333-347.
93. Pichler, H.; Ziesecke, K.-H., <u>Brennstoff-Chem.</u>, 1949, 30,13-22.

RECEIVED December 26, 1979.

Alkyl, Hydride, and Related Bis(trimethylsilyl)amide Derivatives of the 4f- and 5f- Block Metals

RICHARD A. ANDERSEN

Chemistry Department and Materials and Molecular Research Division of Lawrence Berkeley Laboratory, University of California, Berkeley, CA 94720

Metal derivatives of the bis(trimethylsilyl)amido ligand, [(Me$_3$Si)$_2$N], have been extensively investigated for the p- and first-row d-block elements. An exhaustive review by Harris and Lappert, concentrating upon synthetic chemistry, has recently appeared (<u>1</u>). A review of the molecular and electronic structure of three-coordinate and related (Me$_3$Si)$_2$N-derivatives, which reports a number of unpublished results has appeared (<u>2</u>). A rather more general review of the transition metal derivatives also has been published (<u>3</u>).

One area of silylamide (this short-hand abbreviation will be used for (Me$_3$Si)$_2$N) chemistry that has been largely ignored is the f-block element derivatives. The silylamide ligand is potentially a very valuable ligand in this part of the Periodic Table principally due to its size. Association by way of dative bonding (I) is prevented, since in a hypothetical tri-

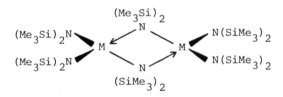

I

valent, binary derivative (with coordination number of four) the steric congestion about the metal atom is far too great. Further, the lone-pair of electrons on the nitrogen atom is considerably less basic relative to an analogous dialkylamide, (a (Me$_3$Si)$_2$N) group is electron-withdrawing relative to a Me$_3$C group, silicon being less electronegative and/or possessing low-lying d$_\pi$-orbitals for electron-delocalization) (II). The resulting decrease in basicity of the nitrogen lone pair minimizes its ability to act as a two electron donor to the metal atom. Thus, the large size and electron-withdrawing

0–8412–0568–X/80/47–131–031$05.00/0

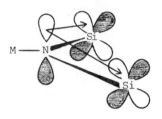

II

ability of the trimethylsilyl group prevents association and
monomeric, hydrocarbon soluble, volatile derivatives are to
be expected. In addition the extreme simplicity of the
nuclear magnetic resonance and vibrational spectra greatly
simplifies analysis.

In light of these considerations it is surprising that
the f-block metal derivatives of this readily accessible
ligand have not been explored. In this review we describe
some results which begin to rectify this deficiency.

Trivalent Lanthanide and Actinide Derivatives

Compounds of the type $[(Me_3Si)_2N]_3M$ have been prepared for
all of the lanthanide elements except Pm, Tb, Dy, Tm, and Er
(4). The synthetic method used in their preparation is nucleo-
philic substitution with three molar equivalents of lithium -
or sodium - bis(trimethylsilyl)amide on the metal trichlorides
in tetrahydrofuran. The compounds are rather high melting
solids (145-170°C) which can be isolated by crystallization
from pentane as long needles or by vacuum sublimation (80-
100°C). The binary silylamides are monomeric in refluxing
benzene solution, in the gas phase (by mass spectrometry), and
in the solid state (by x-ray crystallography, see below). Thus,
these compounds are three-coordinate, a unique coordination
number for the lanthanide elements.

The colors of the silylamides closely parallel those of
the metal ions in aqueous solution, i.e., $Nd[N(SiMe_3)_2]_3$ is
pale blue, $Eu[N(SiMe_3)_2]_3$ is orange, and $Yb[N(SiMe_3)_2]_3$ is
pale yellow. The optical spectra of the praesodymium and neodymium derivatives have been studied in gaseous, solution (CCl_4),
and low temperature (4 K) solid phases (5). The spectra are
very similar (number and intensity of absorptions) which
indicates that the symmetry is identical in all phases.
Further, the silylamide ligand does not greatly perturb the
energy levels of the free ion since the observed spectra are
very similar to the trivalent metal ions in aqueous solution.

Low temperature magnetic susceptibility measurements (the
binary derivatives are all paramagnetic, except those with
closed-shell f^0 or f^{14} electronic configuration) also closely

parallel the trivalent metal ion values in aqueous solution, for those silylamide derivatives which have been examined (Table 1). An exception is $Eu[N(SiMe_3)_2]_3$ which is a temperature independent paramagnet (T.I.P.), the temperature independent susceptibility being ca. 0.4 B.M. consistent with the $^7F_0$ ground state found for Eu (III).

Table I

Magnetic Susceptibilities of $M[N(SiMe_3)_2]_3$

M	μ_B(B.M.)	Temp. range (K)	θ (K)	C_M	Ref.
Nd	3.27	4.2-90	12	1.33	6
Gd	7.94	98-298	--	--	4
Yb	3.10	5.1-46	0	1.09	7
U	2.51	10-70	-10.5	1.64	13

$$X_M = C_M(T+\theta)^{-1}; \quad \mu_{eff} = 2.837\sqrt{X_M}$$

The crystal structure of three representative [Nd (6), Eu (8), and Yb (2)] amides have been determined. The compounds are monomeric and there is no indication of molecular association in the solid state. The $(Me_3Si)_2N$ groups are neither coplanar nor orthogonal to the MN_3 unit but half-way between, such that the dihedral angle (defined by the Si_2NM and MN_3 planes) are ca. 50° (Figure 1). The molecules are thus molecular propellers and chiral. Therefore, two enantiomeric forms may exist which differ only in their sense of twist. This is the conformation that is expected on the basis of steric arguments, the trimethylsilyl groups pack in the crystalline lattice in such a fashion to minimize Van der Waals contact between them.

A curious feature of the solid-state structural results is that the MN_3 unit is not coplanar as found in the first-row transition or Group IIIA series (2) but pyramidal, the metal atom being out of a plane defined by the three nitrogen atoms by ca. 0.4 Å. Since the molecules have no dipole moment in solution (2) the geometry could well be due to packing forces in the solid state.

An interesting feature of the binary, trivalent silyl-amides is the way that they pack in the crystalline lattice. Six $[(Me_3Si)_2N]_3M$ units are orientated about a six-fold rotation axis such that a cylindrical channel, large enough to include a benzene ring, is formed. This phenomenon has been described in some detail (9a). The existence of the hollow cavity accounts for the low density of the solid (ca. 1g cm^{-3}). A single crystal, when viewed under a microscope, appears to contain a hole in the center, the macroscopic structure apparently mirrors the microscopic array (9b). The molecular packing shows that the molecules are three-dimensional molecular sieves.

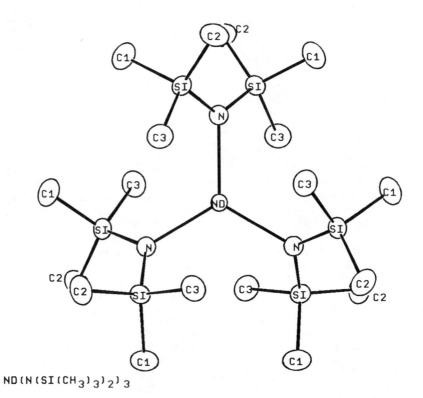

ND(N(SI(CH$_3$)$_3$)$_2$)$_3$

Figure 1.   Molecular structure of Nd[N(SiMe$_3$)$_2$]$_3$

Consideration of metal-nitrogen bond lengths in light of
the ionic-bonding model advanced by Raymond (10) leaves little
doubt that the bonding in the binary silylamide derivatives of
the lanthanide elements is predominantly ionic (11). Indeed,
all of the tris-silylamide derivatives of the p-, d-, and f-
block elements can be viewed as being mainly ionic.

Synthesis of a trivalent uranium silylamide derivative
is of considerable interest. The principal synthetic difficulty is the lack of a suitable, large-scale preparation of
uranium trichloride (12). We have devleoped an in situ preparation of uranium trichloride by reducing commercially available uranium tetrachloride with one molar equivalent of sodium
napthalene in tetrahydrofuran (13). Though we have not characterized this species Moody has recently isolated some
coordination complexes of UCl_3, prepared by sodium hydride
reduction of UCl_4 (14, 15). A trivalent tris-silylamide of
uranium $U[N(SiMe_3)_2]_3$ can be readily prepared in good yield
from UCl_3 in situ and sodium bis(trimethylsilyl)amide in
tetrahydrofuran. The deep red, paramagnetic (Table I) needles
(m.p. 137-140°C) crystallize from pentane. The compound is
volatile (sublimation temperature 80-100°/10^{-3} torr) and
monomeric in gas phase (by mass spectrometry) (13).

The structure of this unique uranium (III) derivative is
of much interest. Unfortunately we have been unable to obtain
crystals satisfactory for an X-ray analysis. The compound is
most likely similar to that of the binary, trivalent lanthanide
derivatives, e.g., pyramidal rather than planar, on the basis
of infrared spectroscopy. Planar $M[N(SiMe_3)_2]_3$ show bands due
to νas $MNSi_2$ at 900 cm^{-1} whereas pyramidal ones absorb at
990 cm^{-1}. Since $U[N(SiMe_3)_2]_3$ has its νas $UNSi_2$ absorption band
at 990 cm^{-1} it is most likely pyramidal in the solid state. Not
surprisingly the analogous thorium (III) derivative cannot be
prepared in a similar fashion.

Coordination Chemistry. The coordinative unsaturation of
the three-coordinate derivatives suggests that these molecules
should have a rich coordination chemistry. This has not been
found, doubtless due to the steric congestion about the metal
atom. Neodymium tris[bis(trimethylsilyl)amide] forms pale
blue 1:1 coordination complexes with the sterically small
Lewis bases Bu^tNC and Bu^tCN. Triphenylphosphine oxide yields
a 1:1 complex with the silylamides of La, Eu, and Lu (16).
A crystal structure of the lanthanum derivative has also been
described. The dissociation pressure in gas phase of Ph_3PO
complex is appreciable as the base-free compound sublimes when
heated in vacuum at 80-100°C. In contrast, the uranium tris-
silylamide does not yield isolable complexes with Bu^tNC,
Bu^tCN, pyridine, or trimethylphosphine oxide (13). The inability to isolate coordination complexes with a variety of
Lewis bases is rather surprising since uranium (III) is larger

than to its cogenic, neodymium (III) (17). Thus, the steric
congestion about the lighter f-metal must be greater than about
its heavier derivative. However, Nd(III) is more electropositive (by ca. 0.5 units) than that of U(III) and the
coordinative affinity of uranium to a reference acid is therefore considerably less.

One attempt to isolate a U(III) complex of trimethyl-
amineoxide resulted in oxidation of [(Me$_3$Si)$_2$N]$_3$U to the
pentavalent oxide [(Me$_3$Si$_2$N]$_3$UO. The oxide could be prepared
rather more simply from the tris-silylamide and molecular
oxygen. The oxide is involatile and is not soluble enough in
suitable solvents for a solution molecular weight determination.
Thus, its degree of association in solution is unknown. It
could be a coordination oligomer (IV), as suggested by its
relative insolubility, or a discrete monomer (V). The U=O

$$[\,(Me_3Si)_2N]_3U\diagup\!\!\!\!\!\overset{\textstyle O}{\underset{\textstyle O}{\diagdown\!\!\!\!\!\diagup}}\!\!\!\!\!\diagdown U[N(SiMe_3)_2]_3 \qquad\qquad [\,(Me_3Si)_2N]_3U{=}O$$

IV   V

stretching frequency (930 cm^{-1}) is similar to that found in
monomeric O$_2$U[N(SiMe$_3$)$_2$]$_2$(thf)$_2$ (938 cm^{-1}) (18a). This suggests
that the uranium-oxygen bond order is the same in both examples
and that [(Me$_3$Si)$_2$N]$_3$UO is a monomer (V). A weak intermolecular
interaction which would account for the poor solubility in
hydrocarbon solvents in solid-state cannot be ruled out.

Divalent Lanthanide Derivatives

The divalent oxidation state of the lanthanide elements
is commonly found for europium and ytterbium, though some
simple salts have been prepared for most of the 4f-series (12).
Only a few molecular compounds have been described and these
have been mainly with cyclopentadienyl and cyclooctateraene
ligands (18b). None have been characterized by X-ray methods
since the compounds are generally insoluble and nonvolatile an
observation that is consistent with a polymeric constitution.
The compounds doubtless polymerize in an attempt to increase
their coordination number to a maximum value while maintaining
satisfactory metal-ligand interactions.

Since the silylamide ligands occupy considerable volume
about a metal atom and the low basicity of the lone pair of
electrons on the nitrogen atom will minimize intermolecular
association, divalent species, [(Me$_3$Si)$_2$N]$_2$M, are likely.
The structural and magnetic properties of these molecular
compounds are of considerable interest.

The most useful preparative method for the europium derivative is shown (7). The chloroamide of europium (III) is

$$2 \; Eu[N(SiMe_3)_2]_3 + EuCl_3 \xrightarrow{\text{thf}} 3 \; ClEu[N(SiMe_3)_2]_2$$

$$ClEu[N(SiMe_3)_2]_2 + Na \; Napthalene \xrightarrow{\text{thf}} Eu[N(SiMe_3)_2]_2(thf)_2$$

not isolated but is reduced in situ. The 1:2 tetrahydrofuran
complex is yellow, paramagnetic ($\mu_B = 7.8$ B.M.), and soluble
in toluene. The tetrahydrofuran can be displaced by pyridine
or 1,2-dimethoxyethane giving 1:2 complexes or by bipyridine
giving a 1:1 complex. The 1:2 complexes with monodentate
ligands and the 1:1 complex with bipyridine are most likely
four coordinate complexes with a distorted tetrahedral geometry.
The complex with 1,2-dimethoxyethane, however, must be six
coordinate. Since the latter complex crystallizes as long,
yellow needles from pentane, a crystal structure analysis was
done (Figure II) (9b).

The amide is six coordinate but it does not conform to
any regular polytopal form since the N-Eu-N bond angle is 135°.
The Eu-N bond length is 2.52 Å.

Tetravalent Actinide Derivatives

The monochlorotris-silylamides, $[(Me_3Si)_3N]_3MCl$, of the
two most readily accessible actinide elements, thorium and
uranium, have been prepared by reaction of three molar equivalents of sodium bis(trimethylsilyl)amide and the metal chloride in tetrahydrofuran. The thorium derivative is a colorless,
diamagnetic, pentane soluble compound which is monomeric in
the gas phase (by mass spectrometry) (19, 20). The tan uranium
analogue is paramagnetic ($\mu_B = 2.8$ B.M.) (20).

The chloride ligand in both derivatives can be replaced
by a methyl group upon reaction with methyllithium in the case
of uranium and dimethylmagnesium in the case of thorium (20).
The reactions of the chloro-derivatives are summarized in
Scheme I. Both methyl derivatives are readily soluble in
pentane from which they can be crystallized. They are thermally
stable to ca. 130°C. Heating the methyls to ca. 20°C above
their respective melting point results, in each case, in elimination of methane and formation of the unique metallocycle VI (21).

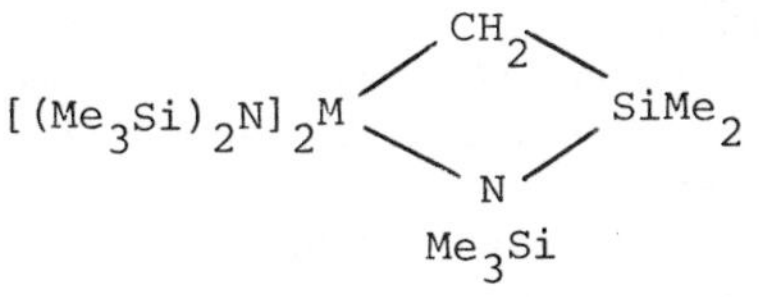

$$[(Me_3Si)_2N]_2M \underset{N}{\overset{CH_2}{\diagup\diagdown}} SiMe_2$$

$$Me_3Si$$

VI

The diamagnetic, monomeric thorium derivative has been completely characterized by its ^1H and ^{13}C nuclear magnetic resonance spectra. Data are shown in Table II. In particular the ^{13}C NMR spectrum, proton-coupled, shows a triplet pattern clearly due to splitting by two equivalent hydrogen atoms. The uranium metallocycle is paramagnetic (μ_B = 2.7 B.M.) and we have only observed its ^1H NMR spectrum. A titanium metallocycle, $Cp_2\overline{TiN(SiMe_3)(SiMe_2)(CH_2)}$, has been previously described (22).

Some interest is attached to the conformation of the metallocycle in solution, as either conformation VII or VIII are possible. In the former (VII), there is no symmetry plane plane containing the MCSiN-four membered ring, the methylene and Me_2Si protons are chemically nonequivalent. In the latter case, (VIII), there is such a symmetry plane and the methylene and Me_2Si protons are chemically equivalent. The room temperature and -80°C ^1H nuclear magnetic resonance spectrum is consistent with VIII. However, conformation (VII) could be present in solution if the molecule were fluxional even at -80°C. A single crystal X-ray examination, which is in progress, will prove which conformation exists in the solid-state (9b).

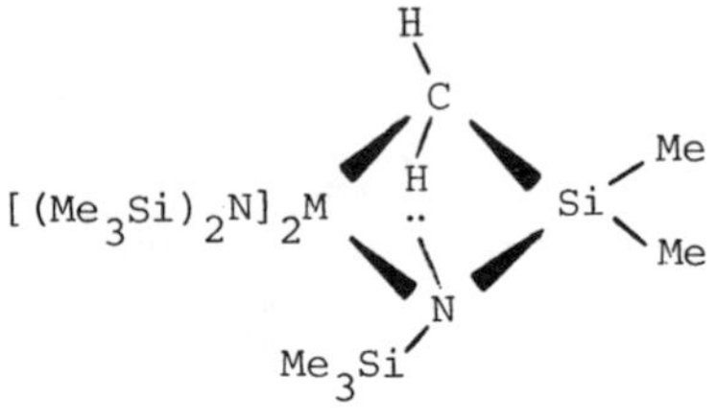

VII 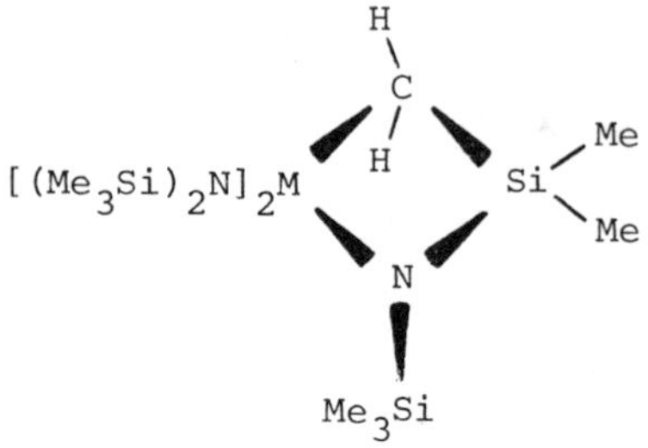   VIII

The high thermal stability of the metal-carbon bond in the actinide methyl derivatives suggests that a series of alkyl derivatives can be made. This does not prove to be the case. Reaction of $ClM[N(SiMe_3)_2]_3$, where M is thorium or uranium, with either ethyllithium or trimethylsilylmethyllithium at room temperature in diethyl ether yields the metallocycle (VI) and ethane or tetramethylsilane. A mechanism for this transforma-tion, which involves a γ-proton abstraction, is shown below. The γ-abstraction process from trimethylsilylmethyl (23) or neopentyl (24) groups is known. The proposed pathway goes by way of the ylide (IX). The ylide could be either a transition state or an intermediate depending upon whether the reaction is concerted or not. A curious facet of the abstraction re-action is that the methyl and hydride derivatives can be isolated and converted into the metallocycle by heating, whereas attempts to prepare alkyls with larger organic groups gives the

Table II
Nuclear Magnetic Resonance Data

$[(Me_3Si)_2N]_2\overline{ThN(SiMe_3)SiMe_2CH_2}$	^1H NMR (δ)	^{13}C NMR (δ)
$(Me_3Si)_2N$	0.37	3.46 q J=117 Hz
Me_3SiN	0.38	4.52 q J=117 Hz
Me_2SiN	0.56	5.55 q J=118 Hz
CH_2	0.46	68.8 t J=120 Hz
$[(Me_3Si)_2N]_3ThMe$		
$(Me_3Si)_2N$	0.57	4.17
Me	0.85	73.6
$[(Me_3Si)_2N]_2\overline{UN(SiMe_3)SiMe_2CH_2}$		
$(Me_3Si)_2N$	-23.3	
Me_3SiN	-9.90	
Me_2SiN	+2.08	
CH_2	-128.6	
$[(Me_3Si)_2N]_3UMe$		
$(Me_3Si)_2N$	-1.49	
Me	-224	

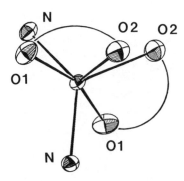

Figure 2. Molecular structure of Eu-[N(SiMe₃)₂]₂[MeOCH₂CH₂OMe]₂: N = N(SiMe₃)₂; O = MeOCH₂CH₂OMe.

$$[(Me_3Si)_2N]_2M \overset{R \cdots H}{\underset{N-Si}{\diagdown}} CH_2 \quad \xrightarrow{-RH} \quad [(Me_3Si)_2N]_2M^+ \diagdown_{N-Si} \bar{C}H_2$$

$$Me_3Si \quad Me_2 \qquad\qquad\qquad Me_3Si \quad Me_2$$

$$[(Me_3Si)_2N]_2M \diagdown \overset{CH_2}{\underset{N}{\diagup}} SiMe_2$$

$$Me_3Si$$

metallocycle straightaway. This is most reasonably ascribed to
a steric effect. In hypothetical $[(Me_3Si)_2N]_3MCH_2SiMe_3$ the
steric congestion about the metal atom is too great and the
molecule relieves itself of the congestion by elimination of
tetramethylsilane.

The chloro-tris(silylamido) derivatives also react with
lithium borohydride giving $[(Me_3Si)_2N]_3MBH_4$. The infrared
spectra and a crystal structure analysis of the thorium derivative shows that the borohydride group is tridentate, that is
the metal atoms are six coordinate as they are bonded to three
nitrogen atoms and three hydrogen atoms. The ^{1}H nuclear magnetic
resonance spectrum of the diamagnetic thorium and paramagnetic
uranium derivative indicate that they are fluxional, a 1:1:1:1
quartet being observed even at -80°C (20).

In an attempt to prepare a tetrasilylamide we allowed
$[(Me_3Si)_2N]_3MCl$ to react with an equivalent of sodium bis(trimethylsilyl)amide in refluxing tetrahydrofuran (25). In each
case, Th or U, we isolated the unique hydride, $[(Me_3Si)_2N]_3MH$.
The colorless, diamagnetic thorium hydride (m.p. 145-147°C)
shows an ThH absorption in the infrared spectrum at 1480 cm^{-1}
which shifts to 1060 cm^{-1} in the deuteride. The paramagnetic
(μ_B = 2.6 B.M.), brown-yellow uranium hydride (m.p. 97-98°C)
shows a similar pattern, νUH = 1430 cm^{-1} and νUD = 1020 cm^{-1}.
The proton nuclear magnetic resonance spectrum of the thorium
hydride shows a resonance at δ 0.90 due to the hydride. The
source of the hydride has been shown to be one of the hydrogen
atoms of tetrahydrofuran since conducting the reaction in
perdeuterotetrahydrofuran yields the monodeuteride. Tetrahydrofuran is essential since boiling $ClM[N(SiMe_3)_2]_3$ with
$NaN(SiMe_3)_2$ in benzene, toluene, isooctane, or diethyl ether
results in isolation of unreacted $ClM[N(SiMe_3)_2]_3$.

The chemical nature of the hydride is clearly demonstrated
by some of its reactions shown in Scheme II. In particular
the hydrogen-atom abstraction with the strong base n-BuLi is
noteworthy. We have not been able, as yet, to isolate the

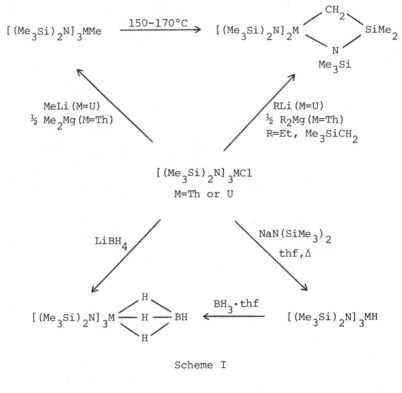

Scheme I

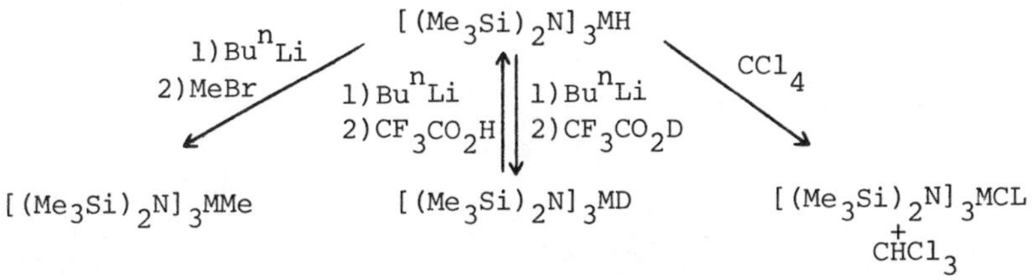

Scheme II

lithium-containing species. The hydridic nature is further illustrated by its reaction with borane in tetrahydrofuran giving the borohydride (Scheme I). In effect, the actinide hydrides act as a soluble source of sodium hydride. This is not unexpected since the actinide elements are quite electropositive.

Possibly the most fascinating reaction of the hydrides so far discovered is their facile exchange of all fifty-five hydrogen atoms with deuterium under one atmosphere of deuterium pressure at room temperature in pentane (21). The conversion

$$[(Me_3Si)_2N]_3MH + D_2 \longrightarrow DM\{N[Si(CD_3)_3]_2\}_3$$

is 92-97% complete. Neither the methyl, chloro, borohydride
derivatives nor the uranium (III) derivative, $[(Me_3Si)_2N]_3U$,
undergo exchange with deuterium under similar conditions.

The mechanism of this unique H-D exchange is of considerable importance. The usual mechanism proposed for H-D exchange
in d-block transition metals involves a series of reductive-
elimination, oxidative-addition cycles. The prerequisite in
this type of process is the ability of the metal atom to shuttle
between two readily available oxidation states. This type of
mechanism could be invoked to explain the exchange reaction in
the uranium hydride since uranium (VI) is a well-known oxidation
state. The observation that the thorium hydride also undergoes
exchange at a comparable rate shows that such a process is not
viable as thorium (VI) is unknown. Scheme III outlines an

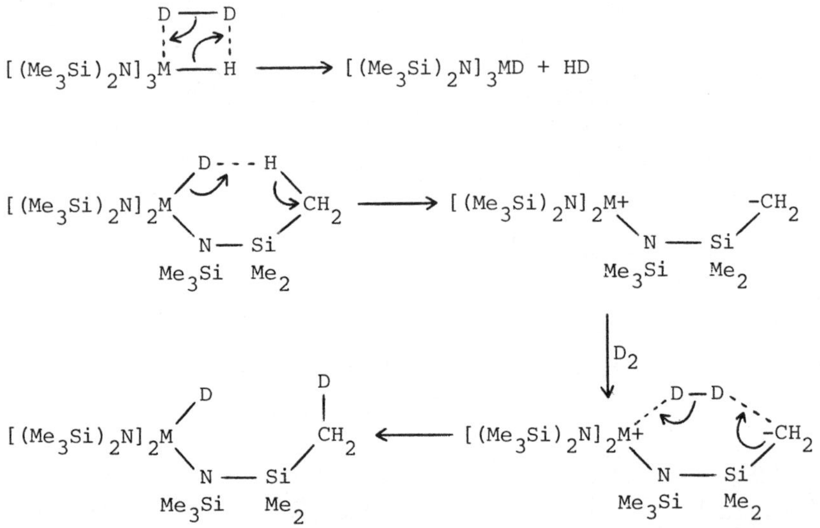

Scheme III

alternative mechanism. The first step is hydride for deuteride
exchange by way of a four-center transition state, so common in
early p- and d-block chemistry (26, 27). Elimination of HD to
form the ylide or its valence tautomer, the fully-developed
metallocycle, then occurs. The ylide reacts or the metallocycle
opens in the presence of deuterium yielding the di- deutero-
compound which then reenters the cycle. The cyclic pathway is
followed until all of the hydrogen atoms are exchanged by
deuterium. The key to the mechanistic scheme is formation and
opening of the metallocycle. Isolation of the metallocycle and

reaction of it with deuterium to yield the perdeutero-derivative strongly implies that the mechanistic scheme is essentially correct. The observation that the hydrides, $[(Me_3Si)_2N]_3MH$, extrude hydrogen and form the metallocycle is added support for the essential validity of the proposed mechanism. Hydrocarbon activation of a related type has been shown for zirconium (28), thorium or uranium (29).

Acknowledgement. I thank Dr. S. J. Simpson, Mr. H. W. Turner, and Mr. T. D. Tilley for providing the experimental results from which this account was constructed. I also thank Dr. A. Zalkin and Dr. D. H. Templeton for the crystal structure analysis, and Dr. N. M. Edelstein for much valuable insight into the electronic and magnetic properties of the f-block elements. This work was supported by the Division of Nuclear Sciences, Office of Basic Energy Sciences, U. S. Department of Energy.

Literature Cited

1. Harris, D. H.; Lappert, M. F. Organomet. Chem. Library: Organosilicon Rev., 1976, 4, 13.
2. Eller, P. G.; Bradley, D. C.; Hursthouse, M. B.; Meek, D. W. Coord. Chem. Rev., 1977, 24, 1.
3. Bradley, D. C.; Chisholm, M. H. Acct. Chem. Res., 1976, 9, 273.
4. Bradley, D. C.; Ghotra, J. S.; Hart, F. A. J. Chem. Soc. Dalton Trans., 1973, 1021.
5. Conway, J.; Edelstein, N. M., personal communication.
6. Andersen, R. A.; Templeton, D. H.; Zalkin, A. Inorg. Chem., 1978, 17, 2317.
7. Tilley, T. D., unpublished results.
8. Ghotra, J. S.; Hursthouse, M. B.; Welch, A. J. J. Chem. Soc. Chem. Comm., 1973, 669.
9. a) Hursthouse, M. B.; Rodesiler, P. F. J. Chem. Soc. Dalton Trans., 1972, 2100. b) Zalkin, A., personal communication.
10. Baker, E. C.; Halstead, G. W.; Raymond, K. N. Struct. and Bonding (Berlin), 1976, 25, 23.
11. Eigenbrot, C.; Raymond, K. N., personal communication.
12. Brown, D. "Halides of the Transition Elements-Halides of the Lanthanides and Artinides"; John Wiley: New York, 1968.
13. Andersen, R. A. Inorg. Chem., 1979, 18, 1507.
14. Moody, D. C.; Penneman, R. A.; Salazar, K. U. Inorg. Chem., 1979, 18, 208.
15. Moody, D. C.; Odom, J. D. J. Inorg. Nucl. Chem., 1979, 41, 533.
16. Bradley, D. C.; Ghotra, J. S.; Hart, F. A.; Hursthouse, M. B.; Raithby, P. R. J. Chem. Soc. Dalton Trans., 1977, 1166.

17. Shannon, R. D. Acta Cryst., 1976, 32A, 751.
18. a) Andersen, R. A. Inorg. Chem., 1979, 18, 209. b) Marks, T. J. Prog. Inorg. Chem., 1979, 18, 209.
19. Bradley, D. C.; Ghotra, J. S.; Hart, F. A. Inorg. Nucl. Chem. Let., 1974, 10, 209.
20. Turner, H. W.; Andersen, R. A.; Templeton, D. H.; Zalkin, A. Inorg. Chem., 1979, 18, 1221.
21. Simpson, S. J.; Turner, H. W.; Andersen, R. A. J. Am. Chem. Soc., in press.
22. Bennett, C. R.; Bradley, D. C. J. Chem. Soc. Chem. Comm., 1974, 29.
23. Andersen, R. A.; Jones, R. A.; Wilkinson, G. J. Chem. Soc. Dalton Trans., 1978, 446.
24. Foley, P.; Whitesides, G. M. J. Am. Chem. Soc., 1979, 100, 2732.
25. Turner, H. W.; Simpson, S. J.; Andersen, R. A. J. Am. Chem. Soc., 1979, 101, 2782.
26. a) Brown, H. C. "Boranes in Organic Chemistry"; Cornell University Press: Ithaca, 1972, p. 265. b) Brown, H. C.; Scouten, C. G.; Wang, K. K. J. Org. Chem., 1979, 44, 2589.
27. Gell, K. I.; Schwartz, J. J. Am. Chem. Soc., 1978, 100, 3246.
28. McAlister, D. R.; Erwin, D. K.; Bercaw, J. E. J. Am. Chem. Soc., 1978, 100, 5966.
29. Marks, T. J., ACS Symposium: Washington, D. C., Sept. 1979, Abstract Number 157.

RECEIVED January 30, 1980.

Organic Derivatives of the *f*-Block Elements: A Quest for *f*-Orbital Participation and Future Perspective

TAKESHI MIYAMOTO (*1*) and MINORU TSUTSUI

Department of Chemistry, Texas A&M University, College Station, TX 77843

Summary

The long-term quest for f-orbital participation in bonding is described along with progress in organometallic chemistry of f-block elements. A wide variety of sigma-bonded organolanth- anide compounds have been synthesized with the expectation of the f-orbital participation in metal-carbon bonding. To our dis- appointment, visible spectra, magnetic properties and x-ray analyses did not show any definite evidence of f-orbital parti- cipation in bonding. However, recent ESCA studies have revealed unique electronic structures of a series of f-block phthalo- cyanine and porphyrin complexes in which f-orbitals play an im- portant role in the core ionization process.

Plets' first attempt to prepare organolanthanide compounds marked the dawn of a new area of chemistry (2). Since then, a wide variety of complexes of f-block elements have been synthe- sized and their structures have been satisfactorily explained in terms of "steric congestion" of ligands around f-metal ions (3,4,5,6). However, the question of whether the 4f valence orbitals participate in bonding has not as yet been answered. The purpose of this article is to review the recent progress in some of the f-element chemistry which is related to the possible f-orbital participation in bonding. Tsutsui and co-workers have continued to explore a series of σ-bonded organic derivaties of 4f-elements in order to search for possible participation of 4f- orbitals in the bonding (7,8). Below their expectation, x-ray analyses, the visible spectra and magnetic susceptibilities have not shown any evidence of f-orbital participation, until a few years ago. However, they have recently found unique electronic structures of known lanthanide complexes by ESCA, which yields

0–8412–0568–X/80/47–131–045$05.00/0

evidence for f-orbital participation in the bonding. This article deals with the current developments of their studies on this subject in addition to their past efforts.

Historical Background

The first π-bonded organometallic compounds of the f-block elements were prepared by Wilkinson et.al. $(Ln(C_5H_5)_3$, Ln=La, Ce, Pr, Nd, Sm. Gd, Dy, Er, or Yb, and $U(C_5H_5)_3Cl)$ in mid-1950 (9,10). The pentahapto coordination of the cyclopentadienyl ring in $U(C_5H_5)_3Cl$ (11) was shown by x-ray analysis. The $U(C_5H_5)_3Cl$ complex is different from $Ln(C_5H_5)_3$ with respect to the reaction with $FeCl_2$. The $Ln(C_5H_5)_3$ complex reacted readily with $FeCl_2$ to give ferrocene while the $U(C_5H_5)_3Cl$ complex did not yield ferrocene. This result might indicate more covalency in η^5-cyclopentadienyl-uranium bonding. A wide variety of Lewis bases also form adducts $Ln(C_5H_5)_3B$, with the $Ln(C_5H_5)_3$ complexes, where B is ammonia, THF, triphenylphosphine or isocyanide (12). The infrared study of $Ln(C_5H_5)_3$ cyclohexylisocyanide complexes shows that C≡N stretching frequencies of the complexes are 60 to 70 cm^{-1} higher in energy than that of the free ligand. This result suggests that the lanthanide ions act predominantly as σ-acceptors and metal to ligand π back-donation is very small. The first tetrakiscyclopentadienylactinides, $An(C_5H_5)_4$, (An=Th or U) were prepared by Fischer and co-workers. Many organolanthanides and organoactinides have been synthesized. These include $Ln(C_5H_5)_2X$, $Ln(C_5H_5)Cl_2$ · 3THF (Ln-Sm, Gd, Dy, Ho, Eu, Er or Yb. X=Cl, I, OCH_3, OCH_5, HCOO CH_3COO or C_6H_5COO (13,14) and An $(C_5H_5)_3$ (An = Pu or Am) (15,16).

In spite of these synthetic efforts, the organometallic chemistry of f-block elements was nearly ignored in comparison with chemistry of d-transition metals until the preparation of bis(cyclooctatetraene) uranium in 1968 (17). From studies of uranocene and its derivatives Streitwieser claimed that the f-electrons of uranium were involved in a covalent bonding between uranium and cyclotetraenyl rings. Following the preparation of uranocene, the cyclooctatetraene derivatives of lanthanides Ln(COT), Ln=Eu, Yb; $K[Ln(COT)_2]$, Ln=La, Ce, Pr, Nd, Sm. Gd, or Tb; Ln(COT)Cl·2THF, Ln=Ce Pr, Nd, Sm; $Ln_2(COT)_3$ · 2THF, Ln-La, Ce, Nd or Er, $Ce(COT)_2$ (18,19,20,21) and trisindenyl lanthanides {Ln (Ind)$_3$, Ln=La, Sm. Gd, Tb, Dy or Yb} (22) have been synthesized and characterized.

Another step in answering the question regarding the mode of bonding in f-block complexes is to synthesize a series of organo σ-bonded derivatives. Tsutsui and Gysling examined the NMR spectrum of (Ind)$_3$La·THF (THF, tetrahydrofuran) (23) and observed an A_2X pattern for the protons of the five-membered ring of the indenyl group. A similar pattern was observed for the π-bonded (Ind)$_2$Fe, (Ind)Fe(Cp), and (Ind)$_2$Ru complexes. However, examination of the NMR spectrum of (Ind)$_3$Sm·THF revealed an ABX pattern

for the analogous protons. The ABX pattern is similar to that
for the σ-bonded CpFe(CO)$_2$(1-indenyl) complex. Therefore, the
authors (23) postulated that the indenyl groups of (Ind)$_3$Sm·THF
were σ-bonded to the Sm(III) ion. If this claim could be substantiated by an X-ray examination, the (Ind)$_3$Sm·THF derivative
might be the first lanthanide derived with three σ- metal-carbon
bonds. The lanthanide contraction and the increased "hardness"
of Sm(III) relative to La(III) were suggested as possible reasons
for the different binding modes.

$$SmCl_3 + 1.5Mg(Ind)_2 \xrightarrow[benzene]{} Sm(Ind)_3 + 1.5MgCl_2$$

Although the structure of the monotetrahydrofuran complex is
unknown at this time, Atwood et.al. reported the preparation
(Eq 1) and the crystal and molecular structure of (Ind)$_3$Sm
(Figure 1) (24). They found that the five-membered rings have an
approximate trigonal configuration about the samarium atom, and
that the distances between Sm and all the carbon atoms of the
five-membered rings show no significant differences. Therefore,
no evidence for enhanced covalent bonding to the particular carbon position is found. Unfortunately, no comparison between
(Ind)$_3$Sm prepared in benzene and that resulting from the removal
of THF in vacuo from (Ind)$_3$Sm·THF has yet been made. The first
well-characterized complexes containing a lanthanide-carbon
σ-bonding were the tetraaryl anions {Li Ln(C$_6$H$_5$)$_4$}]. Subsequently an analogous crystalline lutetium complex, tetrakis
(tetrhydrofuran) lithium tetrakis (2,6-dimethylphenyl) lutetiate
(25) was reported. The structure of the anion is shown in Figure
2. It was observed from the structure that the metal is surrounded by a tetrahedral array of ligands, making this the first known
example of a four-coordinate lanthanide complex. The mean Lu-C$_\sigma$
bond distance of 2.45A is approximately 0.2A shorter than the mean
η^5 metal-carbon distance calculated for the Lu(η^5-C$_5$H$_5$) bonds.
The analogous ytterbium complex was also prepared and found to be
isostructural. However, attempts to prepare σ-bonded derivatives
of the larger lanthanides were unsuccessful.

Three different research groups independently reported the
first successful synthesis of the σ-bonded compound, U(C$_5$H$_5$)$_3$R,
where R may be an alkyl or an alkynyl group (26,27,28). The
isolated compounds are very thermally stable and the crystal
structure of UCp$_3$ (C≡C Ph) was determined by Atwood, Tsutsui and
co-workers (Figure 3) (29).

The U-C σ bond distance of 2.33 (2) A is at least 0.3 A
shorter than the mean uranium-(η^5-C$_5$H$_5$) distance (2.68A). The
shortening can be attributed to the σ-bonded nature of the
phenylacetylide group, supporting evidence for which was reported
by Atwood when he determined the U-C σ bond distance in (C$_5$H$_5$)$_3$
U-C=CH to be 2.36A (30). Unfortunately, crystallographic diffi-
culties limited the accuracy of the observed U-C σ bond distance

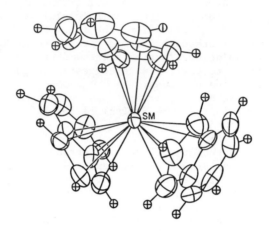

Figure 1.   Structure of Sm(Ind)₃

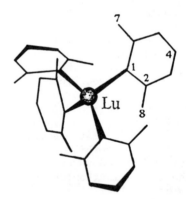

Figure 2.   Structure of the four-coordi-
nated [Lu(C₈H₉)₄]⁻

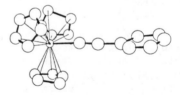

Figure 3.   Structure of tris(cyclopentadienyl)phenylethynyluranium(IV)

in the acetylide derivative to $+0.03$ Å, thus precluding any definitive comparison among the structures of the σ-bonded organo-
uraniums.

NMR spectra of the organouranium complexes $U(C_5H_5)_3R$ have
been examined by Tsutsui et al. (23,26) and Marks et al. (27).
The most outstanding characteristic is that the signals for the
protons of the R moieties have large chemical shifts. In all
cases, the Cp protons appear as a sharp singlet between 9 and 12
ppm upfield from benzene. The isotropic chemical shifts for
these compounds have been separated into contact and pseudo-contact effects. The contact shift reflects the nature of the unpaired spin density (5f electrons for uranium) through the ligand
system and can be related to covalency and bonding. The chemical
shifts of the Cp_3U-R complexes contain large contact contributions. This large contact contribution can be explained by a
delocalization mechanism which transfers the ligand electron
density into empty orbitals on the metal, and such a contact
shift is therefore indicative of a covalent interaction between the
metal and the ligand. This mechanism is the same as that proposed
by Streitwieser for the uranocenes (31).

Raymond et al. reported the synthesis and structural analysis
of an unusual actinide allyl complex, $Cp_3U[CH_3C(CH_2)_2]$ (32).
The most important feature of the structure is that the 2-methyl-
allyl group is σ and not π bonded to the uranium atom. At first
glance the η^1-coordination of the allyl group is surprising,
since in the analogous Cp_4U complex all four Cp rings are bonded
in the η^5- fashion (33). Finally, a series of Cp_3Th-R complexes
have been synthesized in a manner analogous to the Cp_3U-R compounds (34). As expected, the properties of Cp_3Th-R complexes
are similar to the analogous uranium derivatives (thorium (IV) is
diamagnetic).

Tsutsui and co-workers have been successful in synthesizing
the following unusual bimetallic complexes, Cp_3U-fer(Ferrocene) I
and $(Cp_3U)_3$-fer II.

$$Cp_3UCl \ + \ Lifc \ \xrightarrow{\text{THF}} \ \text{[fer-}UCp_3\text{]}$$

(fc = ferrocene)

$$2\,Cp_3UCl \ + \ 1,1'\,Li_2fc \cdot TMED \ \xrightarrow{\text{THF}} \ \text{[}Cp_3U\text{-fer-}UCp_3\text{]}$$

(TMED = tetramethylenediamine)

These complexes have a number of interesting properties.
One of the decomposition products if I and II is ferrocene. This
is viewed as further evidence that these Cp_3U-R complexes decom-
pose by the intramolecular proton abstraction mechanism as seen
in the "monomer" organoactinide (27, 36). The mass spectrum of I
reveals that a cyclopentadienyl ring is lost first (presumably
from Cp_3U) in preference to cleavage of the U-C σ bond.
     The NMR data for I are also in accord with the above form-
ulation. The protons of the η^5-Cp_3U group appear as sharp singlet
at δ -2.33. The five protons of the unsubstituted cyclopentadienyl
ligand on ferrocene appear as a sharp singlet at δ-1.64, a shift
of over 5.7 ppm from free ferrocene. The signals for the four
protons on the substituted ring which came much further upfield
were found but could not be located exactly, coming roughly at δ
-13 and -30 furthest upfield signal is assigned to the two
protons.
     The NMR spectra of the bimetallic complex II would be very
interesting to study for comparative purposes; however, this compound has such poor solubility in common organic solvents (e.g.,
THF) that no spectra have as yet been obtained.
     The compounds were synthesized with the expectation that
magnetic coupling via the organometallic bridge might be observed
for the unpaired 5f electrons.
     Magnetic susceptibility studies of these complexes have
shown some interesting and anomalous results. The complexes con-
tinue to display temperature-dependent paramagnetism below 100 K.
This is in sharp contrast to Cp_3UCl and Cp_4U, as well as other
Cp_3U-R complexes whose paramagnetism is independent of temperature
below 100°K.
     The values of μ_{eff} for these complexes decreased markedly
with decreasing temperature in the range studied. This behavior
is in contrast to other uranium organometallic compounds. The
reasons for this unusual behavior has not as yet been elucidated.
A possible explanation involves spin-spin interactions of the
metals or electron coupling through the ligand.
     Tsutsui and Ely have extended the general method for pre-
paring σ-bonded actinide complexes to the synthesis of compounds
containing lanthanide-carbon σ-bonds (37,38,39). Using the reaction shown in eq.3, they synthesized a number of alkyl, aryl,
and alkynyl derivatives. The metals chosen vary from Gd to Yb.
Like their uranium analogs, they are oxygen and moisture sensitive, but they are surprisingly thermally stable.

$$2 \; NaCp + LnCl_3 \xrightarrow{\text{THF}} Cp_2Ln \; Cl$$

$$Cp = h^5 - C_5H_5 \quad Ln = Sm - Lu$$

$$Cp_2 \; LnCl + RLi \xrightarrow[\text{RMgX}]{\text{THF}} Cp_2 \; Ln \; R \quad Ln = Sm, \; Gd, \; Er, \; Yb$$

$$R = CH_3, \; -C \equiv C \; -\bigcirc, -\bigcirc, allyl$$

By a similar procedure, Ely and Tsutsui synthesized and characterized a monocyclopentadienyl bisphenylacetylide complex (40).
Visible spectra of all the Cp_2Ln-R complexes (except the Yb ones,
which were not studied) show a charge-transfer band which is

$$(\eta^5-C_5H_5)HoCl_2 \cdot 3THF + 2LiC\equiv CC_6H_5 \xrightarrow{THF} (\eta^5-C_5H_5)Ho(C\equiv CC_6H_5)_2$$

(4)

+ 2LiCl

absent in the starting Cp_2LnCl complex. This band shifts to
lower energy as the reducing strength of the R moiety is increased, an effect consistent with its formulation as ligand to metal
charge transfer (41). Also, in the complexes $Cp_2Ho(C=CPh)$ and
$CpHo(C=CPh)_2$, the charge-transfer band is shifted to lower energy
in the spectrum of the complex with two R moieties, indicating
that the charge transfer involves ligand to metal interactions.

A number of the spectra displayed hypersensitive (42,43)
transitions for some of the bands observed (39). Although hypersensitivity has been related to either increased interaction, the
polarizability of ligand or symmetry changes around the metal ion,
the symmetry might remain essentially the same on going from the
Cp_2LnCl complex to the Cp_2Ln-R complex. Therefore, the appearance of hypersensitivity in the spectra of some of the Cp_2Ln-R
complexes may reflect enhanced interaction between metal (Ln) and
ligand (R), or increased polarizability of ligand (R).

Magnetic susceptibility studies of these complexes were performed, and the values of μ_{eff} were found to decrease on lowering
temperature, unlike the magnetically more well behaved chloride
analogues Cp_2LnCl.

The temperature dependence of μ_{eff} in the Cp_2Ln-R complexes
appears to arise from the difference of site symmetry and the
strength of the crystal field interactions.

An $\eta^1(\sigma)$ carbanion such as CH_3- has its electron density concentrated on one carbon atom where it could be more readily
available for some type of localized interaction with the metal
and thus may lead to the unusual optical and magnetic effects.

Structural studies by Baker, Brown, and Raymond (42) have
shown the dimeric nature of lanthanide dicyclopentadienyl halides.
They reported that the molecular structure of $[Yb(C_5H_4CH_3)_2Cl]_2$
consists of two ytterbium atoms, each with two η^5-bound methyl-
cyclopentadienyl rings, which are nearly symmetrically bridged by
the two chlorine atoms. The crystal structure of $Yb(C_5H_5)_2Me$ was
reported by Halton et al. (43). The complex actually has a
dimeric structure, $Cp_2YbMe_2YbCp_2$, remarkably similar to Me_2AlMe_2
$AlMe_2$. The overall molecular geometry is identical with that of
the chloride analogue $Yb(C_5H_4CH_3)_2Cl$.

In an effort to prepare a σ-bonded allyl derivative analogous
to $Cp_3UC_3H_5$, Tsutsui and Ely (38) prepared $Cp_2LnC_3H_5$(Ln=Sm,Er,Ho)
by reacting Cp_2LnCl with allylmagnesium bromide in THF-ether
solution at $-78°C$. Characterization of these new allyl deri-

vatives revealed the formation of an η -allyl-lathanide bond in preference to the η^1-allyl bond observed in the analogous actinide derivatives.

The size of the coordination site available is an important factor which governs the molecular geometry. One example is that sigma-bonded alkyl derivates of the type (Cp$_2$LnR) have been synthesized for only the late lanthanides eléments. Those of early lanthanides series; La, Ce, Pr and Nd have conspicously been absent due to their low thermal stability. The difference between the late and the early lanthanides may be a feature of the lanthanide contraction, and coordination saturation may be the key factor in controlling the stability and/or reaction pattern of the organolanthanindes. John and Tsutsui recently have prepared the stable σ -bonded organometallic compounds of the early lanthanides, using trimethylene bridged biscyclopentadienyl ligand (eq.5) (44) which is much more sterically bulky than the Cp ligand.

$$(CH_2)_3 \quad LnCl + \begin{array}{c} RLi \\ RNa \end{array} \xrightarrow{\text{THF}} (CH_2)_3 \quad LnR$$

Ln = La, Ce

$$R = -\bigcirc \quad , \quad -C\equiv C-\bigcirc$$

Visible spectra of the Ln[Cp(CH$_2$)$_3$Cp]C$\equiv$CPh complexes show a charge transfer band which is absent in the starting LnCp(CH$_2$)$_3$CpCl, again indicating that the charge transfer can be attributable to ligand to metal interactions or the polarizability of C$\equiv$CPh group.

During syntheses and characterization of organolanthanum compounds, the data of magnetic properties, optical spectra and x-ray analyses have been ineffective to claim the existence of f-orbital participation in bonding. The ionic model version satisfactorily explains these data. Accordintly, it is believed for the lanthanide compound, that the 4f orbitals did not extend far enough spatially to enter into covalent bonding or to be split by ligand fields to any great extent. In the actinides, the 5f orbitals are much less shielded than the 4f orbitals and the binding energies are lower than the lanthanides. These factors have been attributed to an increase in covalent bonding for the actinides. Indeed, Raymond and co-workers have presented extensive correlations of the crystallographic data on the organolanthanide and

organoactinide complexes on the basis of an ionic model (7,8).
They concluded that there might be some appreciable f-orbital contribution to the bonding in the early actinide (IV) complexes, but
there is essentially none in the actinide(III) or lanthanide (III)
complexes.

f-Orbital Participation

For the lanthanide complexes, even if the amount of covalent
interaction is very small, we might have a chance to get the evidence of f-orbital participation in bonding (Figure 4).
Recently, Tsutsui and co-workers have shown some interesting results from ESCA studies on a series of Ln(OH)$_3$ (Ln-La, Ce, Pr, Nd,
Sm, Eu, Gd), H[LnPc$_2$], Pc=phthalocyanine and AnPc$_2$ (45,46). From
an investigation of satellite structures of Ln3d5/2 and An4d5/2
peak (Figure 5), a puzzling question arose as to the shake-up
satellite of ligand f-orbitals charge-transfer type. The ligand
4-f shake-up satellite was not observed in Pr(III) (f^2) and Nd
(II)(f^3) complexes which have sufficient vacant f-orbitals to
receive electrons from ligands, whereas La(III)(f^0) complex has
a propensity to show the satellites. (Figure 5) The analysis of
the date shows that f-orbitals (both half-occupied and vacant)
play an important role in the core-ionization process to give the
sharp variation in intensity to the satellites.

The above result may not relate directly to the f-orbital
participation in reactions of f-elements, but is indicative of
important role of f-orbitals (or f-electrons) for bonding significance.

ESCA studies are also effective for the elucidation of structure. The N 1s spectrum of Pc$_2$NdH (Figure 6) shows a sharp single
peak (Figure 7), while that of a phthalocyanine free base has two
types of peak based on aza nitrogen atoms and pyrrole nitrogen atoms.
The date implies that eight central nitrogen atoms are
chemically equivalent to each other, and thereby the acidic hydrogen does not bind strongly to any of nitrogen atoms in the complex
(47).

The nitrogen 1s spectra have been also investigated for a
series of octaethylporphyrin and tetraphenylporphine complexes
of lanthanides Ln(DEP) (OH) and Ln (TPP)(acac) [Ln=Sm,Gd,Er and
Yb; acac = acetylacetove] (48). The profile of N 1s spectrum for
each lanthanide porphyrin showed that the four nitrogen atoms were
equivalent in the complex. No significant change was detected
between the N 1s binding energies of the lanthanide porphyrins. A
good correlation between N 1s line width (FWHM) and a number of
unpaired electrons in the complex was found. This result implies
the presence of unpaired valance electrons on the nitrogen atoms,
which are induced through an interaction between nitrogen valence
orbitals and half-occupied 4f orbitals of lanthanides. Although
the ESCA results clearly demonstrate the evidence for covalency
including f-orbital participation, at the present stage, it is
difficult to estimate their extents.

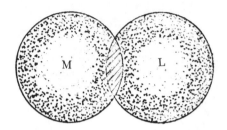

Figure 4.   s, p, d, *[f?]*

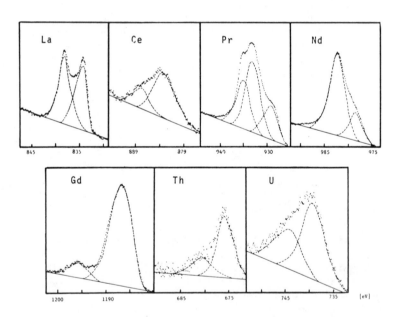

*Figure 5.   Photoelectron spectra of the $Ln3d_{5/2}$ and $An4d_{5/2}$ levels of $H[LnPc_2]$-
(Ln = La, Ce, Pr, and Nd) and $AnPc_2$ (An = Th and U).   Deconvolution of the
satellite structure is given by the dotted line.*

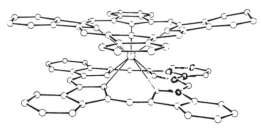

Journal of the American Chemical Society

Figure 6.   Structure of [Pc₂Nd(III)]⁻H⁺ (47)

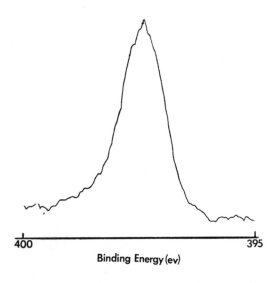

Binding Energy (ev)

Figure 7.   N 1s signals of Pc₂NdH

Future Perspective

Up to the present time, studies on organo-lanthanide complexes have fallen almost exclusively under the category of tri-
valent-lanthanide chemistry. When the oxidation state of organo-
lanthanides is reduced to 0, 1, or 2, the size of coordination
sphere and f-orbital participation in bonding would be altered.
Due to these effects, low-valent organo-lanthanides will show an
increased variety in their reaction facets and catalytic activities.

In spite of this potential for unusual chemistry, investigations of the reductive lanthanide chemistry has just been
started by one group (49). Co-condensation of 1,3-butadiene with
lanthanides (Er, Nd, Sm or La) gave a variety of new organolanthanides; $Er(1,3-butadiene)_3$, $Er(2,3-dimethyl-1,3butadiene)_2$, Nd (1,
3butadiene)$_3$, $Sm(1,3-butadiene)_3$. The optical spectra of these
complexes do not contain the usual sharp absorption bands characteristic of trivalent lanthandies. The room temperature magnetic susceptibilities of these compounds are somewhat different
from susceptibilities previously measured for trivalent lanthanide
species. The most striking differences were observed for $Sm(C_4H_6)_3$
and $La[(CH_3)_2C_4H_4]_2$ where the latter compound was the first reported paramagnetic organo-lathanum complex.

A variety of the metal-metal bonded complexes or clusters
also provide a foothold for the studies of f-orbital participation. Examples of such organo-lanthanide complexes include
cyclopentadienyl lanthanides with lanthanide-to-transition metal
bonding $\{(\eta^5-C_5H_5)_2LnW (\eta^5-C_5H_5)(CO)_3$, $(\eta^5-C_5H_5)_2LnMo(\eta^5-C_5H_5)$
$(CO)_3$, $(\eta^5-C_5H_5)_2LnFe(\eta^5-C_5H_5)(CO)_2$, $(\eta^5-C_5H_5)_2 LnCo(CO)_4$, Ln =
Dy, No, Er, Yb} (39); biscyclopentadienyl erbium-triphenylgermane,
-triphenylstannane and biscyclopentadienyl yetterbium-triphenyl-
stannane (50, 51).

While substantial progress has been made in elucidating the
nature of the f-electron participation in bonding, a number of
problems remain. First, the question of whether 4f valence
orbitals participate in bonding in the ground state has not yet
been answered. More definitive data are needed to resolve the
question. A thorough examination of the spectroscopic data of a
series of lanthanide complexes, together with a well qualified
M.O. approach, might provide the necessary insight by which the
question of f-orbital participation in bonding might ultimately
be resolved.

We are grateful for support by the Robert A. Welch Foundation
(Grant No. A420) for the preparation of this article.

Literature Cited

1. On leave from the University of Tokyo.
2. V.M. Plets, Compt. Rend. Acad. Sci., 1938, 20, 27.
3. T.J. Marks, Prog. Inorg, Chem., 1978, 24, 51.
4. T.J. Marks, Prog. Inorg. Chem., 1979, 25, 224.
5. K.N. Raymond in "Organometallics of the f-elements," T.J. Marks and R.D. Fischer; Ed., Reidel Publishing Co., Dordrecht, Holland 1979, pp. 249–280.
6. E.C. Baker, G.W. Halstead and K.N. Raymond, Struct. Bonding (Berlin), 25, 23 (1976).
7. M. Tsutsui, N. Ely and R. Dubois, Acc, Chem. Res., 1976, 9, 217.
8. H. Gysling and M. Tsutsui, Adv. Organomet, Chem., 1970, 9, 361.
9. J.M. Birmingham and G. Wilkinson, J. Am. Chem. Soc., 1956, 78, 42.
10. L. Reynold and G. Wilkinson, J. Inorg. Nucl. Chem., 1956 2, 246.
11. C. Wong, T. Yen and T. Lee, Acta Cryst., 1965, 18, 340.
12. E.O. Fischer and H. Fischer, J. Organomet. Chem., 1965, 3, 181.
13. S. Manastyrskyj and M. Dubeck, Inorg. Chem., 1964, 3, 1697.
14. M. Tsutsui, T. Takino, and D. Lorenz, Z. Naturforsch, 1965, 21B, 1.
15. F. Baumgartner, E.O. Fischer, B. Kanellakopulos and P. Laubereau, Angew, Chem., 1965, 77, 866.
16. F. Baumgartner, E.O. Fischer, B. Kanellakopulos and P. Laubereau, Angew, Chem. Int. Ed. in English, 1966, 5, 134.
17. A. streitwieser, Jr., and U. Muller-Westerhoff, J. Am. Chem. Soc., 1968, 90, 7364.
18. R.G. Hayes and J.L. Thomas, J. Am. Chem. Soc., 1969, 91, 6876.
19. K.O. Hodgson, F. Mares, D.F. Starks and A. Streitwieser, Jr., J. Am. Chem. Soc., 1973, 95, 8650.
20. C.W. Dekock, S.R. Ely, T.E. Hopkins and M.A. Brantt, Inorg. Chem., 1978, 17, 625.
21. A. Creco, S. Cesca, and G. Bertolini, J. Organomet. Chem., 1976, 113, 321.
22. M. Tsutsui and H.J. Gysling, J. Am. Chem. Soc., 1969, 91, 3175.
23. A. Gebala and M. Tsutsui, J. Am. Chem. Soc., 1975, 14, 78.
24. J.L. Atwood, J.H. Burns, and P.G. Laubereau, J. Am. Chem. Soc., 1973, 95, 1830.
25. S.A. Cotton, F.A. Hart, M.B. Hursthouse and A.J. Welch, J. Chem. Soc., Chem. Commun., 1972, 1225.
26. M. Tsutsui, N. Ely and A. Gebala, Inorg. Chem., 1975, 14, 78.
27. T. Marks, A.M. Seyam and J.R. Kolb, J. Am. Chem. Soc., 1973, 95, 5529
28. G. Lugli, W. Marconi, A. Mazzei, N. Paladino and U. Pedretti, Inorg. Chim. Acta, 1969, 3, 253.

29. J.L. Atwood, C.F. Hains, Jr., M. Tsutsui and A.E. Gebala, J.C.S. Chem. Commun., 1973, 452.
30. J.L. Atwood, M. Tsutsui, N. Ely and A.E. Gebala, J. Coord. Chem., 1976, 5, 209.
31. A. Streitwieser, Jr., D. Dempf, G.N. LaMar, D.G. Karraker and N. Edelstein, J. Am. Chem. Soc., 1971, 93, 7343.
32. G.W. Halstead, E.C. Baker and K.N. Raymond, J. Am. Chem.Soc., 1975, 97, 3049.
33. J.H. Burns, J. Organomet. Chem., 1976, 69, 225.
34. T.J. Marks and W.A. Wachter, J. Am. Chem. Soc., 1976, 98,703.
35. M. Tsutsui and N. Ely, J. Am. Chem. Soc., 1974, 96, 3650.
36. E.C. Baker, K.N. Raymond, T.J. Marks and W.A. Wachter, J.Am. Chem. Soc., 1974, 96, 7586.
37. M. Tsutsui and N. Ely, J. Am. Chem. Soc., 1974, 96, 4042.
38. M. Tsutsui and N. Ely, J. Am. Chem. Soc., 1975, 97, 3551.
39. N. Ely and M. Tsutsui, J. Am. Chem. Soc., 1975, 97, 1280.
40. N. Ely and M. Tsutsui, Inorg. Chem., 1975, 14, 2680.
41. J.C. Burnes, J. Chem. Soc., 1964, 3880.
42. E.C. Baker, L.D. Brown and K.N. Raymond, Inorg. Chem., 1975 14, 1376.
43. J. Halton, M.F. Lappert, D.G.H. Ballard, R. Pearce, J.L. Atwood and W.E. Hunter, J.C.S. Dalton, 1979, 54.
44. J. John and M. Tsutsui, to be submitted to J. Am.Chem. Soc.
45. K. Tatsumi, M. Tsutsui, G.M. Beall, D.F. Mullica and W.O. Milligan, J. Elec. Spec., 1979, 16, 113.
46. K. Tatsumi, K. Kasuga and M. Tsutsui, J. Am. Chem. Soc., 1979, 101, 484.
47. K. Kasuga, M. Tsutsui, R.C. Petterson, K. Tatsumi, N. Van Opdenbosch, G. Pepe and E.F. Meyer, Jr., to be submitted to J. Am. Chem. Soc.
48. K. Tatsumi and M. Tsutsui, J. Am. Chem. Soc., In Press.
49. W.J. Evans, S.C. Engerer and A.C. Neville, J. Am. Chem. Soc., 1978, 100, 331.
50. A.E. Crease and P. Legzdins, J.C.S., Chem. Commun., 1972,268.
51. A.E. Crease and P. Legzdins, J. Chem. Soc., Dalton Trans., 1973, 1501.

RECEIVED December 26, 1979.

Synthesis and Spectroscopy of Novel Mixed-Ligand Organolanthanide Complexes

R. DIETER FISCHER and GUDRUN BIELANG

Institut für Anorganische und Angewandte Chemie der Universität Hamburg, D-2000 Hamburg 13, Germany

In the past few years, considerable development has taken place in the field of f-element organometallics (1, 2, 3), highly representative examples cited being usually complexes involving one singular ligand ("homoleptic" organometallics), e.g. $(C_5H_5)_n^f M$ with n = 2-4, $[(C_8H_8)_2^f M]^q$ (q = 0 or -1), $[Li(tmed)]_3[Ln(CH_3)_6]$ (4) and $[(CH_3)_2P(CH_2)_2]_3^f M$ (5). Although extensive studies of such compounds undoubtedly have their merits apart from purely aesthetic aspects, it is almost exclusively the much wider field of mixed-ligand systems that provides valuable information wherever chemistry with f-organometallics is concerned. Thus it is well-documented that various organouranium compounds can catalyze the stereospecific formation of cis-1,4-polybutadienes from 1,3-butadiene (6) in which homogeneous process complexes with organic ligands are highly superior to classical oxides or halides. Although well-defined organo-uranium complexes such as $(C_5H_5)_3 UX$ and $(C_3H_5)_4 U$/Lewis acid, respectively, have been reported to be most efficient, the (unknown) catalytically active species will undoubtedly be a mixed-ligand system involving the substrate as well as a co-catalyst and/or the solvent.

We can visualise the capability of suitable lanthanide (Ln) compounds (1, 6), e.g. as homogeneous catalysts with respect to olefins, by invoking similar intermediates. Although the series of reportedly catalytically active Ln-complexes spans from the pure trihalide via tris(ß-diketonato)complexes to the organometallic tris(cyclopentadienyl) and tetra(allyl)complexes (8), respectively, no really optimal combination of ligands on a Ln-element has been found so far. Promising aspects are, however, based on some evidence for "reaction steering" in that either cis- or trans-polybutadienes can be obtained from 1,3-dienes, and either polymers or metathesis products from monoolefins, respectively (Table I).

0–8412–0568–X/80/47–131–059$05.25/0

Table I. Catalytic Activity of Lanthanide Compounds towards
Unsaturated Hydrocarbons (according to References
1, 6, 7, 8)

Starting "Catalyst"	Substrate	Product(s)
$SmCl_3/AlEtCl_2$	monoolefin	olefin-metathesis
$LnCl_3/AlEt_3$	"	saturated polymers
Ce-octanoate/AlR_3	"	cis-1,4-polymers
Ln(diket)$_3$/AlR_3		"
$(C_5H_5)_3Sm$	1,3-diene	trans-1,4-polymer
$LnCl_3/Sn(allyl)_4/4LiR$	"	"
$Li[Ln(allyl)_4]\cdot$dioxane	"	"
$(C_5H_5)_3Sm$	$HCCC_6H_5$	triphenylbenzene

General Reaction Pattern

Frequently, the preparation of a distinct mixed-ligand complex, L^CMX_m (L^C = ligand bonded via M-C bonds), is quite a difficult task, in particular if the starting material involves a pure
metal halide. In organolanthanide chemistry, however, an almost
unique and very effective route to arrive at a large variety of
Cp_nLnX_m systems (Cp = η^5-C_5H_5) is offered by eq. (1)

$$Cp_n^fM + mH^{ac}-X \quad \rightarrow \quad Cp_{n-m}^fMX_m + mC_5H_6 \qquad (1)$$

The stepwise substitution of η^5-coordinated Cp-ligands initiated
by the attack of a proton acidic reagent H^{ac}-X is usually possible
under mild conditions in various inert organic solvents, and the
resulting cyclopentadiene is easily removable with the solvent.
Reaction (1) was first adopted by Fischer and Fischer in 1965
(10), and has been extended mainly by Kanellakopulos et al. (11,
12).

Unlike the lanthanide complexes Cp_3Ln, and many degradation
products, $Cp_{3-n}LnX_n$, some actinide, and the majority of d-block
metal, cyclopentadienides are not susceptible to reaction (1). A
reasonably good test for the reactivity of metal-bonded Cp with
H-acids consists in the addition of water or methanol. While all
known lanthanide complexes will immediately be decomposed, many
organo-uranium compounds of the type Cp_3UX either simply add H_2O
and/or undergo substitution of X (13):

$$Cp_3UX + H_2O \xrightleftharpoons{H_2O} [Cp_3UX(H_2O)] \qquad (2)$$

$$[Cp_3UX(H_2O)] \xrightleftharpoons{H_2O} [Cp_3U(H_2O)_x]^+ + X^- \qquad (3)$$

While aqueous solutions of the cation $[Cp_3U(H_2O)_x]^+$ remain stable down to a pH of -1, the presence of fluoride ions causes the immediate rupture of all Cp-U bonds (9). Likewise, ß-diketones have been reported to replace one Cp-ligand (and the halide) from Cp_3UCl (18).

Kanellakopulos et al. have demonstrated the wide applicability of the protic acid NH_4^+ for the elegant preparation of many mixed-ligand metal cyclopentadienides of f-block and main group elements (14):

$$Cp_nM + mNH_4X \xrightarrow[\text{reflux}]{\text{THF}} Cp_{n-m}MX_m + mCpH + mNH_3 \qquad (4)$$

If the acid $H^{ac}-X$ is also furnished with a lone electron pair, the following two-step mechanism appears feasible in case of strongly Lewis-acidic substrates such as $Cp_3{}^fM$:

$$Cp_3{}^fM + :X-H \rightleftharpoons Cp_3{}^fM{\leftarrow}X-H^{ac} \qquad (5a)$$

$$Cp_3{}^fM{\leftarrow}X-H^{ac} \rightarrow 1/2[Cp_2{}^fMX]_2 + CpH \qquad (5b)$$

Usually, the adduct $Cp_3{}^fM{\leftarrow}XH^{ac}$ offers favourable steric conditions for a subsequent intramolecular H-transfer to one of the Cp-ligands. Moreover, the proton acidity of HX will often increase by the coordination. Thus, pure Cp_4U which is surprisingly stable against water has no possibility to form an adduct with H_2O, while the shape of the Lewis base HCN would not completely preclude this possibility. Anhydrous HCN replaces in fact one CpH-molecule (14).

Following the pathway of eqns. (5a) and (5b), even a "protic acid" as weak as NH_3 has successfully been applied to replace CpH (10)

$$Cp_3Yb \xrightarrow{\text{liqu. } NH_3} Cp_3Yb \cdot NH_3 \xrightarrow[-CpH]{250\ ^0C} 1/2[Cp_2YbNH_2]_2 \qquad (6)$$

(green)   (light green)   (bright yellow)

Reaction (6) thus exemplifies, in a sense, the reversal of the well-documented substitution of secondary amines by CpH (15),

$$U(NR_2)_4 + 2CpH \rightarrow Cp_2U(NR_2)_2 + 2HNR_2 \qquad (7)$$

There are various, albeit mainly unpublished, observations suggesting that many organic amines, phosphines, and other Lewis bases carrying at least one H-atom will initiate reaction (6), however, without yielding isolable adducts (16, 17).

Potential Proton Acids of Special Interest

For a better understanding, and further exploration of the

general applicability, of reaction (1), a larger, and more systematic, variation of the protic acid H-X than previously has appeared worthwhile.

Table II. Survey of Proton Acids H-X Subjected to Eq. (1)

Reagent H^{ac}-X	approximate pK_a
carboxylic acids	ca. 5
ß-diketones and	ca. 9
ß-ketoimines	
pyrazole	14
cyclopentadiene	15 - 16
pyrrole and its	16,5
benzo-derivatives	
indene	20 - 21
phenylacetylene and	20 - 21
other alkynes HCCR	
fluorene	25
(ammonia)	33
(toluene)	37 - 39

Table II offers a survey of the classes of compounds, and specific singular compounds, respectively, that have so far been subjected to reactions as specified by eq. (1). Two main objectives
governing this selection have been (a) the question on the relevance of the proton acidity of the reagent H-X, and (b) our increasing interest in the properties of mixed-ligand systems
Cp_nLnX_m (n+m = 3) where X is another polydentate ligand. While
until recently no example of such a complex with a genuine chelate ligand was known, there is, moreover, particular interest
in reaction (1) as a possible route to complexes of the type
$[Cp_2LnX]_n$ of the lighter Ln-elements (Ce-Sm). This sub-class of
organolanthanides has been found to be very unstable (19) the
only well-documented example being the product of the reaction
of Cp_3Nd with HCN, $[Cp_2NdCN]_n$ (12). This complex is probably oligomeric and stabilized by bridging CN-ligands. In this context,
the actinide complex $[Cp_2UCl]_n$ obtained from Cp_3U and anhydrous
HCl (20) may be considered as an outstanding example of a non-
oligomeric $Cp_2{}^fMX$-system where ^fM exhibits an ionic radius comparable to those of the early Ln^{3+}-ions. The pK_a-values of the
acidic reagents in table II vary over more than five units. It
should, however, be kept in mind that the determination of pK-
values in non-aqueous media, and their transferability from one
medium to another, present rather delicate problems (21). Other
important factors to account for in view of the reactivity of an
individual H^{ac}-X system are: (a) its initial Lewis basicity, (b)
the actual increase in proton acidity on coordination, (c) the
total "hapticity" finally displayed by X, and (d) various essential steric conditions.

Table II presents examples of reagents that involve O and N
as well as C as the proton-donating element. Only very few compounds from the table such as formic acid (19), ammonia (10),
and phenylethyne (22) had been reacted with $Cp_3{}^fM$ prior to our
study; it has been claimed that Cp_3Sm only catalyzes the trimeri-
zation of phenylethyne.

Characterization of Reaction Products

For the majority of reactions involving a "heavier" Ln-element Yb was preferred, owing to the relatively good solubility of
Cp_3Yb even in toluene and pentane, and to the very characteristic
colour change (see eq. 6) whenever a reaction of type (1) is taking place.

More detailed information on the nature of an organo-ytterbium system involving Cp-ligands is offered by the optical absorption spectra. The specific differences of f-f-spectra of
Yb^{3+}-complexes of the types Cp_3Yb, Cp_2YbX and $CpYbX_2$ have been
discussed earlier (16, 23). Yb^{3+}-cyclopentadienyl complexes offer
the additional advantage that the lowest-lying charge transfer
corresponding to the intramolecular redox process (23):
[ligands] $4f^{13} \longrightarrow$ [ligands]$^{-1}$ $4f^{14}$ may lie as low as ca.
16.600 cm^{-1} if three η^5-Cp-ligands - or eventually a corresponding
set of sufficiently reducing ligands - are involved.

It has also turned out mainly during our studies of mixed-
ligand organo-ytterbium systems that many Yb^{3+}-complexes present
excellent conditions for the observation of their ^{1}H-NMR spectra
in spite of the strong paramagnetism of this f^{13}-system. Previous
papers dealing particularly with Pr^{3+}- and U^{4+}-systems, respectively, have already pointed out the almost indispensable role
of the NMR spectra of ligands bonded to paramagnetic ^fM-central
ions (24, 25).

Table III exemplifies by some room temperature Cp-proton
shift data of Yb^{3+}-complexes the high diagnostic value even of
one singular NMR shift value per system. More sophisticated
studies are concerned with the complete temperature dependence
of particular resonances over a range of usually 100 degrees
(-70 to + 30 ^{0}C).

Figure 1 presents as an example the ^{1}H-NMR-spectra of two
different Yb^{3+}-organometallics, $[Cp_2YbO_2C(n-C_4H_9)]_2$ and
$[Cp_2YbC_2(n-C_4H_9)]_3$, respectively, both of which contain two Cp-
ligands and one tert-butyl group in the secondary ligand. Although the spectral patterns regarding the numbers, and relative
intensities, of the resonances are compatible, the quite dramatically different isotropic shifts strongly suggest different
electronic and/or steric conditions in the two species.

Figure 2 exemplifies the variation of the "isotropic" proton
shifts of the Yb^{3+}-alkynyl complex with the temperature. It is
noteworthy that the shift of the $\alpha-CH_2$ protons of the tert-butyl

Table III. Cp-ring ^{1}H-NMR shifts of various systems $[Cp_2YbX]_n$. (a): reference signal: internal TMS; (b): in parentheses: weaker satellite resonances; (c): weak "shoulder" present; (d): probably formation of 1:1 adduct; (e): $CH_3COCHC(CH_3)N\emptyset$.

Ligand X	Solvent L	Cp-^{1}H-NMR Shift(s)			Remarks
η^5-C_5H_5	toluene-d_8	-55,2			-
"	"	-54,4			with $P\emptyset_3$[d]
Cl	benzene-d_6	-61,0			probably
"	toluene-d_6	-61,6			all dimeric
"	CD_2Cl_2	-65,6			
Cl	CD_3CN	(-31,9)[b]	-34,0		probably
"	THF-d_8		-35,0		monomeric
"	$(CD_3)_2CO$	(-29,3)	-36,5		adducts
"	pyridine-d_5	(-33,0)	-38,0		$Cp_2YbCl \cdot nL$
NH_2	THF-d_8	-10,0			
"	benzene-d_6	-12,0			all dimeric
n-$C_4H_9CO_2$	toluene-d_8	-19,0			
"ketim"[e]	"	-33,9			
n-$C_4H_9C_2$	toluene-d_6	-57,6	(-64,2)		trimeric (C_6H_6)
"	"	-31,4	(-48,9..	-60,0)	with pyridine-d_5
$C_6H_{11}C_2$	"	-56,6	(-59,6)		
FcC_2	"	-56,3[c]			

a: reference signal: internal TMS
b: in parentheses: weaker satellite resonances
c: weak "shoulder" present
d: probably formation of 1:1 adduct
e: $CH_3COCHC(CH_3)N\emptyset$

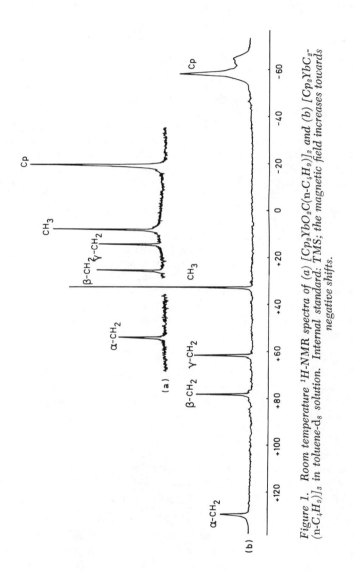

Figure 1. Room temperature ^{1}H-NMR spectra of (a) $[Cp_2YbO_2C(n-C_4H_9)]_2$ and (b) $[Cp_2YbC_2-(n-C_4H_9)]_3$ in toluene-d_8 solution. Internal standard: TMS; the magnetic field increases towards negative shifts.

group is almost comparable in magnitude with the α-CH$_2$ shift of
the $5f^2$-system Cp$_3$U(n-C$_4$H$_9$) (26).

None of the "lighter" Ln-elements offers optimal conditions
for the facile visual indication of a successful reaction of
Cp$_3$Ln with H^{ac}-X. Moreover, the low solubility of both the
starting material Cp$_3$Ln (Ln = Pr, Nd, Sm) and of most reaction
products affords dealing with suspensions rather than with veri-
table solutions. As, moreover, the isotropic ^{1}H-NMR shifts of
Cp$_3$Nd/Lewis base adducts are exceptionally weak (25), the best
spectroscopic method to identify, and characterize, most Nd^{3+}-
organometallics is absorption spectroscopy (either of solutions
or of Teflon pellets) in the NIR/VIS range. The region of the
so-called hypersensitive transitions between crystal field states
of the manifolds ^{4}I$_{9/2}$ (ground manifold) and ^{4}G$_{5/2}$ (ca. 580 –
610 nm) has proved to be of particularly high diagnostic value.
Even at room temperature, and in case of polycrystalline solid
samples (i.e. pellets), various ensembles of fairly sharp f-f-
transitions appear which allow tracing the constituents of mixtures of compound even in cases of rapid interconversion.

Figure 3 gives an impression of the appearance of some typical spectra of Nd^{3+}-complexes in the "hypersensitive region".

The optical spectra of organometallic Pr^{3+}-systems usually
suffer from the lack of suitably hypersensitive transitions as
well as from the widely expanding side wing of a charge-transfer
band. On the other hand, the ^{1}H-NMR spectra of the few sufficiently soluble Pr^{3+}-complexes so far obtained have turned out
too complex to arrive at reliable assessments.

Results and Discussion

Significance of the pK$_a$-Value of H^{ac}-X. The total of results obtained in this study suggests that all reagents H^{ac}-X
with pK$_a$-values lower than that of cyclopentadiene are capable
of readily replacing more than one, and most frequently all, Cp-
ligands of the substrate Cp$_3$Ln. Thus it is only by dropwise
addition of the stoichiometric quantity of the reagent that an
immediate substitution of all Cp-ligands can be avoided. Exem-
plaric products of such careful titration-like procedures have
been the before-mentioned U(III)-complex [Cp$_2$UCl]$_n$ (20), the
monovaleriato complex [Cp$_2$YbO$_2$C(n-C$_4$H$_9$)]$_2$(27), various systems
involving derivatives of ß-diketones, [Cp$_2$Ln(chel)]$_n$ (see below),
and even the (probably oligomeric) reaction product (1:1) of
Cp$_3$Yb and pyrazole (pK$_a$ $\simeq$ 14,2).

Reaction of Cp$_3$Ln with an excess of "strongly acidic"
(pK$_a$ < 15) reagents usually leads to the non-organometallic products LnX$_3$. The light brown tris(pyrazolyl) complex of Yb which
is again very insoluble even in THF is of interest in view of the
nature of the pyrazolyl anion as a versatile polydentate ligand.
By application of an excess of the ß-ketoimine CH$_3$COCHC(CH$_3$)CNHR
= "H-ketim" (see below) on Cp$_3$Ln, the corresponding tris(chelate)

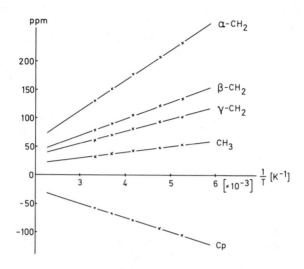

Figure 2. Variation of the ^{1}H-NMR shift values of $[Cp_2YbC_2(n\text{-}C_4H_9)]_3$ with temperature

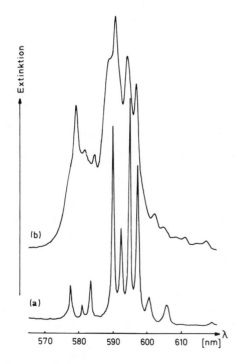

Figure 3. Absorption spectra (toluene solution, room temperature) of (a) Cp_3Nd and (b) "$Cp_2Nd(thd)$" in the region of hypersensitive transitions (thd = 2,2',-6,6'-tetramethylheptane–3,5-dionate)

complexes $Ln(ketim)_3$ are readily accessible. It is worth men-
tioning in this context that the first "classical" lanthanide
complex involving the chelate ligands ketim ($R = C_6H_5$ and
$t-C_4H_9$) have not been described until 1979 (28), e.g.

$$Ln(O-iC_3H_7)_3 + n\ H\text{-ketim} \xrightarrow[\text{reflux}]{C_6H_6} \tag{8}$$

$$(iC_3H_7O)_{3-n}Ln(ketim)_n + n\ i-C_3H_7OH$$

So far, however, it has not been possible to arrive at tris-
(ketimino)complexes by this route.

Somewhat surprisingly, Cp_3Ln-systems may also react with
a wide variety of reagents in cases of considerably larger pK_a-
values than that of cyclopentadiene ($pK_a > 15$). One important
difference is, however, that the "weaker acids" replace   no
more than one Cp-ligand per metal complex. In some instances,
such a reaction is notably improved by reacting the Cp_3Ln-
systems with the solvent-free H-X at elevated temperatures up
to 60 ^{0}C. For example, reaction of Cp_3Yb even with pyrrole
(pyr-H) ($pK_a \simeq 16,5$) only yields the orange product $[Cp_2Yb(pyr)]_n$
the optical spectra of which are devoid of evidence for a η^5-coor-
dination of the pyrrolyl ligand. Somewhat different metal-to-
pyrrolyl bonding conditions may be expected for the deeply green
tris(pyrrolyl) lanthanide complexes which are accessible according to eq. 9 (29). Figure 6 shows the NIR/VIS-spectrum of
$Yb(pyr)_3$.

$$LnCl_3 + 3Na(pyr) \xrightarrow[\text{reflux, 1h, room temp.}]{\text{toluene or THF}} \text{green soln.}$$

$$\text{170 }^0C$$

$$\text{decomp.} \longleftarrow \quad \tag{9}$$

$$Ln(pyr)_3$$

$$\begin{array}{l}\text{colourless} \longleftarrow \quad \text{EtOH} \\ \text{soln.}\end{array}$$

$$\begin{array}{l}Ln = \text{Sm: brownish-green} \\ \text{Ho: dark green} \\ \text{Yb: almost black}\end{array}$$

Reaction with ß-Diketones. Unlike with the other "strong
acids" $Hac-X$ mentioned above, the final products of virtually
stoichiometric reactions (1:1) of Cp_3Ln (Ln = Yb, Ho, Sm, Nd
and Pr) with the ß-diketones

$$\text{H-diket} = \begin{array}{c} R \\ HC \end{array}\begin{array}{c} C-O \\ \quad\ H \\ C-O \end{array} \qquad R = CH_3, \tag{10}$$

$$t-C_4H_9$$

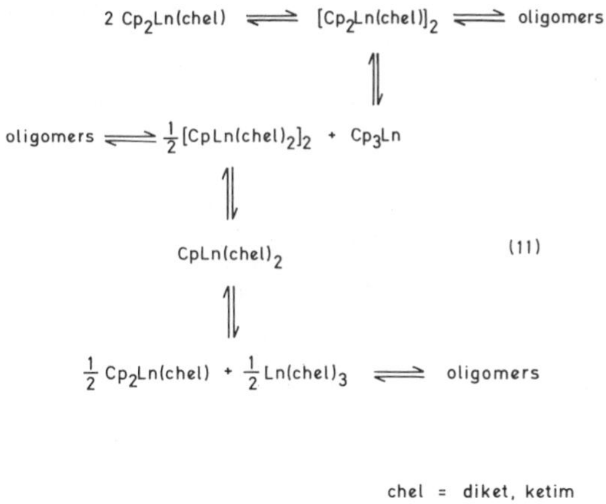

2 Cp2Ln(chel) ⇌ [Cp2Ln(chel)]2 ⇌ oligomers
oligomers ⇌ 1/2 [CpLn(chel)2]2 + Cp3Ln
CpLn(chel)2
(11)
1/2 Cp2Ln(chel) + 1/2 Ln(chel)3 ⇌ oligomers
chel = diket, ketim

Scheme 1

N
O
Nd
Nd
N
O

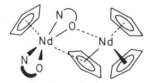

Figure 4. Schematic of $[Cp_2Nd(ketim)]_2$ *in case of one particular phase of unsymmetric coordination*

Extinction
(b)
(a)
580
600
620
[nm]
λ

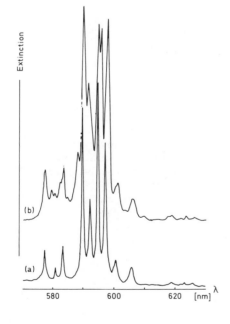

Figure 5. Optical absorption spectra of (a) Cp_3Nd *and (b)* $[Cp_2Nd(ant)]_2$ *(toluene, room temperature) in the region of hypersensitive* f–f-*transitions*

have turned out to be equilibrium mixtures involving at least
some of the different substitution products principally imagi-
nable (30). Such results which are mainly supported by mass-
and NMR-spectroscopic findings (27) are consistent with the
general behaviour of f-metal chelate complexes (31), but differ
from the well-documented apparently higher complex stability of
the corresponding Ti- and even Sc-complexes (32).

The absorption spectrum of the material obtained from Cp_3Nd
and H-diket with R = tert-butyl clearly shows that the starting
compound has been attacked by the acid. The data of Table IV
indicate for diket = 2,2',6,6'-tetramethylheptane-3,5-dionate
(thd) that most fragments apparently involve two or even three
thd-moieties although Cp_3Ln and H-tmd had been reacted in equimolar quantities.

Reaction with ß-Ketoimines. A different situation is met
if the ß-diketone H-diket is replaced by a related ß-ketoimide
(or semi-Schiff base), H-ketim (33):

$$\text{"H-ketim"} = \begin{array}{c} R \\ \overset{|}{C}\text{-}O \\ HC \quad H \\ \overset{|}{C}\text{-}N \\ R \quad R' \end{array} \qquad \begin{array}{l} R = CH_3 \\ R' = C_6H_5, \ t\text{-}C_4H_9 \end{array} \qquad (12)$$

With this modified chelate ligand "ketim" (and Ln heavier than
Nd) it has been possible to arrive at sufficiently kinetically
stable complexes $[Cp_2Ln(ketim)]_n$ which turn out to represent the
first examples of the general type $Cp_2Ln(chel)$. Contrary to the
known ß-diketonato-complexes $Cp_2M(chel)$ (M = Sc, Ti) (32), none
of the formally related Ln-systems occurs as a monomer in solution. The yellow complex $[Cp_2Yb(ketim)]_2$ with R' = C_6H_5 displays
mass-, ^{1}H-NMR- and optical spectra consistent with its formulation
as a bis(η^5-Cp)-complex involving bridging ketim ligands (30).
The corresponding complex with R' = $t\text{-}C_4H_9$ has so far been obtained as an oil which is difficult to purify.

Table V displays the ^{1}H-NMR data of $[Cp_2Yb(ketim)]_2$ at
various temperatures. The rather different isotropic shifts(opposite signs!) of the two methyl groups suggest their location in
fairly different "magnetic environments". It is, however, not
possible to decide if there are N- or O-bridging links. As only
one Cp ring proton resonance appears, some fluxional behaviour of
the bridging ketim ligand is not unlikely.

Contrary to the mass spectra of the ß-diket-complexes,
$[Cp_2Yb(ketim)]_2$ (with R = C_6H_5) gives rise to a spectrum in-
volving the expected fragments: M^+, $M\text{-}Cp^+$, $M\text{-}2Cp^+$, and $M\text{-}ketim^+$.
However, no fragments involving two Yb-atoms are observed.

On reaction of Cp_3Yb with two moles of H-ketim, a yellow
product of the expected composition $[CpYb(ketim)_2]_n$ (R' = C_6H_5)
is obtained. Similarly, the reaction of Cp_3Yb with the potential

Table IV. Mass spectroscopic data of two products
"$Cp_2Ln(thd)$". (a): Ln = Nd, 130 ^{0}C; (b): Ln = Yb, 100 ^{0}C.

Ion	m/e (a)	m/e (b)	rel. intensity(%) (a)	rel. intensity(%) (b)
$Ln(thd)_3^+$	–	723	–	18,3
$\{Ln(thd)_3 - H_2\}^+$	689	–	3,7	–
$\{Ln(thd)_3 - CH_3\}^+$	–	708	–	3,4
$\{Ln(thd)_3 - C_3H_6\}^+$	–	681	–	2,0
$\{Ln(thd)_3 - t \cdot C_4H_9\}^+$	–	666	–	56,5
$Ln(thd)_2^+$	508	540	100,0	100,0
$\{Ln(thd)_2 - C_2H_6\}^+$	478	510	6,4	3,0
$\{Ln(thd)_2 - t \cdot C_4H_9\}^+$	–	482	–	6,7
$Ln(thd)^+$	325	357	11,7	35,2
$\{Ln(thd) - C_2H_7\}^+$	–	326	–	6,0
$\{Ln(thd) - t \cdot C_4H_9\}^+$	–	299	–	2,7
$CpLn(thd)_2^+$	573	–	4,2	–
$Cp_2Ln(thd)^+$	455	–	11,4	–
$CpLn(thd)^+$	390	–	67,7	–
Cp_2Ln^+	272	–	13,5	–
$CpLn^+$	207	–	6,4	–

Table V. ^{1}H-NMR-data of $[Cp_2Yb(ketim)]_2$ (R = C_6H_5) in
toluene-d_8 solution. Internal standard: TMS; in parentheses relative intensities. One signal of intensity
(1) is undetectable.

T/K	C_5H_5 (10)	CH_3 (3)	CH_3 (3)	CH (1)	C_6H_5 (2)	C_6H_5 (2)
300	-33,91	11,88	-11,26	-46,07	36,21	44,43
285	-36,91	13,48	-12,54	-49,16	38,43	47,31
265	-42,21	16,03	-14,63	-54,66	41,80	52,00
245	-48,54	18,68	-17,02	-56,97	45,38	56,19
225	-54,50	22,61	-19,89	-66,30	49,57	61,57
205	-61,90	26,76	-23,18	-72,75	54,01	67,24

"double chelate" "H$_2$-acacen" ($\underline{34}$)

$$
\begin{array}{ccc}
CH_3 & CH_2-CH_2 & CH_3 \\
\diagdown C \doteq N & \diagdown N \doteq C & \\
HC & H \quad H & CH \\
\diagup C \doteq O & O \doteq C & \\
CH_3 & & CH_3
\end{array}
\qquad = H_2\text{-acacen} \qquad\qquad (13)
$$

yields in a clean reaction a species of the composition [CpYb (acacen)]$_n$. Both complexes exhibit optical spectra reminiscent of other CpYbX -systems, whereas the ^{1}H-NMR-spectra turn out too complicated to be reliably interpreted. The complex [CpYb (ketim)$_2$]$_n$ is also obtained by reacting [Cp$_2$Yb(ketim)]$_2$ with one further mole of H-ketim.

Starting from Cp$_2$LnX-systems such as [Cp$_2$YbCl]$_2$, novel organometallics involving three different ligands are obtained, e.g.

$$
\begin{array}{c}
1/2[Cp_2YbCl]_2 + H\text{-ketim} \quad \xrightarrow[\;25\ ^0C\;]{\text{toluene}} \\
\text{(red)}
\end{array}
$$

$$
\begin{array}{c}
[CpYb(ketim)Cl]_x + HCp \\
\text{(red-brown)}
\end{array}
\qquad\qquad (14)
$$

The complex [CpYb(ketim)Cl]$_x$ is non-volatile and sparingly soluble even in THF, suggesting again an oligomeric structure linked via Cl- and/or bridging ketiminato groups. Such complexes involving still one halide ligand are expected to react with alkaliorganic reagents, M-R, thus offering a new route towards CpLn(chel)R-systems involving three different ligands per metal. Although [CpYbCl]$_2$ also reacts readily with ß-diketones, H-diket, all reaction products so far obtained could not be purified enough to confirm the formation of one single compound.

As the central metal Ln is varied from the end of the lan-thanide series towards the first half, various complications arise particularly in the ^{1}H-NMR-, optical and mass spectra indicating the usual reluctance of Ce - Nd to form regular Cp$_2$LnX-systems. Thus the pale blue (1:1) reaction product of H-ketim (R = C$_6$H$_5$) and Cp$_3$Nd suggests by its slight solubility even in pentane, and its much lower thermal stability relative to Cp$_3$Nd, that the Cp$_3$Nd had in fact reacted with the H-ketim. However, the optical absorption spectrum of the product differs only slightly from the spectrum of authentic Cp$_3$Nd, the differences being most pro-nounced in pentane solution and almost non-existent for a solid teflon pellet. These results differ from those obtained on "Cp$_2$Nd(diket)" in that a Cp$_3$Nd/Lewis base adduct seems to occur as a final product in the former case, but apparently a mixture of Cp$_n$Nd(diket)$_m$-systems (n+m = 3) with n < 3 in the latter. In-dependent support for the formation of a Cp$_3$Ln/Lewis base adduct (in spite of the liberation of one equivalent of cyclopentadiene during the initial reaction) is also provided by the ^{1}H–NMR–spectrum of a "Cp$_2$Pr(ketim)"–solution which displays an intense resonance typical of Cp$_3$Pr–systems.

In Table VI the mass spectroscopic behaviour of various representatives of the series "Cp$_2$Ln(ketim)" with Ln = Pr, Nd, Sm and Yb, and R = C$_6$H$_5$, is summarized. It is immediately apparent that the tendency of the primary species to undergo subsequent rearrangements towards Cp$_3$Ln decreases along with the ionic radius of Ln^{3+}.

Our present results suggest that it should be possible to find a suitable chelate ligand (chel) that could stabilize an intermediate of the intramolecular ligand transfer postulated for a dimeric Cp$_2$Ln(chel)–system (Figure 4).

Likewise, it will be a matter of an optimal choice of the secondary ligand to arrive at genuine Cp$_2$LnX–systems even if Ln = Pr or Nd. There is some evidence from the optical spectra that Cp$_3$Nd and anthranilaldehyde (H–ant) give rise to a well-soluble, probably dimeric species, [Cp$_2$Nd(ant)]$_2$ (Figure 5). Unlike H–diket and H–ketim, respectively, the reagent H–ant already belongs to the group of acids that are weaker than C$_5$H$_6$.

<u>Reaction with Weak Proton Acids: 1–Alkynes.</u> While pyrazole (pK$_a$ < 16) is capable of replacing up to three Cp–ligands from Cp$_3$Ln, pyrrole (pK$_a$ > 16) can liberate only one Cp–ligand. Similar observations have been made on the related N–heterocyclic systems indole and carbazole. Although none of the isolated complexes of the type [Cp$_2$LnX]$_n$ is likely to be mononuclear, the optical spectra of the indolyl–system [Cp$_2$Yb(indo)]$_x$ suggest, like for Yb(pyr)$_3$, a hapticity η^n with n between 1 and 5 (Figure 6).

Very surprisingly, even protic "acids" as weak as various 1–alkynes, HCCR, have been found to react with Cp$_3$Ln–complexes, giving rise to substitution products of the type [CpLnCCR]$_n$. So far, six different 1–alkynes with the following substituents R have been adopted (37):

$$R = \text{n–butyl, n–hexyl, cyclohexyl,}$$
$$\text{phenyl and ferrocenyl (= Fc).}$$

With the exception of R = cyclohexyl where reaction (15) occurs already at room temperature the optimal reaction temperature

$$Cp_3Yb + HCCR \rightarrow \frac{1}{x}[Cp_2YbCCR]_x + CpH \qquad (15)$$

is 60 – 80 ^{0}C. Table VII summarizes some characteristic properties of the resulting Yb–alkynyl complexes. All compounds are non–volatile in vacuo and very reluctant to adopt a decently crystallized form. By comparison with published data, the complex with R = C$_6$H$_5$ is identical to the complex Cp$_2$YbCCC$_6$H$_5$ prepared by Tsutsui and Ely (35) from [Cp$_2$YbCl]$_2$ and LiC$_6$H$_5$.

While Tsutsui et al. have not been able to provide any further information on the nature of the homologous products Cp$_2$LnCCC$_6$H$_5$ with Ln = Gd, Ho, Er, Yb the much better solubilities

Table VI. Mass spectroscopic data of various products
"$Cp_2Ln(apo)$" ("apo" = ketim with R = C_6H_5). Source
temperatures between 175 ^{0}C (Yb) and 250 ^{0}C (Sm).

Ln	Pr		Nd		Sm		Yb	
	m/e	%	m/e	%	m/e	%	m/e	%
$Cp_2Ln(apo)^+$	446	1,5	447	14,7	–	–	–	–
$\{Cp_2Ln(apo)-H\}^+$	445	2,4	–	–	455	43,9	478	30,1
$CpLn(apo)^+$	381	6,5	–	–	–	–	–	–
$\{CpLn(apo)-H\}^+$	380	29,5	381	80,4	390	100,0	413	100,0
$\{CpLn(apo)-C_6H_3\}^+$	305	7,6	–	–	–	–	–	–
$Ln(apo)^+$	–	–	–	–	326	69,7	–	–
$\{Ln(apo)-H\}^+$	–	–	–	–	–	–	348	49,0
$LnCp_3^+$	336	62,7	337	14,7	–	–	–	–
$LnCp_2^+$	271	100,0	272	36,3	282	43,2	304	3,7
$LnCp^+$	206	49,1	207	34,3	217	48,5	239	7,4
Ln^+	141	8,2	142	100,0	–	–	–	–

Table VII. Some characteristic properties of the various
products $[Cp_2YbC_2R]_n$ (Fc = ferrocenyl).

R	colour	$\Delta\nu(C\equiv C)$ (cm^{-1})	decomp.temp. (^0C)	solubility (toluene)
$n-C_6H_{13}$	yellow	-57	180	very good
$n-C_4H_9$	orange	-57	150	" "
C_6H_{11}	"	-18	150	good
Fc	"	-54	130	"
C_6H_5	"	-56	180	insoluble

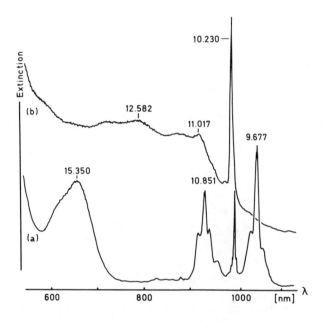

Figure 6.   NIR/VIS-absorption spectra of (a) Cp₃Yb (toluene solution) and (b) of Yb(pyr)₃ (THF-solution). Note the f–f-absorptions above 11.000 cm⁻¹ which suggest ηⁿ-coordination (n > 1) of (pyr).

particularly of the complexes with R = n-C$_4$H$_9$ and n-C$_6$H$_{13}$ have admitted cryoscopic molecular weight studies in benzene according to which these systems are most probably trimeric. Careful thin-layer chromatographic studies in the case R = Fc have proved that no other ferrocene derivatives than "Cp$_2$YbCCFc" occur. By this observation all alternative reaction patterns involving a reaction of the CC-triple bond can be ruled out, leaving a loose adduct of the type (16)

$$\text{Cp}_3\text{Yb} \leftarrow \cdots \begin{matrix} \text{H} \\ | \\ \text{C} \\ \text{\Large III} \\ \text{C} \\ | \\ \text{Y} \\ \text{R} \end{matrix} \qquad\qquad (16)$$

as the most probable intermediate. While reactions of type (15) with acetylenes have not been described before, the observation of a very small paramagnetic shift of solutions of Cp$_3$Nd (Cp = C$_5$H$_5$ and CH$_3$C$_5$H$_4$) in the presence of HCCC$_6$H$_5$ was assigned to the corresponding adduct Cp$_3$Nd·HCCC$_6$H$_5$ (36).

In contrast to these latter observations, we found that Cp$_3$Nd reacts, in spite of its poor solubility, with all 1-alkynes even at room temperature. Rather surprisingly, the resulting products are orange to red in colour (Cp$_3$Nd is pale blue), and their IR-spectra are devoid of a ν(C≡C) vibrational band in the expected region. Nevertheless, the optical absorption spectra due to the f-f-transitions are again identical to the corresponding spectra of authentic Cp$_3$Nd. The red colour of the new products is, however, accounted for by the very extended low-energetic wing of an intense non-f-f-transition. Unfortunately, the low solubility of all products, even in case of R = C$_6$H$_{13}$, has so far hampered all NMR-studies, and attempts to grow crystals for a structural determination.

The ^{1}H-NMR-spectra of [Cp$_2$YbCC(n-C$_6$H$_{13}$)]$_3$ and its homologue [Cp$_2$YbCC(n-C$_4$H$_9$)]$_3$ are extremely similar (see Figure 1) and can be easily assigned in terms of an "internal shift reagent effect" which gives rise to seven and five, widely spaced singlets, respectively, for the Cp ring protons and for the methylene or methyl protons in α- to ω-position. As the singlet in the range characteristic for the Cp ring protons is accompanied by a weaker satellite, the presence of another organometallic by-product in much lower quantities cannot be completely ruled out.

Graphical plots of the observed isotropic shifts versus 1/T give in all cases rise to series of straight lines (Figure 2) which result seems to be somewhat in favour of one singular species rather than of equilibria of throughout very rapidly interconverting species [Cp$_2$YbCCR]$_x$ with different x. All isotropic shifts exhibited by the two n-alkylacetylide complexes are unexpectedly large by comparison with the spectra of e.g. [CpYbO$_2$C(n-C$_4$H$_9$)]$_2$ and Cp$_3$Yb(n-C$_4$H$_9$)$_3$P (38). One possible explana-

tion for these findings is a very "compact" structural arrange-
ment of the oligomers that would allow the exposure of the various
nuclei in R to the magnetic field of more than one paramagnetic
Yb^{3+}-ions.

The ^{1}H-NMR spectra of the Yb-alkynyl complexes with R =
C_6H_{11} and Fc support the assumption of considerable steric congestion. Thus the low-temperature spectra of the cyclohexyl
system do not reflect the expected appearance of axially and
equatorially substituted cyclohexane (37).

For the time being, the best model system to exemplify the
particular type of alkynyl bridging may be the aluminium compound $(CH_3)_2AlC_2CH_3$ which is dimeric both in the crystalline
and in the gaseous state (39). The $\Delta\nu(C\equiv C)$-values known for this
main group metal alkynyl system match nicely those listed in
Table VII.

Reaction with other weakly CH-acidic compounds. Organic
proton acids of an acidity intermediate between cyclopentadiene
and 1-alkynes are represented by various derivatives of cyclopentadiene. We have found that e.g. indene and fluorene do not
react with Cp_3Yb in boiling toluene, but during the careful evaporation of the solvent, as well as under solvent-free conditions,
a reaction takes place, yielding brownish-yellow products of very
weak solubility even in THF (27). The composition $[Cp_2YbL]_x$
(L = indenyl or fluorenyl) is suggested by the elemental analyses.
In view of the fact that a monomeric tris(η^5-indenyl) complex
of Yb of green colour is known (40), the brownish-yellow products might be oligomers in which the ligand L plays a similar
role as in the dimeric methyl complex, $[Cp_2Yb(\mu-CH_3)]_2$ (41).

In a very similar manner, N,N-dimethyl-o-toluidine reacts
with Cp_3Yb to give a very unstable green 1:1-adduct. After refluxing the toluene solution for ca. one hour, its colour turns
orange-brown, and again a very poorly soluble brown powder is
finally isolated. In view of the considerable instability of
many Cp_3Ln/Lewis base adducts it is worth noting that in the
absence of any potentially acidic C-H bond intact adducts even
with rather unusual Lewis bases such as d-metal carbonyls and
metal carbonyl anions, respectively, can be isolated (42).

Acknowledgements

We are greatly indebted to the Deutsche Forschungsgemein-
schaft, Bad Godesberg, and the Fonds der Chemischen Industrie,
Frankfurt (Main), for financial support. We also thank B. Kanellakopulos (Karlsruhe) and T.J. Marks (Evanston) for stimulating
discussions.

Literature Cited

1. Marks, T.J.; Progr. Inorg. Chem., 1978, 23, 51.
2. Marks, T.J.; Fischer, R.D., Eds. "Organometallics of the f-Elements"; D. Reidel Publ. Comp., Dordrecht, Boston, London, 1979.
3. Schumann, H., Nachr. Chem. Tech. Lab.,1979, 27, 393.
4. Schumann, H.; Müller, J., Angew. Chem.,1978, 90, 307.
5. Schumann, H.; Hohmann, S., Chem. Ztg., 1976, 100, 336.
6. Mazzei, A., loc. cit. Ref. 2, p. 379.
7. According to references quoted in Ref. 1, p. 99.
8. Poggio, S.; Brunelli, M.; Pedretti, U.; Lugli, G.; communicated on the N.A.T.O.-ASI on "Organometallics of the f-Elements", Sogesta, Urbino (Italy), Sept. 11-22, 1978.
9. Fischer, R.D.; Landgraf, G., unpublished results (1975).
10. Fischer, H., Dissertation, Technische Universität München, 1965; Fischer, E.O.; Fischer, H., J. Organometal. Chem., 1966, 6, 141.
11. Marks, T.J.; Grynkewich, G.W., Inorg. Chem., 1976, 15, 1302.
12. Kanellakopulos, B.; Dornberger, E.; Billich, H., J. Organometal. Chem., 1974, 76, C42.
13. Fischer, R.D.; Klähne, E.; Kopf, J., Z. Naturforsch., 1978, 33b, 1393.
14. Dornberger, E.; Klenze, R.; Kanellakopulos, B., Inorg. Nucl. Chem. Lett., 1978, 14, 319.
15. Jamerson, J.D.; Takats, J., J. Organometal. Chem., 1974, 78, C23.
16. Bielang, G.; Fischer, R.D., J. Organometal. Chem., 1978, 161, 335.
17. Schumann, H., personal communication.
18. Bagnall, K.W.; Tempest, A.C., loc. cit. Ref. 2, p. 233.
19. Maginn, R.E.; Manastyrskyj, S.; Dubeck, M.,. J. Amer. Chem. Soc., 1963, 85, 672.
20. Kanellakopulos, B., Habilitation Thesis, Universität Heidelberg, 1972; loc. cit. Ref. 2, p. 24 (Fig. 5.12).
21. Schlosser, M., "Struktur und Reaktivität polarer Organometalle", Springer-Verlag Berlin, Heidelberg, New York, 1973, p. 43.
22. Tsutsui, M.; unpublished observations (1965) cited in Gysling, H.; Tsutsui, M., Adv. Organometal. Chem., 1970, 9, 365.
23. Pappalardo, R.; Jørgensen, C.K., J. Chem. Phys., 1967, 46, 632.
24. Fischer, R.D. in "NMR of Paramagnetic Molecules", LaMar, G.N.; Horrocks, W. DeW.; Holm, R.H., Eds., Academic Press, New York and London, 1973, p. 521.
25. Fischer, R.D., loc. cit. Ref. 2, p. 337.
26. Marks, T.J.; Seyam, A.M.; Kolb, J.R., J. Amer. Chem. Soc., 1973, 95, 5529.
27. Bielang, G., Dissertation, Universität Hamburg, 1979.

28. Agarwal, S.K.; Tandon, J.P., Monatsh. Chem., (1979), 110, 401.
29. The ytterbium complex was also independently prepared by E. Mastoroudi and B. Kanellakopulos; personal communication.
30. Bielang, G.; Fischer, R.D., Inorg. Chim. Acta, 1979, 36, L 389.
31. Bagnall, K.W., loc. cit. Ref. 2, p. 231; Siddall III, T.; Stewart, W.E., J. Chem. Soc. Chem. Communic., 1969, p. 922.
32. Coutts, R.S.P.; Wailes, P.C., J. Organometal. Chem., 1970, 25, 117; Austr. J. Chem., 1969, 22, 1547.
33. Combes, M.A., Bull. Soc. Chim. France, 1888, [2], 49, 89; Holtzclaw, Jr., H.F.; Collman, J.P.; Alire, R.M., J. Amer. Chem. Soc., 1958, 80, 1100.
34. McCarthy, P.J.; Hovey, R.J.; Uena, K.; Martell, A.E., J. Amer. Chem. Soc., 1955, 77, 5820.
35. Tsutsui, M.; Ely, N.M., J. Amer. Chem. Soc., 1975, 97, 1280 and 3551; Inorg. Chem., 1975, 14, 2680.
36. Crease, A.E.; Legzdins, P., J. Chem. Soc. Dalton Trans., 1973, 1501.
37. Fischer, R.D.; Bielang, G., J. Organometal. Chem., in press.
38. Marks, T.J.; Porter, R.; Kristoff, J.S.; Shriver, D.F., in "Nuclear Magnetic Resonance Shift Reagents" Academic Press, New York, 1973, p. 247.
39. Fries, W.; Schwarz, W.; Hausen, H.-D.; Weidlein, J.; J. Organometal. Chem., 1978, 159, 373, and further references therein.
40. Tsutsui, M.; Gysling, H.J., J. Amer. Chem. Soc.. 1968. 90, 6880; ibid., 1969, 91, 3175.
41. Holton, J.; Lappert, M.F.; Ballard, D.G.H.; Pearce, R.; Atwood, J.L.; Hunter, W.E., J. Chem. Soc. Chem. Communic., 1976, 480.
42. Onaka, S.; Furuichi, N., J. Organometal. Chem., 1979, 173, 77.

RECEIVED January 30, 1980.

Cyclooctatetraeneactinide(IV) Bis-borohydrides

JEFFREY P. SOLAR and ANDREW STREITWIESER, JR.

Materials and Molecular Research Division, Lawrence Berkeley Laboratory, Berkeley, CA 94720 and Department of Chemistry, University of California, Berkeley, CA 94720

NORMAN M. EDELSTEIN

Materials and Molecular Research Division, Lawrence Berkeley Laboratory, Berkeley, CA 94720

Although bis(η^8-cyclooctatetraene)actinide(IV) (Figure 1a) complexes have been extensively studied (1) since the synthesis of uranocene in 1968 (2,3), mono-COT actinide "half-sandwiches" (Figure 1b) were unknown until recently (4,5). The proposed covalent bonding, involving overlap between filled ligand e_{2g} orbitals with empty metal 5f orbitals (3), could also apply to the bonding in mono-ring complexes. Whereas uranocene has been compared to ferrocene, COT half-sandwiches could show the varied reactivity exhibited by mono-cyclopentadienyl transition metal complexes such as $CpFe(CO)_2Cl$. Thus, reactions such as ligand substitution and reactions of coordinated ligands might be observed along with the usual reactions of uranocene. Mono-COT actinide complexes could also show chemistry similar to the $(C_5Me_5)_2MX_2$ compounds studied by Marks and coworkers (6).

The possibility of COT half-sandwich complexes was established by the observation of an intermediate COT signal in the nmr spectrum of the preparation of thorocene from $ThCl_4$ and K_2BuCOT (4). LeVanda and Streitwieser were able to isolate a white crystalline solid from the reaction of equimolar amounts of $ThCl_4$ and K_2COT (4) and an x-ray crystal structure determination (7) confirmed the product as $(COT)ThCl_2(THF)_2$, 1a. A better synthesis for 1a and complexes with substituents on the cyclooctatetraene ring is reaction of a thorocene and $ThCl_4$ in refluxing THF.

$$ThCl_4 + K_2COT \xrightarrow[\Delta]{THF} (COT)ThCl_2(THF)_2 \qquad (1)$$

$$\tfrac{1}{2}ThCl_4 + \tfrac{1}{2}Th(RCOT)_2 \xrightarrow[\Delta]{THF} (RCOT)ThCl_2(THF)_2 \qquad (2)$$

$$\text{1a, R=H; b, R=n-Bu; c, R=1,3,5,7-Me}_4$$

Interestingly, disproportionation, e.g.,

$$2(RCOT)ThCl_2 \longrightarrow (RCOT)_2Th + ThCl_4 \qquad (3)$$

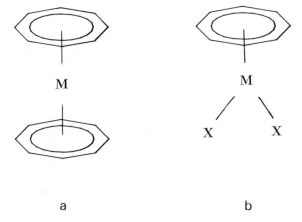

Figure 1. Cyclooctatetraeneactinide complexes

does not occur in this case or for $(C_5Me_5)_2MCl_2$ ($\underline{8}$), whereas $(C_5H_5)_2UCl_2$ is apparently unstable ($\underline{9}$). The increased stability of the pentamethylcyclopentadienyl system has been attributed to steric hindrance. The cyclooctatetraene complexes acquire two molecules of solvent (THF) to complete the coordination sphere.

In order to take advantage of the known volatility of actinide borohydride complexes ($\underline{10}$), we sought to prepare half-sandwich borohydrides $(RCOT)M(BH_4)_2$. Several routes have been developed for the synthesis of the thorium complexes in 65 to 90% yield ($\underline{5}$). $Th(BH_4)_4(THF)_2$ ($\underline{11}$) has been found to react with an equimolar quantity of a cyclooctatetraene dianion in THF to give 2 as a microcrystalline white solid. In contrast to the preparation of 1, 2 is readily prepared at ambient temperature without prolonged reflux. 2 can also be prepared in a manner analogous to the synthesis of 1, using $Th(BH_4)_4(THF)_2$ and $Th(RCOT)_2$.

$$Th(BH_4)_4(THF)_2 + Th(RCOT)_2 \xrightarrow[\Delta]{THF} 2(RCOT)Th(BH_4)_2(THF)_2 \quad (4)$$

$$\text{2a, R=H}$$
$$\underset{\sim}{b}\text{, R=n-Bu}$$

The preparation of $\underset{\sim}{2}$ directly from $ThCl_4$ is a modification of the recent synthesis of $Cp_2U(BH_4)_2$ ($\underline{12}$).

$$ThCl_4 + 2LiBH_4 \xrightarrow{THF} \xrightarrow{K_2COT} (COT)Th(BH_4)_2(THF)_2 \quad (5)$$

This synthesis is complicated, however, by the need to remove the LiCl formed in the reaction. In addition, metathesis of 1 with $LiBH_4$ gives the borohydride half-sandwich.

$$(RCOT)ThCl_2(THF)_2 \xrightarrow[THF]{LiBH_4} (RCOT)Th(BH_4)_2(THF)_2$$

A 1,2-dimethoxyethane complex of $\underset{\sim}{2}$, $(COT)Th(BH_4)_2(DME)$, $\underset{\sim}{3}$, was prepared by the reaction of $Th(BH_4)_4(THF)_2$ and K_2COT in DME. In contrast to the THF complex, the DME complex is only sparingly soluble in toluene.

Although satisfactory elemental analyses have not been obtained, spectroscopic characterization of 2 is straightforward. The ^1H nmr spectrum of 2a listed in Table 1 contains signals from one cyclooctatetraene, two equivalent borohydride groups, and two molecules of coordinated THF. The borohydride signal appears as a quartet due to coupling to the ^{11}B nucleus and indicates that the terminal and bridging protons are rapidly exchanging. The borohydride signal of $Cp_3Th(BH_4)$ also appears as a quartet, at δ 3.30 ($\underline{13}$). At low temperature, the quartet of 2a collapses symmetrically to a broad singlet at δ 3.0. Marks and Shimp have observed this effect in the spectrum of $Zr(BH_4)_4$ and have interpreted the process as a temperature-dependent loss of B-H coupling rather than a slowing of the H_t-H_b exchange ($\underline{14}$).

Table 1. Infrared Spectra (Nujol, cm^{-1})

Compound	$B-H_t$	$B-H_b$			COT		Other
2a (COT)Th(BH$_4$)$_2$(THF)$_2$	2482(s)	2282(w)	2220(s)	2150(m)	722(sh)	714(s)	1180(s) 1163(s)
3 (COT)Th(BH$_4$)$_2$(DME)$_2$	2464(s)	2375(w)	2282(s)	2215(s)	730(s)	716(m)	1165(s) 1120(s) 1095(m) 1038(s)
4 (COT)U(BH$_4$)$_2$(THF)$_x$	2472(s)	2339(w)	2210(s)	2142(m)	722(s)		1170 (br,s)
5 (COT)UCl$_2$(THF)$_x$					734(s)	719(s)	1135(sh) 1011(s)
6 (COT)Th(BH$_3$Et)(THF)$_2$		2310(s)	2185(s)	2128(s)			

The infrared spectra of 2 (Table 2) provide information regarding the binding of the borohydride broup to the metal ($\underline{13}$). Both bidentate (A) and tridentate (B) forms are known for actinide borohydrides ($\underline{13,15}$).

$$
\begin{array}{cc}
\text{M}\diagdown\!\!\!\diagup\text{B}\diagup\!\!\!\diagdown\text{H} & \text{M}\!\!-\!\!\text{H}\!\!-\!\!\text{B}\!\!-\!\!\text{H} \\
\text{(A)} & \text{(B)}
\end{array}
$$

The infrared spectra of 2 are in accord with that expected for a tridentate form (B) and are similar to that from $Th[N(SiMe_3)_2]_3$ (BH_4) for which tridentate bonding has been established by x-ray crystal structure determination ($\underline{16}$).

The infrared spectra of 2 also contain a strong band at $714\ cm^{-1}$; 1 gives similar spectra in this region. In thorocene, a strong absorption at $698\ cm^{-1}$ has been assigned to a perpendicular C–H deformation ($\underline{17}$) or, more recently, to an asymmetric ring–metal–ring vibration ($\underline{18}$). Bands at 742 and $775\ cm^{-1}$ in thorocene have been interpreted as C–H or C–C–C perpendicular bending modes ($\underline{18}$). The absence of an absorption at ca. $700\ cm^{-1}$ and the presence of strong bands at 15–$30\ cm^{-1}$ higher frequency appear to be characteristic of the half-sandwich complexes.

The preparation of mono-COT complexes of uranium was also of interest. Marquet-Ellis has reported an ^{1}H nmr spectroscopic study of the reaction between UCl_4, $LiBH_4$, and K_2COT ($\underline{19}$). He observed signals from both uranocene and a second product assigned as $(COT)U(BH_4)_2$. We have studied this reaction and found that it is not straightforward. A high yield of uranocene was formed when cyclooctatetraene dianion, either solid or in THF solution, was added to a solution formed from the reaction of UCl_4 and two equivalents of $LiBH_4$. A mixture of equimolar amounts of $U(BH_4)_4$ and K_2COT also yielded uranocene upon addition of THF. However, a second product, 4, could be obtained by the slow dropwise addition of a THF solution of K_2COT to the UCl_4–$2LiBH_4$ reaction mixture. The brown product is soluble in benzene and was separated from the insoluble uranocene byproduct by filtration.

$$UCl_4 + 2LiBH_4 \xrightarrow{\ THF\ } \xrightarrow{K_2COT} (COT)U(BH_4)_2(THF)_x + U(COT)_2$$

$$\underset{\sim}{4}$$

The visible spectrum of 4 (Table 3) contains a strong band at 396 nm and tails off to longer wavelength with a series of shoulders much like a uranocene. The infrared spectrum of the compound is nearly identical to that of the thorium analog 2a and is consistent with tridentate bonding. We are, however, unable to reproduce the nmr spectrum reported by Marquet-Ellis.

Table 2. NMR Spectra (δ vs TMS)

Compound	Solvent	COT	BH_4	Other
2a $(COT)Th(BH_4)_2(THF)_2$	C_6D_6	5.69(s,8H)	3.11(q,J=85Hz,8H)	3.34,0.90(br,8H,8H,THF)
3 $(COT)Th(BH_4)_2(DME)_2$	C_6D_6	6.7(s)	3.1 (q,J=85Hz)	3.0, 2.8(DME)
5 $(BuCOT)UCl_2(THF)_x$	THF-d_8	-21.7(t,1H), -29.0(t,2H), -33.0(d,2H), -44.7(t,broad,2H)		

Table 3. Visible Spectra (λ_{max}, nm)

4 $(COT)U(BH_4)_2(THF)_2$	396, 457, 492, 525
5 $(BuCOT)UCl_2(THF)_x$	402, 456, 499, 541

Another route to mono-cyclooctatetraene compounds is cleavage of one ring from the bis-COT complexes by protonation. Kablitz and Wilke have prepared $(COT)ZrCl_2$ and its THF adduct in this manner (20).

$$(COT)_2Zr + 2HCl \text{ (g)} \xrightarrow{C_6H_5CH_3} [(COT)ZrCl_2]_n + C_8H_{10}$$

$$(COT)_2Zr + 2HCl \text{ (g)} \xrightarrow{THF} (COT)ZrCl_2(THF)_2 + C_8H_{10}$$

Although the decomposition of uranocene in the presence of aqueous acids has been noted (3), we have found that the reaction of $(BuCOT)_2Th$ with two equivalents of dry HCl gives 1b.

1,1'-Dibutyluranocene also reacts with HCl. The product is a dark green powder that gives a red-brown solution in THF with a visible spectrum similar to that of 3.

$$U(RCOT)_2 + 2HCl \text{ (g)} \xrightarrow{THF} (RCOT)UCl_2(THF)_x$$

$$5, \text{ R=n-Bu}$$

The ^1H nmr spectrum of 5 clearly shows a COT ring distinct from that of $(BuCOT)_2U$ and, unlike a substituted uranocene, coupling is observed between the nonequivalent protons on the COT ring. Although the signals from 5 are strongly shifted by the paramagnetic uranium, the signals are much sharper than those of a uranocene. The infrared spectrum of 5 contains absorptions from the COT ring at 734 and 719 cm^{-1}, in the region expected for a half-sandwich complex. 5, however, is much less soluble than the thorium analog 2b, and unreacted uranocene can be removed by washing the mixture with benzene. LeVanda has also observed this product from the prolonged reflux of UCl_4 and $U(BuCOT)_2$ (21).

Several other reactions of these complexes have been investigated. Although $Cp_2Zr(BH_4)_2$ reacts readily with tertiary amines, converting metal-BH_4 groups to metal hydrides by formation of BH_3-amine complexes (22), $(COT)Th(BH_4)_2(THF)_2$ was unchanged on treatment with excess NEt_3 at room temperature. 1a was observed to react slowly with excess BEt_3 in refluxing benzene.

$$(COT)Th(BH_4)_2(THF)_2 + BEt_3 \xrightarrow[\Delta]{C_6H_6} (COT)Th(BH_3Et)_2(THF)_x$$

$$6$$

Marks and Kolb have carried out the analogous reaction of $Cp_3U(BH_4)$ and have observed complete substitution of the terminal proton in 36 hours, In contrast, 1a was only about 50% reacted after 40 hours and a trace remained after four additional days with a fresh charge of BEt_3. The infrared spectrum of 6 is much like that of 1a but the B-H terminal absorption is absent. Neither 1a nor 6 sublime on heating in vacuo and 6 is slowly

decomposed to thorocene (and other unidentified products) at
ca. $100°C$ at 10^{-6} torr. Thorocene is also formed when 1a is
allowed to react with K_2COT at room temperature in THF.~~

$$(COT)Th(BH_4)_2(THF)_2 \xrightarrow{\frac{K_2COT}{THF}} (COT)_2Th$$

Further studies of the synthesis and reactivity of these
half-sandwich compounds are in progress.

Experimental

General. Due to the air-sensitive nature of the compounds
in this study, all work was carried out in an argon atmosphere
using a glove box or standard Schlenk techniques. Solvents were
distilled under argon from sodium-benzophenone ketyl. $LiBH_4$ was
purified by Soxhlet extraction with diethyl ether before use and
other reagent grade chemicals were used as received. Uranocenes
(3,23) and thorocenes (24) were prepared by published procedures.
$Th(BH_4)_4(THF)_2$ was prepared by a modification of the method of
Ehemann and Noth (11) using THF in place of diethyl ether and
was shown by a powder pattern to be isostructural with
$U(BH_4)_4(THF)_2$ (25). The synthesis (5) and x-ray crystal structure
(7) of $(RCOT)ThCl_2(THF)_2$ have been published.
    Infrared spectra were run of Nujol mulls prepared in a glove
box and sealed between salt plates. Samples could then be transferred to a Perkin-Elmer Model 283 spectrometer without decomposition. Visible spectra were run in stoppered glass cells on
a Cary 118 spectrometer. NMR spectra were run at 180 MHz on
a FT instrument.

    Preparation of $(RCOT)Th(BH_4)_2(THF)_2$ (2). (A) $Th(BH_4)_4(THF)_2$
(0.51 g, 1.2 mmole) and K_2COT (0.21 g, 1.2 mmole) were mixed
and ca. 75 ml of THF was added. The solution was stirred in the
glove box for 16 hrs until the yellow color faded. The white
precipitate was removed by centrifugation and the solids were
washed with 10 ml of THF. The solutions were combined and the
solvent was pumped off to give 0.45 g (76%) of $(COT)Th(BH_4)_2(THF)_2$
(2a).
    ~~ (B) $Th(BH_4)_4(THF)_2$ (0.37 g, 0.85 mmole) and $Th(BuCOT)_2$
(0.44 g, 0.80 mmole) were mixed and 50 ml of THF was transferred
onto the solids in vacuo. The flask was refilled with argon
and the solution was heated at reflux for 5 days. A visible
spectrum indicated no further trace of the thorocene. Removal
of the solvent in vacuo gave a light yellow solid. The residue
was rinsed with hexane and dissolved in ca. 30 ml of benzene.
A small amount of solid was removed by centrifuge and the solvent
was removed to give 0.60 g (67%) of $(BuCOT)Th(BH_4)_2(THF)_2$ (2b)
as a white solid. ~~

(C) THF (60 ml) was added to a mixture of 1.25 g (3.3 mmole) of ThCl₄ and 0.16 g (7.3 mmole) of LiBH₄ and the solution was stirred for 16 hrs in a glove box. A solution of 0.54 g (3.0 mmole) of K₂COT in 40 ml of THF was added and the mixture stirred an additional 16 hrs. The precipitate was removed by centrifuge and washed with two 10 ml portions of THF. The solutions were combined and the solvent removed. The residue was dissolved in benzene and the insoluble material was removed by centrifuge. Removal of the solvent gave an oily white product. Recrystallization from benzene–hexane gave 1.38 g (91%) of (COT)Th(BH₄)₂(THF)₂.

(D) (BuCOT)ThCl₂(THF)₂ (211 mg, 0.347 mmole) and LiBH₄ (36 mg, 1.6 mmole, excess) were mixed and 15 ml of THF was added. The solution was stirred in the glove box for two days. The solvent was removed and the residue was extracted with benzene. Removal of the solvent gave (BuCOT)Th(BH₄)₂(THF)₂ (2b), identified by its infrared spectrum.

Preparation of (RCOT)Th(BH₄)₂(DME) (3). Following synthesis A above, 0.326 g (0.725 mmole) of Th(BH₄)₄(THF)₂ was dissolved in 2 ml of DME and a solution of 0.13 g (0.71 mmole) of K₂COT in 20 ml of DME was added. The solution was stirred for 40 hrs in a glove box. The white solids were removed by centrifugation and washed with 10 ml of DME. The solutions were combined and the solvent was pumped off to give 0.26 g (67%) of (COT)Th(BH₄)₂ (DME).

Preparation of (COT)U(BH₄)₂(THF)ₓ. A solution of 0.22 g (10 mmole) of LiBH₄ in 10 ml of THF was added to a solution of 1.63 g (4.3 mmole) of UCl₄ in 25 ml of THF. The solution was stirred for 24 hrs in a glove box. K₂COT (0.75 g, 4.1 mmole) was dissolved in 100 ml of THF and added dropwise to the UCl₄– LiBH₄ solution. The resultant brown solution was stirred for 16 hrs and the solvent was pumped off. The residue was extracted with benzene and the insoluble material removed by centrifuge. The solvent was removed in vacuo leaving a brown tacky residue. Infrared and visible spectra indicated the product to be (COT)U(BH₄)₂THFₓ. Recrystallization from benzene–hexane gave a tacky brown solid.

Reaction of (COT)Th(BH₄)₂(THF)₂ with BEt₃. (COT)Th(BH₄)₂ (THF)₂ (0.34 g, 0.67 mmole) was dissolved in benzene and 1.7 ml (1.7 mmole) of 1 M BEt₃ in THF was added by syringe. The solution was heated at reflux for 40 hrs. The solvent was removed and an infrared spectrum indicated that the reaction was ca. 50% complete. Benzene and 3.0 ml of BEt₃ in THF were again added and the solution was heated at reflux for four days. The solvent was removed to give a tacky residue of (COT)Th(BH₃Et)₂(THF)₃.

Reaction of $(COT)Th(BH_4)_2(THF)_2$ with K_2COT. $(COT)Th(BH_4)_2$ $(THF)_2$ (0.253 g, 0.496 mmole) and K_2COT (0.089 g, 0.49 mmole) were mixed and 50 ml of THF was added. The solution was stirred for three days and the solids were removed by centrifuge. The solids were washed with additional THF (ca. 25 ml) until no further yellow color was extracted. The solutions were combined and the solvent was removed to give 0.178 g (83%) of $Th(COT)_2$.

Acknowledgments

We thank Rodney Banks for a sample of $U(BH_4)_4$ and Wayne Luke for assistance with the nmr spectra. This work was supported in part by The Division of Nuclear Sciences, Office of Basic Energy Sciences, U.S. Department of Energy under Contract No. W-7405-Eng 48.

Literature Cited

1. Streitwieser, A., Jr. "Organometallic Compounds of the f-Elements," Marks, T.J.; Fischer, R.D., Eds. Reidel Publishing Co., Amsterdam (1979), 149-177.
2. Streitwieser, A., Jr.; Muller-Westerhoff, U. J. Am. Chem. Soc. 1968, 90, 7364.
3. Streitwieser, A., Jr.; Muller-Westerhoff, U.; Sonnichsen, G.; Mares, F.; Morrell, D.G.; Hodgson, K.O. J. Am. Chem. Soc. 1973, 95, 8644-8649.
4. LeVanda, C.; Streitwieser, A., Jr. Abstracts, 175th National Meeting American Chemical Society, Anaheim, California, March 1978, INOR No.68.
5. LeVanda, C.; Solar, J.P.; Streitwieser, A., Jr. J. Am. Chem. Soc., submitted.
6. Fagan, P.J.; Manriquez, J.M.; Marks, T.J. "Organometallic Compounds of the f-Elements," Marks, T.J.; Fischer, R.D., Eds. Reidel Publishing Co., Amsterdam (1979), 113-148; and references therein.
7. Zalkin, A.; Templeton, D.H.; LeVanda, C.; Streitwieser, A.,Jr. Inorg. Chem., to be submitted.
8. Manriquez, J.M.; Fagan, P.J.; Marks, T.J. J. Am. Chem. Soc. 1978, 100, 3939-3941.
9. Kanellakopulos, B.; Aderhold, C.; Dornberger, E. J. Organometal. Chem., 1974, 66, 447-451.
10. Schlesinger, H.I.; Brown, H.C. J. Am. Chem. Soc. 1953, 75, 219-221.
11. Ehemann, M.; Noth, H. Z.Anorg. Allg. Chem. 1971, 386, 87-101.
12. Zanella, P.; DePaoli, G.; Bombieri, G. J. Organometal. Chem. 1977, 142, C21-24.
13. Marks, T.J.; Kennelly, W.J.; Kolb, J.R.; Shimp, L.A. Inorg. Chem. 1972, 11, 2540-2546.
14. Marks, T.J.; Shimp, L.A. J. Am. Chem. Soc. 1972, 94, 1542-1550.

15. Bernstein, E.R.; Hamilton, W.C.; Kiederling, T.A.; LaPlaca, S.J.; Lippard, S.J.; Mayerle, J.J. Inorg. Chem. 1972, 11, 3009–3016.
16. Turner, H.W.; Andersen, R.A.; Zalkin, A.; Templeton, D.H. Inorg. Chem. 1979, 18, 1221–1224.
17. Karraker, D.G.; Stone, J.A.; Jones, E.R., Jr.; Edelstein, N. J. Am. Chem. Soc. 1970, 92, 4841–4845.
18. Goffart, J. "Organometallic Compounds of the f-Elements," Marks, T.J. and Fischer, R.D., Eds. Reidel Publishing Co., Amsterdam (1979), p. 489.
19. Marquet-Ellis, H. NATO Advanced Study Institute, Sogesta, Italy, 1979.
20. Kablitz, H.J.; Wilke, G. J. Organometal. Chem. 1973, 51, 241–271.
21. LeVanda, C. Unpublished results.
22. James, B.D.; Nanda, R.K.; Wallbridge, M.G.H. Inorg. Chem. 1967, 6, 1979–1983.
23. Streitwieser, A., Jr.; Harmon, C.A. Inorg. Chem. 1973, 12, 1102–1104.
24. LeVanda, C.; Streitwieser, A., Jr. Inorg. Chem., submitted.
25. Reitz, R.R.; Edelstein, N.M.; Ruben, H.W.; Templeton, D.H.; Zalkin, A. Inorg. Chem. 1978, 17, 658–660.

RECEIVED December 26, 1979.

Nuclear Magnetic Resonance Studies of Uranocenes

WAYNE D. LUKE and ANDREW STREITWIESER, JR.

Materials and Molecular Research Division, Lawrence Berkeley Laboratory,
Berkeley, CA 94720 and Department of Chemistry, University of California,
Berkeley, CA 94720

I. Introduction and Historical Background

In the past several years a substantial amount of work has
been devoted toward evaluation of the contact and pseudocontact
contributions to the observed isotropic shifts in the H nuclear
magnetic resonance (NMR) spectra of uranium(IV) organometallic
compounds (1-15). One reason for interest in this area arises
from using the presence of contact shifts as a probe for covalent
character in the uranium carbon bonds in these compounds. Several
extensive ^{1}H NMR studies on Cp_3U-X compounds (10-13) and less
extensive studies on uranocenes have been reported (5,6,14,15).
Interpretation of these results suggests that contact shifts contribute significantly to the observed isotropic shifts. Their
presence has been taken as indicative of covalent character of
metal carbon bonds in these systems, but agreement is not
complete (2). In this paper we shall review critically the work
reported on uranocenes in the light of recent results and report
recent work on attempted separation of the observed isotropic
shifts in alkyluranocenes into contact and pseudocontact components.

A. Theory. A detailed derivation of the theory behind paramagnetic shifts in the NMR of paramagnetic compounds, or a complete
review of the literature concerning separation of observed isotropic shifts into contact and pseudocontact components is well
beyond the scope of this paper. Several books and reviews of
these subjects are available (16-21).
The presence of a paramagnetic metal in organometallic
compounds significantly influences the NMR spectrum of ligand
nuclei. Changes in nuclear relaxation times and changes in resonance frequency are the two principal effects arising from interaction between the unpaired electrons on the metal and ligand
nuclei. Nuclear relaxation times are shortened due to increased
spin-spin relaxation and result in increased linewidths of the
resonance signals. In some compounds this broadening of the

0–8412–0568–X/80/47–131–093$12.00/0

resonance signals is large enough to preclude their observation.

The coupling of the unpaired electrons with the nucleus being observed generally results in a shift in resonance frequency that is referred to as a hyperfine isotropic or simply isotropic shift. This shift is usually dissected into two principal components. One, the hyperfine contact, Fermi contact or contact shift derives from a transfer of spin density from the unpaired electrons to the nucleus being observed. The other, the dipolar or pseudocontact shift, derives from a classical dipole-dipole interaction between the electron magnetic moment and the nuclear magnetic moment and is geometry dependent.

Expressions for the contact shift vary depending on the assumptions made. One common form is (22)

$$\delta_{CONTACT} = \frac{A_i g_e^2 \beta_e^2 S(S+1)}{3g_N \beta_N kT} \tag{1}$$

where A_i is the hyperfine coupling constant, g_e is the rotationally averaged electronic g value, β_e is the Bohr magneton, g_N and β_N are the corresponding nuclear constants and S is the spin of the unpaired electrons. For actinide organometallics in which crystal field splitting is small compared the separation between electronic states and characterized by quantum number J, but large compared to kT, the contact shift may be expressed as (12):

$$\delta_{CONTACT} = \frac{A_i(g_J-1)\chi}{Ng_J \beta_e g_N \beta_N} \tag{2}$$

in which χ is the magnetic susceptibility.

The pseudocontact shift may be expressed as (20, 23-26):

$$\delta_{PSEUDOCONTACT} = \frac{\chi_z - 1/2(\chi_x + \chi_y))}{3N} \frac{3\cos^2\theta - 1}{R^3} +$$

$$\frac{(\chi_x - \chi_y)}{2N} \frac{\sin^2\theta \cos 2\psi}{R^3} \tag{3}$$

in which χ_x, χ_y and χ_z are components of the magnetic susceptibility, and the coordinate system is shown in Fig. 1.

The total isotropic shift is the sum of the two components:

$$\delta_{ISOTROPIC} = \delta_{CONTACT} + \delta_{PSEUDOCONTACT} \tag{4}$$

In this paper we define all shifts upfield from TMS as negative and all shifts downfield from TMS as positive. This is the modern accepted convention.

B. NMR of Uranium(IV) Organometallic Compounds. Current
interest in the NMR of U(IV) organometallic compounds has been
concerned with the relative contributions of contact and pseudocontact shifts to the observed isotropic shifts. Much of this
interest arises from the possible presence, and relative role of
covalency in ligand metal bonds in organoactinide compounds.
Ideally, if the isotropic shifts in U(IV) compounds can be
factored into contact and pseudocontact components, the contact
shift can be correlated with electron delocalization and bond
covalency.

From an experimental point of view, the ^{1}H NMR spectra of
U(IV) compounds are ideally suited to such analysis. In general,
the isotropic shifts are less than $\pm$ 100 ppm, which is small
compared to shifts observed in many transition metal complexes.
The linewidths for protons on carbons directly bonded to uranium
are less than 50 Hz and rapidly decrease for protons on carbons
not directly bonded to the metal atom, so that J-J coupling is
often observed. A review of early work on the H NMR spectra of
U(IV) compounds apepared in 1971 (7).

1. Triscyclopentadienyl uranium(IV) compounds. The ^{1}H NMR
resonance of the cyclopentadienyl ligand in Cp_4U is shifted -19.27
ppm upfield from the corresponding resonance in diamagnetic Cp_4Th
at room temperature. The interpretation of this shift involved
some early controversy (1,2,4,10). Moreover, a wide variety of
Cp_3U-X compounds has been prepared and extensive studies on their
^{1}H NMR spectra have been reported. Some confusion exists in
comparing the isotropic shifts reported in the literature. Some
shifts are reported referenced relative to various solvents while
others are referenced relative to the corresponding thorium
compound instead of the universal standard TMS. To facilitate
comparison the reported shifts have been referenced to TMS and
are recorded in Table I.

Assuming axial symmetry along the U-X bond, the isotropic
shifts for compounds 3, 7-16 and 25 have been factored into contact and pseudocontact components. In the rigid cholesteroloxy
ligand, 25, Fischer and co-workers (11) showed the ratio of the
geometry factors $(3\cos^2-1)/R^3$ for the A-ring protons in the β and
γ-positions to be equal to the ratio of the isotropic shifts,
whereas gross deviations occurred when the α-positions were
compared. This implies that all of the isotropic shifts except
those in the α-position arise purely from pseudocontact-type
interactions whereas both pseudocontact and contact interactions
contribute to the α-proton isotropic shifts. The isotropic shifts
were factored into contact and pseudocontact components. Taking
the average geometry factor, (G_r), for the ring protons as -5.49
x 10^{-22} cm^{-3}, the calculated pseudocontact and contact shifts at
room temperature are -6.4 ppm and -17.6 ppm, respectively. The
approximate invariance of the ring proton isotropic shifts in all

of the alkoxy substituted compounds suggests that there is no
great fluctuation in the molecular anisotropy throughout this
series.

Marks and co-workers (12) have studied the alkyl substituted
compounds 7-16. Assuming that INDO/2 molecular orbital calculations on alkyl radicals can reasonably predict experimental elec-
tron-nuclear hyperfine coupling constants, a_i, they have calculated the a_i values for each of the alkyl substituents. Taking
the ratio of the contact shifts of the ortho positions in 7 and
vinylic position in 16 as equal to the ratio of calculated a_i
values and the ratio of the geometry factors as equal to the ratio
of pseudocontact shifts, Marks and co-workers could solve for the
contact and pseudocontact shifts in 7 and 16. Factoring the

Table I

The ^{1}H NMR Resonances of Cp_3U-X Compounds

δ ppm from TMS

	X		Temp	Solvent	Ref.
1	Cp	−13.96	25	THF	1,7,10
2	F	−6.46	25	Benzene	7,10,27
3	Cl	−3.40	25	Benzene	7,10,13,27
4	Br	−3.65	25	Benzene	7,10,27
5	I	−4.28	25	Benzene	7,10,27
6	BH_4	−6.53	25	Benzene	3,7
7	C_6H_5	−3.26	25	Benzene	12,28
8	CH_3	−2.76 −194.76 (CH_3)	25	Benzene	12
9	$n-C_4H_9$	−3.06	25	Benzene	12
10	$i-C_3H_7$	−3.66 −11.46 (CH_3) −20.36 (γ) −26.36 (β) −192.76 (α)	25	Benzene	12
11	$t-C_4H_9$	−4.16 −15.96 (CH_3)	25	Benzene	12
12	neopentyl	−4.36 −14.86 (CH_3) −184.76 (CH_2)	25	Benzene	12
13	allyl	−2.76 −30.96 (CH) −118.76 (CH_2)	25	Benzene	12
14	vinyl	−2.06 31.64 $(\beta$-trans$)$ −9.76 $(\beta$-cis$)$ −156.36 (α)	25	Benzene	12

Table I (cont.)

X		Temp	Solvent	Ref.
15	cis-2-butenyl −3.36 −12.56 (α-CH$_3$) −15.36 (H) −35.06 (β-CH$_3$)	25	Benzene	12
16	Trans-2-butenyl −3.46 30.74 (H) −25.76 (β-CH$_3$) −26.36 (α-CH$_3$)	25	Benzene	12
17	C$_6$F$_5$ −3.66	25	Benzene	12
18	OCH$_3$ −17.06 52.54 (CH$_3$)	RT	Benzene	7,29
19	OC$_2$H$_5$ −18.36 59.04 (OCH$_2$) 16.84 (CH$_3$)	RT	Benzene	7,29
20	O-n-C$_4$H$_9$ −17.86 57.84 (OCH$_2$) 17.04 (β) 8.94 (γ) 4.58 (CH$_3$)	RT	Benzene	7,29
21	O-i-C$_3$H$_7$ −18.56 121.94 (OCH$_2$) 17.84 (CH$_3$)	RT	Benzene	7,29
22	O-t-C$_4$H$_9$ −19.36 19.64 (CH$_3$)	RT	Benzene	7,29
23	n-hexyloxy −17.6 56.9 (α) 16.69 (β) 8.80 (γ) 5.08 (δ) 2.57 ( ) 1.67 (ε)	RT	Benzene	11
24	cyclohexy-loxy −18.3[a] 122.0 (α) 19.0 (β) 17.0 (β) 10.2 (γ) 9.8 (δ) 7.5 (ε)	30	Benzene	11
25	cholester-yloxy −17.7[a,b]	30	Benzene	11

[a] extrapolated from spectrum reported in ref.11

[b] See ref. 11 for substituent proton resonances

isotropic shifts in the remaining members of the series was effected by assuming that the pseudocontact shifts are all proportional to the corresponding geometry factors. Agreement between the calculated shifts and a_i values was fair, and, in general, the contact shifts were less than 50 ppm. For the ring protons an average geometry factor of -7.97×10^{-22} cm^{-3} was used to calculate a pseudocontact shift of 19.1 ppm and a contact shift of −28 ppm. While this contact shift is similar in magnitude to

that calculated for the alkoxy compounds, the calculated pseudocontact shifts for the two series are opposite in sign. This
implies that the replacement of -OR by -R caused a reversal of
sign in the magnetic anisotropy term of eq. 3 (i.e., $\chi_{||}-\chi_{\perp}$).

Recently, Amberger (13) has assigned the bands in the absorption spectrum of 3. In this analysis a set of first-order crystal
field functions was derived which models the known temperature
dependence of the magnetic susceptibility. From these parameters,
the isotropic ^1H NMR shifts of the ring protons were factored
into contact and pseudocontact components. Using the geometry
factor of Marks ($-7.97 \times 10^{-22} cm^{-3}$) or that of Fischer ($-5.49 \times 10^{-22} cm^{-3}$) the calculated pseudocontact shifts at 25°C are 2.38
and 1.64 ppm and the calculated contact shifts are -11.58 and
-10.84 ppm, respectively.

Interestingly, all of the calculated contact shifts for the
ring protons in these Cp_3U-X compounds are of the same sign and
of the same order of magnitude as the isotropic shift in Cp_4U,
suggesting that the ring metal bonding in all of these compounds
is quite similar. Replacement of one Cp in Cp_4U by any other
ligand lowers the symmetry of the complex leading to magnetic
anisotropy and pseudocontact contributions to the isotropic ^1H
NMR shifts. Lower symmetry alone does not completely control the
magnetic anisotropy. The substituent has a profound effect which
can serve to change the sign of the magnetic anisotropy term in
eq. 3 and hence, the sign of the pseudocontact shift.

The temperature dependence behavior of the ring proton isotropic shifts also reflects the effects of lower symmetry. While
the ring proton shift in Cp_4U shows a linear dependence on T^{-1}
from -106°C to 133°C, the ring proton shifts of Cp_3U-X compounds
2-5, 8-10, 12, 17-18, and 23-25 all show marked deviations from
linearity. The alkyl-substituted systems show linear behavior
from ca. -150°C to room temperature but deviate from linearity
above room temperature. The alkoxy compounds show apparent
linearity from ca. 200°C to 400°C but deviations from linearity
below 200°C. All of the halides except for the fluoride display
a slight curvature from 200°C to 400°C. The variable temperature
behavior of the fluoride is solvent dependent and reflects the
formation of dimers.

The presence of the paramagnetic center in Cp_3U-X compounds
also serves as an internal shift reagent and as such has been
used as a conformational probe. In a variable temperature ^1H NMR
study, Marks and co-workers (30) have observed line broadening of
the borohydride proton resonances in 6. The broadening was not a
result of temperature dependent changes in boron quadrupolar
relaxation but instead was interpreted as indicative of slowing
of the chemical exchange process between bridging and terminal
protons. Estimation of the coalescence temperature as -140 ± 20°C
leads to a calculated $\Delta G^{\neq}$ for the process of 5.0 ± 0.6 kcal mole$^-$.
Similarly, the energy barrier to rotation of the isopropyl group
in $Cp_3U-i-C_3H_7$ has been estimated to be $E_a = 10.5 \pm 0.5$ kcal

mole^{-1} from computer simulated line shape analysis of variable
temperature spectra (12). From the coalescence temperature for
fluxionality between monohapto- and trihapto-bonding of the allyl
group in Cp$_3$U-allyl of 43°C, a value of 8.0 kcal mole^{-1} for $\Delta G^{\neq}$
for the process was calculated (12), while in cyclohexyloxy-UCp$_3$,
a lower limit for $\Delta G^{\neq}$ for ring inversion of the cyclohexyl ring
has been estimated to be 2.3 kcal mole^{-1} (11).

2. Uranocenes. Edelstein and co-workers (5) proposed that
the ^{1}H isotropic shift in uranocene can be approximated by

$$\delta_{ISOTROPIC} = \frac{\chi_{||}-\chi_{\perp}}{3} \frac{3\cos^2\theta-1}{R^3} + \frac{Ai}{3} \frac{16g_J'\beta e}{5kT} \qquad (5)$$

The pseudocontact term is simply the axially symmetric form of
eq. 3. The contact term is eq. 2, where β and g_J have been
evaluated using a crystal field model for bis-cyclooctatetraene-
actinide sandwich compounds proposed by Karraker (31).
    The ground state term for U^{+4} is ^{1}H$_4$. In a crystal field of
D$_{8h}$ symmetry this ninefold degenerate state is split into four
doublets (J$_z$ = ±4, ±3, ±2, ±1) and one singlet (J$_z$ = 0). Analysis
of bulk magnetic susceptibility data led to selection of the
ground state as J$_z$ = ±4, provided that an effective orbital
reduction factor of k = 0.8 was included in the crystal field
calculations to correct covalent contributions to metal ligand
bonding (32). This model successfully predicts the magnetic
behavior of uranocene, neptunocene, and plutonocene assuming:
1) only the lowest crystal field state is populated in the temperature range T << 400 K; 2) there is no mixing of J states by the
crystal field; 3) the effects of intermediate coupling are small
and can therefore be neglected (31).
    A direct result of the J$_z$ = ±4 ground state is in the limit
of kT << D, the total crystal field splitting, $\chi_{||}$ = 3χ_{av} and
$\chi_{\perp}$ = 0 where

$$\chi_{av} = 1/3\, \chi_{||} + 2/3\, \chi_{\perp} \qquad (6)$$

Thus, the magnetic susceptibility component of the pseudocontact
shift was evaluated from bulk susceptibility measurements. Using
geometric data from the x-ray structure of Raymond and Zalkin (33)
and a magnetic moment of 2.4 B.M., Edelstein and co-workers
calculated the pseudocontact shift for uranocene ring protons,
(entry 1, Table II). These authors used the Curie Law to relate
X and μ_{eff}, while the magnetic data obeyed the Curie-Weiss Law,
with μ_{eff} = 2.4 B.M. and θ = 9.6°K. Neglect of the Weiss constant, (i.e., the Curie Law instead of the Curie-Weiss Law) under-
estimates the value of X$_{av}$ resulting in smaller values for the
pseudocontact shift. This underestimation amounts to about 3.5%
for the ring ^{1}H resonances in uranocene (entry 2, Table II).

Since the calculated pseudocontact shifts are smaller in magnitude than the observed isotropic shift, Edelstein, et.al., concluded that an upfield contact component contributes to the total isotropic shift, indicative of covalency in the ligand metal bonds of uranocene.

TABLE II

Earlier Analyses of Isotropic ^{1}H Shifts of Uranocene

Proton	$3\cos^2\theta - 1/R^3$ x 10^{21} cm^{-3}	Temp °C	μ_{eff} B.M.	Iso- tropic shift (ppm)	Pseudo- contact shift (ppm)	Contact shift (ppm)
urano-cene[a] ring	-3.55	29	2.4	-41.9	-14.0	-27.9
urano-cene[b] ring	-3.55	29	2.4	-41.9	-14.5	-27.4
octa-methyl[c] ring	-2.0	25	2.38	-41.3	-7.9	-33.4
octa-methyl[c] ring	-5.9	25	2.38	-6.0	-23.6	+17.6
urano-cene[c] ring	-2.0	25	2.38	-42.6	-7.9	-34.7

(a) Ref. 5.   (b)   Correction for Curie-Weiss Law; see text.
(c) Ref. 6.

The plot of shift vs T^{-1} was linear in accord with Curie-Weiss magnetic behavior and in agreement with the linearity predicted by eq. 5. The intercept, however, was ca. 7 ppm instead of zero as predicted by eq. 5.

Subsequently, the ^{1}H NMR of 1,1',3,3',5,5',7,7'-octamethyl-uranocene was analyzed in a similar manner (6). The contact shifts for the ring and α-protons were found to be similar in magnitude, but opposite in sign, implying spin density in a π-MO, and transfer of spin density via a spin polarization type mechanism (entries 3 and 4 in Table II). In this paper, a new, significantly smaller, value for the pseudocontact shift in uranocene

was reported (entry 5, Table II). This value was calculated
using better geometric data from the refined x-ray structure of
uranocene by Raymond and co-workers (34).

These results led to a simple model for the contact shifts
in uranocenes shown in Fig. 2 (35). In the ground state, orbital angular momentum dominates so the two f-electrons on the metal have their magnetic moments opposed to the applied field.
Electron density donated from filled ligand molecular orbitals to
vacant metal orbitals will be spin-polarized so the net spin density in the ligand π-MO gives rise to a magnetic moment aligned
with the applied field. Relay of spin density via a spin polarization mechanism affords an upfield shift to the ring protons,
and via hyperconjugation, a downfield shift to the α-carbons.
Subsequent spin transfer results in an alternating upfield, downfield shift pattern, which decreases substantially the greater
the number of sigma bonds between the observed nucleus and the
ring carbons.

Separation of the isotropic shifts in uranocenes into
pseudocontact and contact components is certainly an appealing
method of attributing covalent character to bonding in uranocene.
However, Hayes and Thomas (7) have advised caution in making deductions about covalency from NMR data on actinide complexes. In
these compounds J is assumed to be a good quantum number and thus,
both spin and orbital angular momentum contribute to the observed
magnetic moment. In actinide complexes, the spin magnetic moment
may not be parallel to the net magnetic moment, which is aligned
with the applied field. In fact, it is opposed if the 5f shell
is less than half full as in uranocene. Hence, direct transfer
of spin density to a ring proton will give rise to a downfield
shift.

Second and more importantly, the ligand metal interaction in
organometallic complexes involves only certain orbitals on both
the ligand and the metal. The electronic states giving rise to
shifts in an NMR experiment may not involve these orbitals.
Hence, little if any direct information on covalency can be derived from NMR experiments. In general, one must consider the
occupancy of the relevant orbitals in the crystal field states
populated over the temperature range of the NMR experiment in
attempted correlation of contact shifts with specific modes of
bonding.

Nevertheless, a model with spin polarization of ligand electrons donated to empty metal orbitals gives rise to positive spin
density in the ligand system and the observed upfield shift to
the ring protons. Such electron donation to metal orbitals does
relate to bonding. Moreover, it appears that contact shifts do
contribute to both the ring and α-proton isotropic shifts in uranocene and 1,1',3,3',5,5',7,7'-octamethyluranocene. Because both
ring and α-positions experience contact and pseudocontact shifts
it is impossible to test if the assumptions used in factoring the
observed shifts are valid. Of particular interest are the assump-

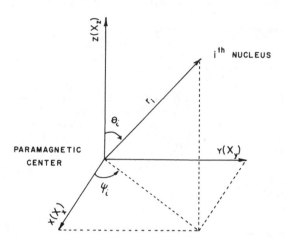

Figure 1.   Coordinates R_i, Θ_i, *and* ψ_i *of a nucleus* i *in the coordinate system* x, y, z, *with the three principal components* X_x, X_y, X_z *of the magnetic susceptibility*

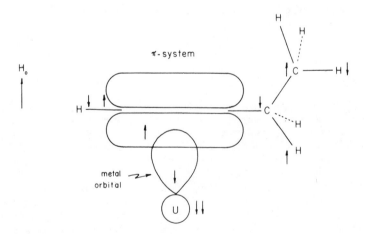

Figure 2.   Spin polarization in uranocenes. Arrows shown refer to magnetic moments.

tions concerning the magnetic anisotropy term $(\chi_{\parallel} - \chi_{\perp})$. Typically, contact shifts are effectively zero if at least three atoms (i.e., four sigma bonds) separate the observed nucleus from the paramagnetic center (15,35). Ideally, in a 1,1',3,3',5,5',7,7'-octaalkyluranocene, where the alkyl groups have β or γ protons, the observed isotropic shifts for these positions would be solely pseudocontact in nature. Unfortunately, none of these systems is known and attempts to prepare the t-butyl compound have not been successful (36).

Numerous substituted uranocenes are now known and could, in principle, provide useful tests. Other factors now, however, become involved and need to be evaluated. The lower symmetry of these compounds means that χ_x and χ_y are no longer constrained to be equal and the eq. 3 needs to be considered in its entirety. Moreover, the substituent could have an effect on magnetic anisotropy. Finally, some substituents have more than one possible conformation which would need to be considered.

If the magnetic moment of a paramagnetic molecule obeys the Curie or Curie-Weiss Law, variable temperature ^{1}H NMR can serve as a conformational probe. Conformationally rigid nuclei or those rapidly oscillating between conformations of equal energy, will exhibit a linear shift dependence on T^{-1} while those which undergo exchange between conformations differing in energy will show a non-linear dependence. Equation 3 shows that the slope of these plots will depend upon the sign of A_i and the sign of the geometry factor.

In the remainder of this paper we will present NMR results for a variety of uranocenes as a function of temperature. The results will be analyzed in terms of the component contact and pseudo-contact contributions with due regard to the foregoing considerations.

II. The Variable Temperature ^{1}H NMR of Uranocene and Substituted Uranocenes

In this section we summarize the experimental results for a number of substituted uranocenes. The compounds studied are listed in Table III and Fig. 3.

The spectra were run on the Berkeley 180 MHz FT NMR spectrometer equipped with a variable temperature probe. All spectra were run in toluene-d_8. In general, spectra were taken at 10° intervals from at least the range -80°C to 70°C. The temperature of the probe was monitored by a pre-calibrated thermocouple 5 mm from the sample tube, and could be held to ±0.3°C over the dynamic temperature range. Shifts were measured relative to the methyl group of toluene rather than stopcock grease; the latter shifted ca. 0.2 ppm over the temperature range. The shifts are reported relative to TMS by assigning the toluene methyl resonance as 2.09 ppm. This resonance differs from that in protio-toluene (2.31 ppm). Often this resonance is erroneously assigned

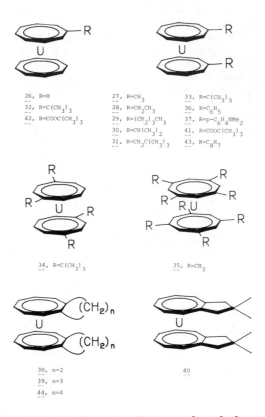

26, R=H
32, R=C(CH$_3$)$_3$
42, R=COOC(CH$_3$)$_3$

27, R=CH$_3$
28, R=CH$_2$CH$_3$
29, R=(CH$_2$)$_3$CH$_3$
30, R=CH(CH$_3$)$_2$
31, R=CH$_2$C(CH$_3$)$_3$

33, R=C(CH$_3$)$_3$
36, R=C$_6$H$_5$
37, R=p-C$_6$H$_4$NMe$_2$
41, R=COOC(CH$_3$)$_3$
43, R=C$_8$H$_7$

34, R=C(CH$_3$)$_3$

35, R=CH$_3$

38, n=2
39, n=3
44, n=4

40

Figure 3.   Structures of compounds studied

TABLE III

Uranocenes Analyzed by Variable Temperature ^{1}H NMR

26	Uranocene
27	1,1'-Dimethyl-
28	1,1'-Diethyl-
29	1,1'-Di-n-butyl-
30	1,1'-Diisopropyl
31	1,1'-Dineopentyl-
32	Mono-t-butyl
33	1,1'-Di-t-butyl
34	1,1',4,4'-Tetra-t-butyl-
35	1,1',3,3',5,5',7,7'-Octamethyl-
36	1,1'-Diphenyl-
37	1,1'-Bis(p-dimethylaminophenyl)-
38	Dicyclobuteno-
39	Dicyclopenteno-
40	Bis(dimethylcyclopenteno)-
41	1,1'-Di(t-butoxycarbonyl)-
42	Mono-(t-butoxycarbonyl)-
43	1,1'-Di(1,3,5,7-cyclooctatetraenyl)-

the same value as in the protio-compound; however, we have experimentally verified the difference which is a recognized secondary deuterium isotope effect in ^{1}H NMR spectroscopy (37,38).

A. Diamagnetic Reference Compounds. Analysis of the isotropic shifts requires referencing the observed shifts to their positions in the spectrum of a corresponding hypothetical diamagnetic uranocene. The diamagnetic thorocenes are probably the closest analogy to such a model uranocene and several of these compounds have now been reported (39,40). The difference between the ^{1}H resonances in the thorocenes and the corresponding cyclooctatetraene dianions is small (Table IV); hence, without important error isotropic shifts in all of the uranocenes discussed in this chapter can be referenced to the ^{1}H shifts in the corresponding cyclooctatetraene dianions. For those cyclooctatetraenes where the dianion has not been isolated and characterized by ^{1}H NMR, the shifts have been estimated by comparison with other

cyclooctatetraene dianions. The error resulting from such reference is probably no more than 1-2 ppm.

TABLE IV

The ^{1}H NMR Resonances of Cyclooctatetraene Dianions
and Thorocenes in THF (ref. 39, 40)

	δ ppm from TMS			
	ring	substituent		
COT$^{=a}$	5.9			
thorocene	6.5			
n-butylCOT$^{=a}$	5.7	2.9	1.3	0.9
1,1'-di-n-butylthorocene	6.5	3.2	1.6	1.0
methylCOT$^{=a}$	5.6	2.8		
1,1'-dimethylthorocene	6.5	3.1		
t-butylCOT$^{=a}$	5.7	1.5		
1,1'-di-t-butylthorocene	6.5	1.7		

[a] as the dipotassium salt

B. The Temperature Dependent ^{1}H NMR of Uranocene and
Octamethyluranocene. Our initial interest was in verifying the
temperature dependence of the ^{1}H isotropic shift in uranocene and
the reported non-zero intercept at $T^{-1}=0$. Recent laser Raman
studies by Spiro and co-workers (41) have established that the
first excited state in uranocene is 466 cm^{-1} above the ground
state. Thus, the isotropic shift may not vary linearly with the
inverse of the temperature from -100°C to 100°C. Indeed, below
100°K some controversy exists concerning the temperature dependence of the magnetic moment in uranocene (42,43).

The temperature dependence of the isotropic shift in uranocene was measured on two independent samples from -80°C to 100°C.
At the same nominal temperature slight differences in the shift
between the two samples are undoubtedly due to slight differences
in the true temperature of the samples and provide an estimate of
the error in temperature measurement or measurement of the resonance frequency in this study.

The plot of shift vs T^{-1} (fig. 4, Table V) is strictly linear
with an extrapolated intercept at $T^{-1}=0$ of zero within experimental error. The difference between this result and that reported
by Edelstein et al. (5), appears to arise entirely from uncertainty in measurement of the temperature. In the earlier work the
uncertainty in the temperature at both the high and low extremes
was +3.0°C while in this study it is +0.3°C. In fact, if one
takes into account the reported error in the temperature measurements in the earlier work, the data can be fitted with a straight
line which intercepts zero at $T^{-1}=0$. (Fig. 5).

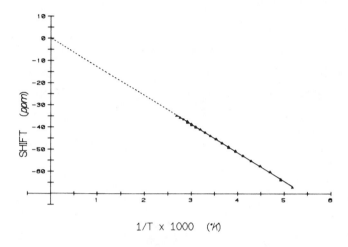

Figure 4. Isotropic shift vs. T⁻¹ for uranocene

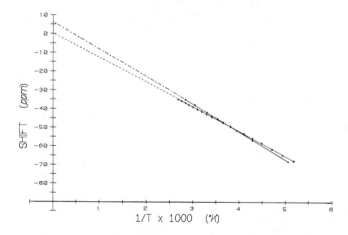

Figure 5. Comparison of older data ((· — ·), Ref. 5) with present results (– – –) for uranocene ring protons

TABLE V

Least Squares Linear Regression Analysis of VT ^{1}H NMR Data for
Ring Protons in Uranocene, Octamethyluranocene and the
Unsubstituted Ring in Monosubstituted Uranocenes.

Compound	Slope	Intercept	r^2
26, Uranocene Run #1	-12.83±0.07	-0.32±0.32	0.9992
Uranocene Run #2	-12.94±0.06	0.21±0.21	0.9997
Uranocene (ref.6)	-14.70±0.17	6.96±0.64	0.9991
32, Mono-t-butyl[a]	-12.62±0.04	-0.32±0.19	0.9998
41, Mono-t-butoxy-carbonyl[a]	-13.54±0.12	1.88±0.47	0.9989
35, Octamethyl	-13.12±0.03	2.45±0.14	0.9999

(a)   Unsubstituted ring; the substituted ring data are in
      Table IX.

Octamethyluranocene, 35, has effective 4-fold symmetry and
χ_x and χ_y are constrained to be equal on the nmr time scale. The
temperature dependence of the ring protons of this compound is
compared with uranocene in Fig. 6 and Table V. The non-zero intercept is probably due to referencing the isotropic shift to the
tetramethylCOT dianion; note in Table IV that the ring protons
of dimethylthorocene differ from methylCOT dianion by almost 1
ppm.
      The near-identity of the slopes of the lines in Fig. 6 has
important implications. The geometry factor for the ring protons
of octamethyluranocene is essentially identical to that for uranocene itself; hence, according to eq. 5, any significant change
in $\chi_\perp$ would be expected to produce a significant change in slope.
The fact that methyl substitutents have little effect on the
slope means either that $\chi_\perp$ does not change significantly by
methyl substitution or that the effect of a change in $\chi_\perp$ is almost exactly balanced by an opposing change in the contact shift.

      C.   Monosubstituted Uranocenes. Some monosubstituted urano-
cenes are known, compounds with one COT and one substituted COT
ligands.   The mono-t-butoxycarbonyluranocene, 42, was prepared
by reaction of one mole of the corresponding COT dianion with one
mole of COT dianion itself and UCl_4 (44).   It could be separated
from the disubstituted compound, 41, also formed, by its greater
stability towards hydrolysis.   Mono-t-butyluranocene, 32, was
obtained and measured as a 1.8:1 mixture with the disubstituted
compound, 33.   A separate preparation of pure 33 allowed complete
analysis of the mixture.   Mono-(di-t-butylphosphino)uranocene
has also been reported by Fischer, et al(45).

The importance of these compounds for nmr interpretations is that we can look at the unsubstituted ring in systems where χ_x and χ_y are not constrained by symmetry to be equal. In both of the monosubstituted uranocenes investigated, the proton resonance of the unsubstituted ring is a singlet.

At 30°C, the protons of the unsubstituted ring in mono-t-butyluranocene resonate at 0.51 ppm lower field and those in the mono-ester resonate at 0.43 ppm higher field than the ring protons in uranocene. These differences are small but real and were established independently by observing the spectrum of mixtures of these compounds.

The temperature dependence of the unsubstituted ring proton resonances are linear functions of T^{-1} and the slopes of shift vs. T^{-1} are identical within experimental error to that of uranocene (fig. 7, Table V). The slight difference in intercepts at $T^{-1}=0$ undoubtedly result from using the proton resonance of cyclooctatetraene dianion as a diamagnetic reference for all the compounds.

Changes in the linewidths at half heights of the unsubstituted ring resonances as a function of temperature parallels that of uranocene and results from the known change in paramagnetic relaxation times as a function of temperature rather than the onset of coalescence (Table VI) (19). This implies that ring rotation in monosubstituted uranocenes is rapid on the NMR time scale or that rotation is slow and the differences between the resonance frequency of the non-equivalent protons is smaller than the linewidths of the observed signals. Bis(1,4-di-t-butylcyclooctatetraene)-uranium, 34, does show coalescence of all of the proton resonances at low temperature corresponding to a barrier to rotation of 8.4 kcal mol^{-1}(46). Substituents smaller than t-butyl should show smaller barriers. We conclude that uranocene and the monosubstituted uranocenes are freely rotating on the nmr scale at our temperatures.

TABLE VI

Linewidth at Half Height of ^{1}H NMR Resonances of Uranocene

(H_z)

	-70°C	30°C	70°C
26, uranocene	102	90	76
32, mono-t-butyl[a]	45	33	30
41, mono-t-butoxycarbonyl-	50	38	32

(a) Unsubstituted ring.

For complete rotation, the final term in eq. 3 averages to zero; hence, if $\frac{1}{2}(\chi_x + \chi_y)$ differs seriously from $\chi_\perp$ of uranocene, we would expect a significant change in slope. The near constancy of the observed slopes for all of the unsubstituted

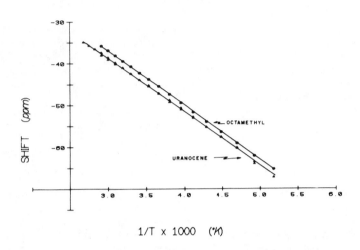

Figure 6. Isotropic shift vs. T^{-1} *for uranocene and the ring protons in 1,1',3,3',-5,5',7,7'-octamethyluranocene, 35*

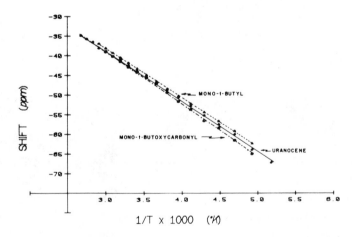

Figure 7. Isotropic shift vs. T^{-1} *for uranocene and the ring protons in the unsubstituted ring of mono-t-butyl, 32, and mono-t-butoxycarbonyluranocene, 42*

rings together with the ring protons of $\underset{\sim}{35}$ provides highly sug-
gestive, albeit not rigorous, evidence that $\chi_x = \chi_y = \chi_\perp$ for all
of these compounds. These approximations cetainly make a strong
working hypothesis.

Recently, Fischer (15,45) has independently arrived at the
same conclusion based on the temperature dependence behavior of
the ^{1}H NMR resonances of the two monosubstituted uranocenes,
$(C_8H_8)(C_8H_7R)U$, $R = P(t-C_4H_9)_2$ and $Sn(t-C_4H_9)_3$. In both of these
compounds the unsubstituted ring resonances is reported to be
identical with that in uranocene.

D. Magnetic Susceptibility of Substituted Uranocenes. We
examine further implications of the potential effects of substituents on magnetic anisotropy. In the limit of rapid ring rotation the final term in eq. 3 averages to zero; in the limit of
frozen rotations this term can contribute and result in non-linearity. The ^{1}H NMR data on $\underset{\sim}{34}$ provide a test (46). At temperatures above coalescence the rings are freely rotating and the
three pairs of equivalent ring protons on each ring are linear
functions of T^{-1}. Below coalescence the three ring proton resonances split into six and all six resonances are again linear
functions of T^{-1}. Moreover, the average of appropriate pairs of
resonances is close to the value extrapolated from three resonances above coalescenece. Thus, even in the "frozen rotation"
region, the last term in eq. 3 makes little contribution, a result that implies $\chi_x = \chi_y$.

We conclude that substitution of the uranocene skeleton, although formally lowering the symmetry of the complex, exerts
only a small perturbation on the crystal field around the uranium. The magnetic behavior remains primarily an atomic property
and from the point of view of the uranium atom, it still experiences a C_{8v} crystal field as in uranocene. Thus, to a good
first approximation, substituted uranocenes can be viewed as having effective axial symmetry regardless of the rate of ring
rotation.

We next inquire whether this result is consistent with other
physical properties of uranocenes. Bulk magnetic susceptibility
measurements at low temperature on several substituted uranocenes
appear to suggest that within experimental error the magnetic
properties of all uranocenes are essentially identical and equal
to 2.4±0.2 B.M. (Table VII). This result is consistent with the
idea confirmed by χ_α Scattered Wave (47) and Extended Hückel
MO (48) calculations that the magnetic properties of uranocenes
are determined principally by the two unpaired electrons that are
primarily metal electrons.

E. ^{1}H NMR of Substituted Uranocenes. Table VIII summarizes
the chemical shifts relative to TMS for a number of uranocenes
at a common temperature (30°C). The results are summarized for
ring and substituent protons for convenience.

F. The Temperature Dependence of Proton Resonances in Substituted Uranocenes. In substituted cyclooctatetraene dianions where substitution lifts the symmetry imposed equivalency of the ring protons, the difference in resonance frequency of the magnetically non-equivalent protons is sufficiently small that the observed resonances appear as a broadened singlet even in high field NMR experiments. Likewise in corresponding substituted thorocenes, the non-equivalent ring proton resonances appear as a broadened signal with no assignable features. However, in substituted uranocenes the non-equivalent ring proton resonances all appear as well resolved singlets for all of the uranocenes whose ^{1}H NMR has been reported.

The structure of a sufficient number of substituted uranocenes has been determined by single crystal X-ray diffraction to establish that both the uranium-ring distance and the $C_{ring}-C_{ring}$ bond distance are invariant, within experimental error, regardless of substituents on the uranocene skeleton. Assuming that the geometry factor for all of the ring protons is the same, and if $\chi_x = \chi_y$ as shown above, then the pseudocontact shift for each will be the same and the observed differences in resonance frequency must arise from differences in the contact shift at the magnetically non-equivalent ring positions. For comparison, differences in the isotropic shifts of the non-equivalent ring protons in substituted bisarenechromium complexes have been attributed to differences in the contact shift (52).

For purposes of convenient identification, the ring proton resonances in the NMR of substituted uranocenes will be labeled alphabetically starting with the lowest field resonance. This does not imply that the "A" resonances in two different uranocenes correspond to the same ring position. We shall discuss below the assignment of the individual ring proton resonances.

The temperature dependence of the ring proton resonances of the uranocenes listed in Table III were determined and plotted as shifts vs. T^{-1}. In all, 60 individual ring proton resonances in 17 different uranocenes were observed. In all cases except for one position in dicyclobutenouranocene, 38, the shifts are linear functions of T^{-1} from at least the range -70°C to 70°C. The non-linearity of the 3-position in dicyclobutenouranocene, 38, probably reflects a temperature dependent geometry change of the ring proton resulting from conformational changes in this strained ring system and will not be discussed further.

Some typical examples of the linear behavior found is summarized in Figs. 8-11. The complete set of plots is given in ref. 53 and the linear regressions are summarized in Table IX.

Note in these results that the total difference between the highest and lowest field resonance of the non-equivalent ring protons in all of the uranocenes increases as the temperature decreases. Moreover, the relative pattern of the ring proton resonances in each uranocene remains constant as a function of temperature except for the two phenyl-substituted uranocenes and 1,1'-

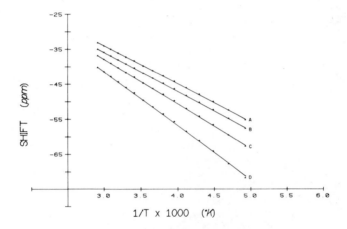

Figure 8.   Isotropic shift vs. T^{-1} *for the ring protons in 1,1'-dimethyluranocene, 27*

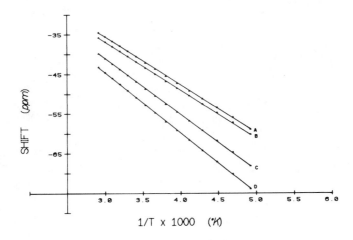

Figure 9.   Isotropic shift vs. T⁻¹ for the ring protons in the substituted ring of
mono-t-butyluranocene, 32

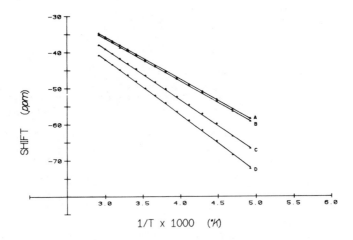

Figure 10.   Isotropic shift vs. T⁻¹ for the ring protons in 1,1'-di-t-butyluranocene,
33

TABLE VII

Magnetic Properties of Uranocenes

	Substituent[a]	Temp. Range °K	μ_{eff} B.M.	Weiss Constant °K	Ref.
26	H	4.2–4.5	2.43	9.56	42
	H	4–10	3.33	9.4	31
	H	10–42	2.3	0.9	31
	H	12–72	2.42	2.9	43
	H	180–300	2.62	3	43
	H	10–300	2.6		5
27	CH_3	14.5–81.5	2.26±0.2	17	49
28	CH_2CH_3	3–8	2.86	14.9	31
	CH_2CH_3	10–27	1.9	0.4	31
	CH_2CH_3	14.7–79.6	2.13±0.2	−7	49
29	$(CH_2)_3CH_3$	3–10	2.85	5.8	31
	$(CH_2)_3CH_3$	10–50	2.3	2.6	31
36	C_6H_5	14–100	2.65±0.2	12.2±3	50
38	cyclobuteno	15–100	2.35±0.2	8.5	14
39	cyclopenteno	15–95.6	2.4±0.2	16.1	49
44	cyclohexeno	14.4–97.8	2.65±0.2	23	49
35	1,3,5,7-tetramethyl	1.9–73.7	2.2±0.2	11.3±3	50
	1,3,5,7-tetraphenyl	4.2–100	2.5±0.1	6.7±1	50
41	$CO_2\text{-}t\text{-}C_4H_9$	30–100	2.64±0.2	10.4	44

(a) Both rings substituted

TABLE VIII

^{1}H NMR Resonances of Substituted Uranocenes

δ ppm from TMS

Substituent[a]	Shift[b] at 30°C
26, H	-36.63
35, 1,3,5,7-tetramethyl	-35.15, -4.21 (CH_3)
27, CH_3	-31.70, -33.67 (H5), -36.10,-40.39 -7.20 (CH_3)
28, CH_2CH_3	-32.89, -34.45 (H5), -36.33, -39.7 -17.47 (CH_2), -1.20 (CH_3)
30, i-C_3H_7	-35.50, -35.98, -36.00 (H5), -36.40 -14.47 (CH), -9.89(CH_3,d,J=4.4 Hz)
29, n-C_4H_9	-32.64, -34.10 (H5), -36.22, -39.74 -19.03 (α-CH_2), 0.22 (β-CH_2) 0.98 (q-CH_2,m), 0.36 (CH_3,t,J= 6.3 Hz)
33, t-C_4H_9	-33.43, -33.80, -37.30, -40.54 (H5) -11.49 (CH_3)
32, t-C_4H_9	-33.41, -34.74, -39.51, -43.37 (H5) -36.02 (8H, unsubstituted ring), -10.82 (CH_3)
34, 1,4-di-t-butyl	-25.23, -39.66, -42.23 -10.25 (CH_3)
31, neo-C_5H_{11}	-32.84, -33.42 (H5), -36.26, -41.07 -23.97 (CH_2), 3.86 (CH_3)
$(CH_2)_3N(CH_3)_2$[d]	-31.5, -32.9 (H5), -34.9, -38.1, -18.3 (α-CH_2,t,J=7.5 Hz) 0.63 (β-CH_2,m), 113 (CH_3) 2.80 (γ-CH_2,t,J=7.0 Hz)
36, C_6H_5	-34.29, -36.15, -36.45, -37.13 (H5) 0.76 (p,d,J=7.2 Hz) 0.85 (m,t,J=7.6 Hz) -13.95 (o,d,J=7.3 Hz)
37, p-$(CH_3)_2NC_6H_4$	-34.29, -36.15, -36.46, -37.13 (H5) -14.10 (o,d,J=7.6 Hz) o.13 (m,d,J=7.6 Hz), -0.04 (CH_3)
38, cyclobuteno	-27.70, -35.90, -43.80 -26.75 ($α_{endo}$) 19.65 ($α_{exo}$) (J=9.64 Hz)

TABLE VIII (cont.)

Substituent[a]	Shift[b] at 30°C
39, cyclopenteno	-32.12, -34.20, -41.15 -32.58 (β_{endo},m) -8.28 (β_{exo},m) -18.78 (α_{endo},m) 24.43 (α_{exo},m)
40, dimethyl cyclopenteno	-32.43, -33.26, -39.83 -12.91 (CH_{3endo}) 5.39 (CH_{3exo}) -22.90 (α_{endo}) 8.28 (α_{exo}) (J=14.5 Hz)
44, cyclohexeno	-30.64, -32.53, -38.70 -22.35 (β_{endo},m) -2.94 (β_{exo},m) -16.42 (α_{endo},m) 6.56 (α_{exo},m)
$C(C_6H_5)_3$[e]	-21.35, -34.87, -49.50, -52.01(H5) 4.88 (o,d,J=6.8 Hz) 4.95 (m,t,J=6.6 Hz) 5.44 (p,t,J=6.6 Hz)
OCH_3[d]	-27.5, -30.2 (H5), -35.6, -43.7 -3.73 (CH_3)
$O-t-C_4H_9$[d]	-28.1, -28.7 (H5), -36.2, -45.7 2.08 (CH_3)
$OCH_2CH CH_2$[d]	-27.9, -30.5 (H5), -35.5, -43.5 -0.33 ($\alpha-CH_2$,d,J=5.0 Hz) 0.70 (trans-H, d,J=17.5 Hz) 1.75 (β-CH,m), 2.60 (cis-H,d,J=10.5 Hz)
41, $CO_2-t-C_4H_9$	-30.51, -32.65, -36.01 (H5),-42.45 -6.07 (CH_3)
42, $CO_2-t-C_4H_9$[c]	-29.42, -33.69, -36.0 (H5), -40.06 -37.06 (8H, unsubstituted ring) -6.27 (CH_3)
$CO_2CH_2C_6H_5$	-29.81, -32.08, -36.23 (H5),-43.16 -2.98 (CH_2), -0.56 (o) 4.09 (m) 5.20 (p)
$CO_2CH_2C_6H_5$[c]	-28.51, -32.40, -32.98 (H5),-40.63 -2.99 (CH_2), -36.06 (8H, unsubstituted ring) -1.16 (o) 3.94 (m) 5.30 (p)

TABLE VIII (cont.)

Substituent[a]	Shift[b] at 30°C
$CO_2CH_2CH_3$	-29.93, -32.69, -35.78 (H5),-42.14 -6.05 (CH_3), -4.23 (CH_2)
$CO_2CH_2CH_3$ [c]	-28.84, -32.93, -36.14 (H5),-40.27 -6.57 (CH_3), -4.45 (CH_2), -37.07 (8H,unsubstituted ring)

(a) Substituent on each 8-membered ring.

(b) In monosubstituted cyclooctatetraene ligands the
     ring H5 could be identified by integration relative
     to the other ring proton resonances.

(c) Monosubstituted.

(d) Data from ref. 51 at 39°C.

(e) At 26°.

biscyclooctatetraenyluranocene. In these latter cases the substituent ^{1}H NMR spectra show slowing of rotation and coalescence
phenomena to be discussed below; these phenomena may also affect
some of the ring protons.

     The high degree of linearity in the temperature dependence of
the ring proton shifts is evident from the correlation coefficients of the least squares regression lines (Table IX). The
slopes of the lines are all negative and similar in magnitude to
that of uranocene. However, the standard deviations of the extrapolated intercepts at $T^{-1}=0$ indicate that a number of the intercepts are non-zero. Ideally, eq. 3 predicts that all of the intercepts should be zero at $T^{-1}=0$.

     Considering all of the ring proton resonances together, there
is no apparent correlation between the non-zero intercepts and
the magnitude of the isotropic shifts at a given temperature, say
30°C. However, for some individual uranocenes, it appears that a
correlation does exist such that the intercept increases the larger the isotropic shift at a given temperature. This seems to
suggest that the non-zero intercepts are in some way associated
with the contact shift.

     The linear dependence of the isotropic shifts on T^{-1} over the
observed temperature range can imply one of two things: 1) both
the contact and pseudocontact shifts are linear functions of T^{-1};
2) the contact shift is a linear function of T^{-1} while the
pseudocontact shift is a function of both T^{-1} and higher orders
of T^{-1}, where the combined contact and pseudocontact T^{-1} dependence is large relative to the higher order terms of the pseudocontact shift. In principle, these two possibilites can be dif-
ferentiated by observing the temperature dependence of α and β

protons whose geometry factor is invarient with temperature. The
contact shift for α and particularly for β-protons should be substantially smaller than for ring protons. Hence, their temperature
dependence should be linear in T^{-1} if the former is true, but non-
linear if the latter is true.

In fact, studies of a number of substituent protons in substituted uranocenes provide linear correlations with T^{-1} (54).
The temperature dependence of the substituent proton resonances in
1,1',3,3',5,5',7,7'-octamethyl-, 35, mono-t-butyl-, 32, and 1,1'-
di-t-butyluranocene, 33, are all linear. Similarly, both the
methylene and methyl protons of 1,1'-dineopentyluranocene 31 are
linear. For this case, the results imply a relatively fixed conformation with the t-butyl group swung away from the central uranium (conformation A in Figure 12; R=t-OBu). The non-linearity of
the methyl protons of 1,1'-diethyluranocene 28 is interpreted as
an effect of temperature on the populations of different conformations having different pseudo-contact shifts. Conformation A in
Figure 12 (R=CH$_3$) predominates but other conformations also contribute. We have no simple interpretation of the non-linearity of
1,1'-dimethyluranocene, 27, at this time. Some of the results are
summarized in Table X.

An interesting special case is that of 1,1'-di(cycloocta-
tetraenyl)uranocene, 43. Both Miller (55) and, recently, Spiegel
and Fischer (56) have reported that the number of substituent
and ring proton resonances vary as a function of temperature indicative of a dynamic process which is slow on the NMR time scale.
Above 90°C, the spectrum consists of four ring proton resonances
in an area ratio of 2:2:2:1 similar to that of other 1,1'-disub-
stituted uranocenes. At 30°C, six broad ring proton resonances
are present and determination of relative areas is extremely difficult. Initially, we had hoped that monitoring coalescence of
the ring protons in this system would provide a method of assigning individual ring proton resonances. However, interpretation of
the temperature dependent changes was not straightforward and no
assignment could be made.

Initially, the B ring resonance begins to broaden at 80°C,
followed by the A resonance at ca. 70°C, and both merge into a
single peak at 50°C. Below this temperature, they rapidly separate into three broad peaks at 40°C and to at least six peaks at
30°C. At 40°C, the C resonance also begins to coalesce followed
by the D resonance at ca. 30°C. Below 30°C, it is not clear
which of the peaks in the 'low temperature' spectrum are associated with peaks in the 'high temperature' spectrum. From 0°C
to -80°C, eleven ring proton resonances are discernible; however,
relative peak areas indicate that not all of the individual resonances are resolved.

Similar temperature dependence behavior is observed for the
substitutent proton resonances. At 90°C, all of the resonances
have coalesced into the baseline, while at 80°C a resonance appears at 1.8 ppm, followed at 70°C by the appearance of two broad

TABLE IX

Least Squares Linear Regression Lines For

Alkyl Uranocene Ring Proton Data

Fig. no.	Substitutent	Proton Resonance	Slope	Intercept	r^2
8	methyl	A	-10.89±0.05	-1.35±0.19	0.9997
		B	-11.20±0.05	-2.30±0.18	0.9997
		C	-12.80±0.06	0.57±0.24	0.9996
		D	-15.59±0.09	5.48±0.36	0.9994
9	t-butyl[a]	A	-12.12±0.04	0.75±0.17	0.9998
		B	-12.08±0.04	0.71±0.16	0.9998
		C	-14.03±0.04	0.92±0.17	0.9999
		D	-15.22±0.05	1.00±0.18	0.9999
10	t-butyl	A	-11.80±0.03	-0.37±0.12	0.9999
		B	-11.89±0.05	-0.51±0.18	0.9998
		C	-14.24±0.08	3.77±0.32	0.9995
		D	-15.59±0.11	4.96±0.41	0.9993
	ethyl	A	-10.95±0.06	-2.43±0.20	0.9994
		B	-10.85±0.06	-4.31±0.20	0.9994
		C	-13.12±0.08	1.17±0.31	0.9991
		D	-16.04±0.13	7.25±0.46	0.9986
	n-butyl	A	-10.78±0.02	-2.77±0.08	0.9999
		B	-10.73±0.02	-4.37±0.08	0.9999
		C	-12.85±0.03	0.39±0.11	0.9999
		D	-15.69±0.07	6.14±0.25	0.9996
	neopentyl	A	-11.27±0.05	-1.68±0.17	0.9998
		B	-11.15±0.05	-2.68±0.19	0.9997
		C	-13.01±0.05	0.58±0.19	0.9998
		D	-15.93±0.07	5.31±0.25	0.9998
	isopropyl	A	-13.06±0.05	1.68±0.19	0.9999
		B	-12.79±0.05	0.39±0.19	0.9998
		C	-13.12±0.05	1.40±0.19	0.9998
		D	-13.44±0.08	2.04±0.30	0.9995
	cyclobuteno	A	non-linear		
		B	-12.81±0.12	0.65±0.48	0.9984
		C	-17.58±0.20	8.45±0.77	0.9979
	cyclopenteno	A	-11.07±0.25	-1.50±1.02	0.9912
		B	-13.20±0.08	3.67±0.31	0.9994
		C	-16.84±0.23	9.02±1.04	0.9960
	dimethylcyclo- cyclopenteno	A	-10.29±0.05	-4.11±0.21	0.9996
		B	-12.26±0.06	1.52±0.22	0.9997
		C	-16.84±0.11	9.93±0.45	0.9993

TABLE IX (cont.)

Fig.no.	Substituent	Proton Resonance	Slope	Intercept	r^2
11	phenyl	A	-12.04±0.08	-0.71±0.32	0.9992
		B	-12.03±0.10	-2.58±0.38	0.9989
		C	-13.95±0.09	3.41±0.34	0.9994
		D	-12.05±0.10	-3.52±0.40	0.9988
	p-dimethyl-aminophenyl-	A	-11.23±0.10	-3.49±0.41	0.9985
		B	-12.20±0.13	-0.49±0.53	0.9979
		C	-11.17±0.09	-5.21±0.38	0.9987
		D	-14.93±0.19	5.74±0.76	0.9971
	t-butoxy-carbonyl[a]	A	-11.02±0.08	0.19±0.31	0.9993
		B	-12.47±0.12	2.89±0.47	0.9987
		C	-13.01±0.08	1.30±0.31	0.9995
		D	-14.62±0.09	0.19±0.36	0.9994
	t-butoxy-carbonyl	A	-4.63±0.02	1.16±0.08	0.9997
		B	-10.52±0.07	-0.37±0.29	0.9993
		C	-12.60±0.12	2.36±0.47	0.9987
		D	-14.05±0.11	0.73±0.43	0.9991

[a] substituted ring of monosubsituted uranocene

resonances at -9.0 ppm and -14.9 ppm and a sharper resonance at
ca. 0.0 ppm. Labeling these resonances as K (1.8 ppm), L (0.0
ppm), M (-9.0 ppm) and N (-14.9 ppm), the L resonance separates
into two peaks at ca. 50°C, while the other resonances remain
fairly sharp. At 30°C, the N resonance begins to broaden and
separates into two peaks at 20°C, followed by broadening of the M
resonance. At 10°C, the M and N regions each consist of two resonances while the two resonacnes of the L region are broadened.
The behavior of the K resonance is obscured by the TMS/grease signal. At 0°C, the L region consists of four resonances. At -80°C,
the M and N signals are both well separated sets of two resonances
each while the K and L regions consist, respectively, of four and
five sets of double resonances of essentially equal area.
     Of the substituent resonances, only the M and N signals can
be definitely assigned to the α position of the uncomplexed ring.
At low temperature, the α position protons are equally distributed in four magnetically different environments.
     A combination of slowing or effective stopping of several
dynamic exchange processes could give rise to the observed changes
in the spectrum: 1) tub-tub interconversion of the uncomplexed
cyclooctatetraene ring; 2) double bond reorganization in the uncomplexed cycloocatetraene ring; 3) rotation about the $C_{ring}-C_\alpha$
bond; 4) ring-ring rotation in the uranocene moiety. The presence of four different α position resonances in the 'low temperature' spectrum requires that double bond reorganization be slow

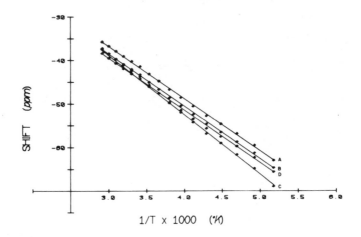

Figure 11.   Isotropic shift vs. T⁻¹ for the ring protons in 1,1′-di-phenyluranocene

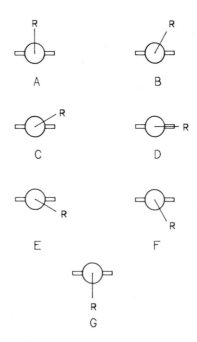

Figure 12.   Conformations of the substituent in substituted uranocenes shown in Newman projection form with the uranium atom below the plane of the ring in each figure

TABLE X

Least Squares Regression Data for Alkyl Uranocene
Substituent Proton Data vs T^{-1}

Compound	Proton Resonance	Slope	Intercept	r^2
α–Protons				
27	methyl	non-linear		
35	octamethyl	-5.53±0.04	11.19±0.14	0.9993
28	ethyl	-12.65±0.17	21.84±0.61	0.9962
29	n-butyl	-12.22±0.10	18.10±0.37	0.9995
31	neopentyl	-12.5 ± 0.73	14.02±0.28	0.9995
30	isopropyl	-8.58±0.05	11.47±0.20	0.9994
β–Protons				
32	mono-t-butyl	-5.42±0.03	4.42±0.10	0.9996
33	t-butyl	-5.60±0.07	5.32±0.25	0.9979
34	tetra-t-butyl	-5.09±0.06	5.03±0.22	0.9980
29	n-butyl	non-linear		
28	ethyl	non-linear		
30	isopropyl	-4.69±0.04	4.51±0.14	0.9991
γ and δ Protons				
31	neopentyl t-butyl	1.14±0.01	-0.97±0.03	0.9992
29	n-butyl $-CH_2$	non-linear		
29	n-butyl CH_3	non-linear		

relative to the NMR time scale. This implies that in the 'high
temperature' spectra, where double bond reorganization is rapid,
four rather than seven substituent resonances should be observed.
Unfortunately, due to solvent and instrumental limitations, we
could not obtain spectra above 100°C to confirm this. The data
do not permit further differentiation between the other possible
dynamic exchange processes.

III.   Identification of Ring Proton Resonances in Substituted
        Uranocenes.

        In all of the mono- and 1,1'-disubstituted uranocenes prepared
to date, the ^1H NMR resonances of the non-equivalent protons in

the substituted rings are all well resolved singlets, three of
area 2 and one of area 1. From Table VIII the total difference
between the highest and lowest field resonances at 30°C in such
uranocenes varies from 0.9 ppm to 30.6 ppm. It seems likely in
most cases that the difference in ring proton resonances arises
from differences in the contact shift at each of the non-equivalent positions in the 8-membered ring. One might therefore expect
a correlation between the contact shift and the spin density at
the various ring positions. Attempting such correlation requires
assigning all of the ring proton resonances in 1,1'-disubstituted
uranocenes.

    Integration readily differentiates the 5 position, of area 1,
from the remaining three postions, of area 2. Inspection of
Table VIII shows that there is no apparent correlation between
the electron-donating or withdrawing character of the substituent
and the position of the 5 proton resonance relative to the other
proton resonances. Figure 13 shows the patterns of ring proton
resonances for some 1,1'-disubstituted uranocenes in a more schematic form. The pattern of the results strongly suggests that for
primary alkyl substituents the assignments of the A,B,C, and D
resonances are all the same. In all of these cases the B resonance is identified with the 5-position. Important changes do
occur, however, for isopropyl and t-butyl substituents. For isopropyl, the ring proton resonances are closely bunched together.
For t-butyl, the 5-position is now the D-resonance for both the
mono- or disubstituted uranocenes.

    A tentative assignment of the other ring protons in the t-
butylCOT ligand may be made in the following way. The barrier to
rotation in tetra-t-butyluranocene, $\underset{\sim}{34}$, suggests that conformations of 1,1'-di-t-butyluranocene with the t-butyl rings close
(Figure 14a) will be relatively unpopulated compared to populations with the bulky t-butyl groups farther apart, Figure 14b, c,
and d. Next, we note that the substituted ring protons in mono-t-
butyluranocene, $\underset{\sim}{32}$, show some significant differences from 1,1'-
di-t-butyluranocene, $\underset{\sim}{33}$. The $\Delta\sigma$ values for the A,B,C, and D
resonances are, respectively, -0.02, 0.94, 2.21, 2.83 (H5) ppm.
The largest change is associated with the known position H5 for
which conformation (d) in Figure 14 has a high population. This
suggests that the presence nearby of a t-butyl group in the other
ring has a perturbing effect to shift the ring proton resonance to
lower field. On this basis, H2, which rarely has such a "de-
shielding" perturbation, may be assigned resonance A. Similarly,
conformation (c) in Fig. 14 is probably more highly populated than
(b); hence, H4 is assigned to resonance C and H3 to B. That is,
this argument provides assignments of resonances A,B,C, and D
to positions 2,3,4, and 5, respectively. Although Figure 14 is
based on eclipsed conformations the same approach applies to analogous staggered conformations.

    This approach finds confirmation in the effect of the t-butyl
group of one ring on the unsubstituted ring of mono-t-butyl-uran-

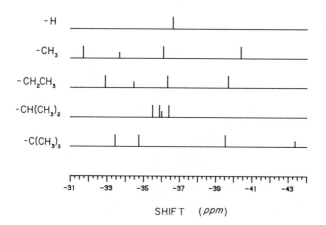

Figure 13. Pattern of ring proton resonances of 1,1'-dialkyluranocenes. H5 is indicated by its reduced intensity.

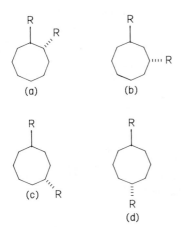

Figure 14. Several conformations of 1,1'-disubstituted uranocenes about the central axis. The 1'-substituent is shown with the dotted line.

ocene, 32. In this compound, each hydrogen is equally likely to have a t-butyl group nearby in the other ring. The result is an average shift of 0.6 ppm having the appropriate direction and approximate magnitude.

A more rigorous approach to assigning resonances to structure is by deuterium labeling. Methylcyclooctatetraene-4-d was prepared via sulfone chemistry pioneered by Pacquette and co-workers (57), (Fig. 15). Lewis acid catalyzed addition of SO_2 to COT gave the sulfone which was dilithiated with butyllithium and quenched with D_2O. The dideuterio compound was mono-metallated with butyllithium and quenched witn methyl iodide. Attempted prior alkylation followed by introduction of deuterium was found to be far less successful. Pyrolysis of the deuteriosulfone by slow sublimation through a pyrex tube packed with glass helices at 400° gave the desired methylcyclooctatetraene-4-d in 87% yield. Reduction to the dianion with potassium metal and reaction with UCl_4 gave 1,1'-dimethyluranocene-4·,4'-d_2. The nmr spectrum showed incorporation of 1.5d. Only the A resonance was affected and can be rigorously assigned the 4-position.

We also prepared deuterated butylcyclooctatetraenes by bromination of butylcyclooctatetraene followed by dehydrobromination, metallation with butyllithium and quenching with D_2O. Location of the deuterium in the product is, however, not straightforward. Pacquette has studied the bromination of methylcyclooctatetraene and has identified different bromination products on different occasions (58,59). It appears from his work that bromination-dehydrobromination of methylcyclooctatetraene can lead to all four possible methylbromocyclooctatetraenes.

In our case with the butyl compound, various workup procedures were applied to the butylbromoCOT product; e.g., reduced pressure short path distillation in one run, silica gel chromatography in another. The deuterio-products were converted to the corresponding deuterated 1,1'-dibutyluranocenes giving the nmr results in Table XI. Included are the total deuterium incorporations by mass spectral analysis. The A resonance is identified by analogy to dimethyluranocene (vide supra) as the 4-position. The B resonance is established by integration to be the 5-position. Of the two remaining positions only the D resonance is undeuterated in all preparations. It seems most likely by consideration of steric hindrance effects in the reaction mechanism for dehydrobromination that the undeuterated position must be the 2-position. Accordingly, the most probable assignment of the ring resonances in primary alkyl uranocenes is that A, B, C, D correspond to positions 4, 5, 3, 2, respectively. The corresponding $\Delta\delta$ in dimethyluranocene relative to uranocene itself is, therefore, 2, -3.8 ppm; 3, +0.5 ppm; 4, +4.9 ppm; 5, 3.0 ppm. Only the 2-position, adjacent to the alkyl group, suffers an upfield shift.

Some comparisons suggest that these effects may be additive. For example, in octamethyluranocene, 35, a given ring proton is 1,2 with respect to two methyls and 1,4 with respect to two more.

TABLE XI

Proton NMR of Deuterated 1,1'-Dibutyluranocenes

Run	Total d-incorporation	% Deuterium Incorporation in Ring Resonances			
		A	B	C	D
1	1.59	69	20	0	0
2	0.93	28	5	20	0
3[a]	1.74	61	0	14	0

(a) Prepared earlier with somewhat different conditions
     by Dr. C. LeVanda.

An additive effect would give $\Delta\delta=2$ (-3.8) + 2 (4.9)=2.2 ppm.
The actual $\Delta\delta$ relative to uranocene is 1.5 ppm (Table VIII) in
good agreement. Further development of this approach may prove
useful in other assignments. For example, if the $\Delta\delta$ values for
1,1'-di-ethyluranocene (-3.1, +0.3, 3.7, 2,2 for the 2,3,4,5
positions, respectively) are applied to the three ring positions
of bis-cyclohexenouranocene, 44, we can assign the ring resonances
A, B, C to positions 5,4,3, respectively, and obtain the following
experimental and calculated $\Delta\delta$ ppm, respectively, relative to uranocene: 3-, -2.1, -2.8; 4-, +4.1, +4.0; 5-, +6.0, +5.9.
    These correspondences help confirm the assignments made above.
But now we can inquire why the 2-position in primary alkyl uranocenes is furthest upfield whereas in t-butyluranocene it is furthest downfield. This marked difference suggests a significant
difference in structure. In all uranocenes whose structures have
been established by X-ray analysis so far, ring-carbon substituent bonds are tilted towards the central uranium by several degrees. This effect probably occurs to provide better overlap between ligand π and central metal orbitals. With the t-butyl group,
however, even for a ring-carbon bond coplanar with the ring,
methyl hydrogens approach within van der Waals distance of the
other ring. We suggest, therefore, that in t-butyluranocenes the
t-butyl group is tilted away from the uranium with a consequent
perturbation of the C_8 ring that shows up in the nmr spectra. We
hope to test this prediction by X-ray structure analysis of suitable compounds.

IV. Factoring the ^{1}H Isotropic Shifts in Alkyluranocenes.

    The discussions above have shown that the pseudocontact com-
ponent of the isotropic shift in 1,1'-dialkyluranocenes is accurately given by the axially symmetric form of eq. 3 and thus, these
systems can be used in evaluating both the assumptions employed in
deriving, and the value of the anisotropy term $(\chi_{||} - \chi_{\perp})$ used,
by previous workers in factoring isotropic shifts in uranocenes.

In this section we present such an analysis comparing pseudo-contact shifts calculated assuming $\chi_{\parallel} - \chi_{\perp} = 3\chi_{av}$ and assuming values of $\chi_{\parallel} - \chi_{\perp}$ derived from isotropic shift and geometric data for protons which experience little or no contact shift. However, prior to such analysis it is important to be cognizant of the accuracy of calculated pseudocontact and contact shifts. Irrespective of the method or the equation(s) used to calculate pseudocontact shifts, three factors limit their accuracy: a) errors in measurement of the isotropic shift; b) errors in the assumed geometry; c) errors in the magnetic anisotropy. For uranocenes, the uncertainty associated with the isotropic shifts is small, larger for the assumed geometries and largest for the assumed anisotropy difference. In calculating shifts assuming $\chi_{\parallel} - \chi_{\perp} = 3\chi_{av}$, Table VII shows that to a good first approximation, $\chi_{av} = 2.4 \pm 0.2$ B.M. for all uranocenes. As a result of the 10% uncertainty in this value, calculated pseudocontact shifts will have an uncertainty of at least 10%. Similarly, in using a value of $\chi_{\parallel} - \chi_{\perp}$ derived from isotropic shift and geometric data, the uncertainty associated with calculated pseudocontact shifts will depend upon the reference compound chosen and will undoubtedly be of the same order of magnitude. Thus, the factored shifts in the following section will have an error of at least 10%.

In the following discussion, all calculated shifts are derived assuming a temperature of 30°C. For numerical convenience, the anisotropy term $\chi_{\parallel} - \chi_{\perp}$ will be expressed in terms of $\mu_{\parallel}^2 - \mu_{\perp}^2$.

Fischer has proposed useful and important methods for factoring the isotropic shifts of uranocenes into contact and pseudocontact components (15); values were reported for uranocene, 1,-1',3,3',5,5',7,7'-octamethyluranocene, and 1 1'-bis(trimethylsilyl)uranocene using a non-zero value of $\chi_{\perp}$. Fischer arrived at values of $\mu_{\parallel}^2$ and $\mu_{\perp}^2$ at several temperatures from the ratio of the geometry factor and the isotropic shift for methyl protons in bis(trimethylsilyl)-uranocene, and bulk magnetic susceptibility data, assuming no contact contributions to the isotropic shift of the methyl protons. From the published data of Fischer, the value of $\mu_{\parallel}^2 - \mu_{\perp}^2$ at 30°C is 8.78 BM^2.

His results show that $\mu_{\perp}$ is small but not zero. The non-zero $\mu_{\perp}$ component has the effect of reducing the magnitude of pseudo-contact shifts. There seems little doubt that Fischer's result is qualitatively correct but the several assumptions required, especially of geometry, make them quantitatively suspect. For example, 1,1'-bis(trimethylsilyl)uranocene shows the same pattern of ring proton resonances as 1,1'-di-t-butyluranocene; hence, the structure may involve a trimethylsilyl group bent away from the ring plane. Such a distortion would change the calculated geometry factors and the derived value of $\mu_{\parallel}^2 - \mu_{\perp}^2$.

In our approach we have determined $\mu_{\parallel}^2 - \mu_{\perp}^2$ by another approach involving dicyclobutenouranocene, 38, and have compared the results for 1,1'-di-t-butyluranocene, 33, and 1,1'-dineo-pentyluranocene, 31. These three test systems contain α, β and γ-

protons constrained in relatively known geometric configurations
relative to the uranium center. In the latter two compounds, contact contributions to the t-butyl isotropic shift must be vanish-
ingly small, whereas in the first compound, the fixed geometric
relationship of the methylene group relative to the 8-membered
ring suggests that both hyperconjugation and the contact shift
must be effectively the same for the exo and endo protons, if
the contact shift results from hyperconjugation transfer of spin
density.

The average geometry factor of the t-butyl group in 1,1'-di-
t-butyluranocene was taken as $1/6$ (A + 2C + 2E + G) (Fig. 12,
$R=CH_3$) in Table XII and for the t-butyl group in the neopentyl
substitutent it was taken as conformation A (Fig. 12, R=t-Bu)
(Table XIII). While the methylene protons in dicyclobutenouranocene are conformationally mobile, as evidenced by their temperature dependent ^1H NMR spectra, we assume that their average
position in solution is given adequately by the average position
of the methylene groups in the X-ray crystal structure. Although
atomic coordinates are reported for all of the atoms in the X-ray
structure, geometry factors calculated from these data are probably in error for two reasons: 1) the reported coordinates are
not thermally corrected, and thus, they reflect an average
$C_{ring} - C_{ring}$ bond length of 1.39 A rather than a thermally corrected value of 1.41 A; 2) the two reported $H_{exo} - C_\alpha - H_{endo}$
bond angles of 104° and 106° are certainly too small and reflect
the large uncertainty associated with the location of hydrogen
atoms by X-ray diffraction.

TABLE XII

Calculated Geometry Factors $\dfrac{3\cos^2\theta-1}{R^3}$

For β Methyl Group
($R=CH_3$ in Fig. 12)

Conformation[a]	Planar $G_i \times 10^{21} cm^{-3}$	5° Tip[b] $G_i \times 19^{21} cm^{-3}$
A	2.563	1.793
B	1.756	0.9977
C	-0.7736	-1.557
D	-5.081	-6.082
E	-10.32	-11.83
F	-14.64	-16.65
G	-16.45	-18.74

(a)   Figure 12.

(b)   Towards uranium.

Formally, the fused 4-membered ring is similar to the 4-membered ring of cyclobutene or benzocyclobutene, and the methylene bond angle should be similar to the methylene bond angle in these compounds. Gas phase electron diffraction of cyclobutene gives this angle as 110° (60), whereas $J_{13_{C-H}}$ coupling constants yield a value of 114° (61). Similarly, $J_{13_{C-H}}$ coupling constant analysis predicts a bond angle of 112° in benzocyclobutene (62). Thus, 112° is certainly a more realistic value for the $H_{exo} - C_\alpha - H_{endo}$ bond angle. In calculating geometry factors for the exo and endo protons, we have used the idealized geometry in Fig. 16, which more accurately describes the location of the methylene protons, rather than the coordinates of the atoms from the published X-ray crystal structure. The geometry factors for the exo and endo protons calculated with these data are -0.7097×10^{21} cm^{-3} and -16.97×10^{21} cm^{-1}, respectively.

Considering first the Edelstein, et al. (5), proposal that $\chi_{\parallel} - \chi_{\perp} = 3\chi_{av}$, the average μ_{eff} of 2.4 ± 0.2 B.M. for uranocene and substituted uranocenes affords a value of 17.28 BM^2 for μ_{av}^2. The calculated pseudocontact shifts for a t-butyl group and the t-butyl protons in a neopentyl group, assuming coplanarity of the $C_{ring}-C_\alpha$ bond and the 8-membered ring, are -23.7 ppm and 14.3 ppm, respectively. With a tipped substituent the values are, respectively, -28.8 ppm and 6.35 ppm. Comparison with the experimental isotropic shifts of -13.29 ppm and 2.76 ppm shows that the calculated values overestimate the magnitude of the pseudocontact shift. Factoring the isotropic shifts of the methylene protons in the cyclobuteno group further reinforces this result. The calculated pseudocontact shifts are: exo -2.80 ppm, endo -67.0 ppm. By difference from the experimental isotropic shifts of 15.19 ppm (exo) and -31.20 ppm (endo), the corresponding contact shifts are 18.0 ppm (exo) and 35.8 ppm (endo). The large difference in the contact shifts cannot result from slight differences in hyperconjugation, which may arise from the ca. 6° decrease in the 180° dihedral angle between the 4- and 8-membered rings, but clearly results from overestimation of the pseudocontact shift.

Reducing the magnitude of the calculated pseudocontact shifts requires smaller values of the anisotropy term $\chi_{\parallel} - \chi_{\perp}$, which can only result if $\chi_{\perp} \neq 0$. This result provides independent confirmation of the same result of Fischer cited above. We noted also that both the electronic structure of uranocene proposed by Warren (63), assuming a $J_z = \pm 4$ ground state, and a recent model proposed by Fischer (15), assuming a $J_z = 3$ ground state, show that $\chi_{\perp}$ is non-zero, and less than $\chi_{\parallel}$, at 30°C.

Using Fischer's value of $\mu_{\parallel}^2 - \mu_{\perp}^2 = 8.78$ BM^2 the calculated pseudocontact shifts for the t-butyl groups in 1,1'-di-t-butyl- and 1,1'-dineopentyl uranocene are -12.1 ppm and 7.28 ppm, respectively, for coplanar substituents, and -14.6 ppm and 3.22 ppm, respectively, for tipped substituents. Agreement between the calculated pseudocontact shifts and the observed isotropic shifts is rather good. Calculation of the pseudocontact shifts for the cyc-

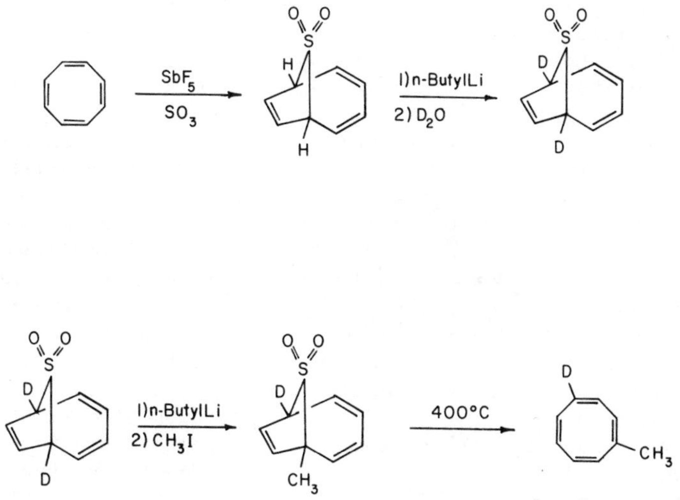

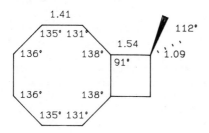

Figure 15.   *Preparation of methylcyclooctatetraene-4-*d

Figure 16.   *Assumed structure of cyclo-butenouranocene, 38*

lobuteno substituent, however, gives values of -1.42 ppm (exo) and
-34.0 ppm (endo) with corresponding contact shifts of 16.6 ppm
(exo) and 2.80 (endo). Again the difference in contact shifts for
the two positions is too large to be theoretically justifiable.
It can not arise from the difference between the reported atomic
coordinates from the X-ray crystal structure data and our 'idealized' geometry for cyclobutenouranocene. The calculated pseudocontact shifts using geometry factors derived from the average
position of the methylene groups in the X-ray structure are -2.36
ppm (exo) and -28.5 ppm (endo), with corresponding contact shifts
of 17.5 ppm (exo) and -2.7 ppm (endo). Moreover, increasing the
dihedral angle between the fused rings in the cyclobuteno ligand
from 173° to 180° results in a larger discrepancy between the calculated contact shifts for the two positions.

The value of the contact shift for the exo and endo protons
can be derived indirectly from the calculated contact shift for
the methyl groups in 1,1'-dimethyluranocene. Assuming a geometry
factor for the methyl group (Table XIV) as $1/6(A + 2C + 2E + G)$
(R = H in Fig. 12) and $\mu_{||}^2 - \mu_{\perp}^2 = 8.78$ BM^2, the calculated
pseudocontact shifts for the coplanar and tipped substituent are
-11.8 ppm and -12.8 ppm, respectively. From the experimental isotropic shift of -10.00 ppm, the corresponding contact shifts are
1.8 ppm and 2.8 ppm. Contact shifts for α-protons are assumed to
arise from hyperconjugative transfer of spin. Hyperconjugation
between a carbon p orbital and a carbon-hydrogen bond is a function of the dihedral angle between the two. When unpaired spin is
transferred by hyperconjugation, the magnitude of both the hyperfine coupling constant in ESR, and the contact shift in NMR, can
be expressed by

$$B = B_o \cos^2(\phi) \tag{7}$$

where ϕ is the dihedral angle and B_o is the magnitude of the hyperfine coupling constant or the contact shift when $\phi=0$ (64-67).
Evaluation of B_o from the contact shift for the methyl group affords values of 3.54 ppm and 5.54 ppm, respectively, for a planar
and a tipped substituent.

The fixed orbital angle between the p-orbitals of the 8-mem-
bered ring and the methylene C-H bonds in the cyclobuteno substituent permit evaluation of the contact shift for the exo and endo
protons from B_o. In our idealized structure of the ligand, the
dihedral angle is 25° which compares favorably with the average
value of 22° from the X-ray data. With $\phi=25°$, the calculated contact shift is 2.91 ppm when $B_o=3.54$ ppm and 4.5 ppm when $B_o=5.54$
ppm. These values are significantly different from those obtained
by difference from the isotropic shifts for the exo and endo prot-
ons and the calculated pseudocontact shifts assuming $\mu_{||}^2 - \mu_{\perp}^2 = 8.78$ BM^2. The discrepancy between the calculated contact shifts
for the exo and endo protons in 1,1'-dicyclobutenouranocene
using Fischer's value of $\mu_{||}^2 - \mu_{\perp}^2$ can only arise from

TABLE XIV

Calculated Geometry Factors $\dfrac{3\cos^2\theta - 1}{R^3}$

For α Protons

(R = H in Fig. 12)

Conformation[a]	Planar Gi x 10^{21}cm^{-3}	5° Tip[b] Gi x 10^{21}cm^{-3}
A	1.388	0.3599
B	0.6717	-0.3653
C	-1.503	-2.604
D	-5.053	-6.357
E	-9.424	-11.14
F	-13.23	-15.46
G	-14.76	-15.55

(a) Figure 12.
(b) Toward Uranium

underestimation of the pseudocontact shifts resulting from underestimation of $\mu_{\parallel}^2 - \mu_{\perp}^2$.

The known geometry of the methylene protons in the cyclobuteno permits an independent calculation of $\mu_{\parallel}^2 - \mu_{\perp}^2$. Assuming that the contact shift for both methylene protons is equla, the relationship between $\chi_{\parallel} - \chi_{\perp}$, the isotropic shift, δ, and the geometry factor G for the exo and endo protons is given by

$$\frac{\chi_{\parallel} - \chi_{\perp}}{3} = \frac{\delta_{exo} - \delta_{endo}}{G_{exo} - G_{endo}} \tag{8}$$

This equation leads to a value of 12.5 BM^2 for $\mu_{\parallel}^2 - \mu_{\perp}^2$ with corresponding pseudocontact shifts of -2.03 (exo), -48.5 ppm (endo) and a contact shift of 17.2 ppm.

This value of $\mu_{\parallel}^2 - \mu_{\perp}^2$ yields respective pseudocontact shifts of -17.2 ppm and 10.4 ppm for the t-butyl groups in 1,1'-di-t-butyl- and 1,1'-dineopentyluranocene, assuming coplanar substituents, and -20.8 ppm and 4.59 ppm, assuming tipped substituents. Although agreement between the calculated and experimental shifts for the neopentyl t-butyl group, assuming a tipped substituent is good, agreement between the t-butyl calculated and experimental data is poor for the coplanar and worse for the tipped substituent. If we assume that the difference in the observed and the calculated shifts for the t-butyl substituent is contact in nature, neither its sign nor its magnitude are consistent with the predicted sign based on transfer of spin by spin polarization, or the magnitude limits established from analysis of the temperature dependence of the methyl resonance in 1,1'-diethyluranocene. Thus, if the difference in calculated and observed shift does not arise from the anisotropy term, it must result from inaccuracies in the assumed geometry factor.

We can now return to our conclusion in the last section where we deduced from the pattern of ring proton resonances and from steric considerations that t-butyl substituents in uranocenes must be tilted <u>away</u> from uranium. This argument does not apply to the neopentyl group which is a normal primary alkyl substituent for which the ring-CH_2 bond can be tilted towards uranium without difficulty.

Tipping the substituent away from the uranium center leads to better agreement between the calculated and observed shift for the t-butyl group in 1,1'-di-t-butyluranocene. With $\mu_{||}^2 - \mu_{\perp}^2 = 12.5$ BM^2, a tip of 5° away from uranium affords a calculated pseudo-contact shift of -13.7 ppm, in excellent agreement with the experimental isotropic shift of -13.29 ppm.

To further demonstrate the difficulties associated with selecting an appropriate reference compound from which $\mu_{||}^2 - \mu_{\perp}^2$ can be derived, we shall derive $\mu_{||}^2 - \mu_{\perp}^2$ from the geometry factor and the isotropic shift of the t-butyl group in 1,1'-dineopentyl-uranocene. Our conformational analysis showed that the substituent is locked in conformation A in Fig. 12. For a coplanar substituent, the derived value of $\mu_{||}^2 - \mu_{\perp}^2$ is 3.33 BM^2, while tipping the substituent 5° toward the uranium leads to a value of 7.51 BM^2. However, relaxing the restriction of exclusive population of conformation A, and assuming an extremely small population of any other conformation where the geometry factor and the pseudocontact shift are negative, will greatly increase the derived value of $\mu_{||}^2 - \mu_{\perp}^2$.

Thus, evaluation of the geometry factor is extremely important in deriving a value of $\mu_{||}^2 - \mu_{\perp}^2$ from geometric and isotropic shift data. Two factors favor our approach to deriving a value of $\mu_{||}^2 - \mu_{\perp}^2$ from the methylene protons in dicyclobutenouranocene: 1) the single crystal X-ray data and the variable temperature ^1H NMR data provide an excellent estimate for the geometry factors for the two methylene protons; 2) calculation involves using the isotropic shift and geometry factor of two magnetically non-equivalent protons rather than one.

From the contact shift of the exo and endo protons in dicyclobutenouranocene, derived using $\mu_{||}^2 - \mu_{\perp}^2 = 12.5$ BM^2, a value of B_o, the maximum contact shift for an α-proton, can be calculated from eq. 7. Taking $\phi=25°$ leads to a value of 20.9 ppm for B_o.

Assuming a geometry factor of 1/6 (A + 2B + 2D + C) for the methyl group in 1,1'-dimethyluranocene, the calculated pseudocontact shifts are -16.8 ppm and -18.2 ppm, respectively, for a coplanar and a tipped substituent. By difference from the isotropic shift, the contact shifts are 6.76 ppm and 8.17 ppm, while calculation of the contact shift from B_o affords a value of 8.71 ppm. Agreement between the contact shifts calculated by both methods is excellent, particularly for the tipped substituent.

Considering the α-protons in 1,1'-dineopentyluranocene, if A is the only populated conformation of the neopentyl substi-

tuent, then the conformation of the α-protons is EE (Fig. 12).
The calculated pseudocontact shift is -26.9 ppm (coplanar), -31.8
(tipped) and by difference from the experimental isotropic shift
of -23.97 ppm, the contact shifts are 2.93 ppm (coplanar), and
7.83 ppm (tipped). Calculation of the contact shift from B_O
affords a value of 5.23 ppm.

Comparison of the calculated pseudocontact shifts for the
neopentyl t-butyl resonances with the isotropic shift showed that
the tipped geometry affords better agreement between the two
values, but still the value of calculated shift was approximately
twice that of the experimental isotropic shift. However, an
extremely small population of any conformation other than A
will readily decrease the magnitude of the calculated pseudocontact shift for the t-butyl resonance. Assuming an extremely
small population of conformations other than A, how does this
affect the factored shifts of the α-protons?

From the geometry factors in Table XIV and eq. 7, the
pseudocontact shift for any conformation other than EE will be
less negative than that for EE, while the contact shift will be
smallest in magnitude for EE and larger for any other conformation.
The magnitude of these changes is such that the isotropic shift
will be less negative as the population of conformations other
than EE increase. Thus, assuming a tipped substituent and an
extremely small population of conformations other than A for the
neopentyl substituent in 1,1'-dineopentyluranocene, leads to
better agreement between the calculated pseudocontact and contact
shifts for both the α and t-butyl resonance, than assuming either
exclusive population of conformation A or a coplanar substituent.

This analysis also accounts for the observed trend in the
isotropic shifts of the α-protons in 1,1'-diethyl-, 1,1'-di-n-
butyl-, and 1,1'-dineopentyluranocene, respectively, -17.47 ppm,
-19.03 ppm, and -23.97 ppm. The increase in magnitude of the
isotropic shift directly parallels the increasing stability of
the preferred conformation of the substituent, (i.e., conformation
A in Figure 12).

V. Summary.

Previous attempts at factoring the isotropic ^{1}H NMR shifts
in uranocene and substituted uranocenes have assumed that these
systems can be viewed as having effective axial symmetry. The
temperature dependent ^{1}H NMR spectra of uranocene and a variety
of substituted uranocenes clearly verify this assumption and
show that eq. 9 can be used to evaluate the pseudocontact contribution to the total isotropic shift in uranocenes. In this
equation $\chi_x \cong \chi_y$ for substituted uranocenes and are replaced by $\chi_\perp$.

$$\delta_{PSEUDOCONTACT} = \frac{\chi_{||} - \chi_\perp}{3N} \quad \frac{3\cos^2\theta - 1}{R^3} \tag{9}$$

Early attempts to factor the isotropic shifts in alkyl-uranocenes using eq. 9 were not completely successful because of failure to correctly assess the conformation of the substituent in solution and overestimation of the value of the anisotropy term $\chi_{\parallel} - \chi_{\perp}$ (5,6,14).

In alkyl-substituted uranocenes, our conformational analysis shows that a primary alkyl substituent populates principally conformations in which the dihedral angle between the substituent $C_{\alpha} - C_{\beta}$ bond and the ring plane is close to 90° on the side of the ring away from the metal. X-ray structure analyses have shown generally that substituents have ring-C_{α} bonds tilted several degrees towards uranium. The pattern of ring proton resonances and steric considerations suggest that t-butyl and related substituents are tilted away from uranium.

Another important result of this study is the confirmation of Fischer's demonstration that $\chi_{\perp}$ is not equal to zero in uranocene. Early attempts to factor isotropic shifts in uranocene have generally assumed that $\chi_{\perp}=0$, and leads to overestimation of the anisotropy term. A precise value of $\chi_{\perp}$ is difficult to determine rigorously from analysis of available NMR data. We have found that $\mu_{\parallel}^2 - \mu_{\perp}^2 = 12.5$ BM^2 leads to the best internal consistency factored isotropic shifts for a wide variety of 1,1'-dialkyluranocenes. Assuming $\mu_{av}^2 = 5.76$ BM^2 and $\mu_{\parallel}^2 - \mu_{\perp}^2 = 12.5$ BM^2, at 30°C, the corresponding values of $\mu_{\parallel}^2$ and $\mu_{\perp}^2$ are 14.09 and 1.59 BM^2, respectively. This implies that $\chi_{\parallel}/\chi_{\perp} \simeq 8$ in uranocene, a value substantilly larger than Fischer's ratio of $\chi_{\parallel}/\chi_{\perp} \simeq 2.8$ (15).

As a result of $\chi_{\perp} \neq 0$, early work on factoring the isotropic shift of the ring protons in uranocene underestimated the magnitude of the contact shift. Using our value of $\mu_{\parallel}^2 - \mu_{\perp}^2 = 12.5$ BM^2, the pseudocontact and contact shifts for uranocene ring protons are -8.30 ppm and -34.2 ppm, ($G_i = -2.34 \times 10^{21}$ cm^{-3}), respectively. Thus, this study confirms that both contact and pseudocontact interactions contribute to the observed isotropic shifts in uranocenes. The contact component is dominant for ring protons, but rapidly attenuates with increasing number of σ-bonds between the observed nucleus and the uranium such that the contact shift is effectively zero for β-protons.

The value of the contact shift for ring protons in uranocene is of the same sign but about 10 to 15 ppm larger in magnitude than the contact shift for ring protons in CP$_3$U-X compounds. If a direct correlation exists between the magnitude of the contact shift and the degree of covalency in ligand-metal bonding in these systems, then the NMR data suggest a higher degree of covalency in the ligand-metal bonds in uranocene.

Acknowledgement

This research was supported in part by the Division of Nuclear Sciences, Office of Basic Energy Sciences, U.S. Department of Energy under contract no. W-7405-Eng-48. We are indebted to Dr. Norman Edelstein and Professor R.D. Fischer for valuable discussions and suggestions.

Literature Cited

1.   von Ammon, R.; Kanellakopulos, B.; Fischer, R.D. Chem. Phy. Let., 1968, 2, 513.

2.   Siddall, T.H. III; Stewart, W.E.; Karraker, D.G. Chem. Phy. Let., 1969, 3, 498.

3.   von Ammon, R.; Kanellakopulos, B.; Fischer, R.D.; Laubereau, R. Inorg. Nucl. Chem. Let., 1969, 5, 219.

4.   von Ammon, R.; Kanellakopulos, B.; Fischer, R.D. Chem. Phy. Let., 1970, 4, 553.

5.   Edelstein, N.; La Mar, G.N.; Mares, F.; Streitwieser, A. Jr. Chem. Phy. Let., 1971, 8, 399.

6.   Streitwieser, A. Jr.; Dempf, D.; La Mar, G.N.; Karraker, D.G.; Edelstein, N. J. Am. Chem. Soc., 1971, 93, 7343.

7.   Hayes, R.G.: Thomas, J.L. Organomet. Chem. Rev. A., 1971, 7, 1.

8.   Paladino, N.; Lugli, G.; Pedretti, U.; Brunelli, M.; Giacometti, G. Chem. Phy. Let., 1970, 5, 15.

9.   Brunelli, M.; Lugli, G.; Giacometti, G. J. Mag. Res., 1973, 9, 247.

10.   Fischer, R.D.; von Ammon, R.; Kanellakopulos, B. J. Organomet. Chem., 1970, 25, 123.

11.   von Ammon, R.; Fischer, R.D.; Kanellakopulos, B. Chem. Ber., 1972, 45, 105.

12.   Marks, T.J.; Seyam, A.M.; Kolb, J.R. J. Am. Chem. Soc., 1973, 95, 5529.

13.   Amberger, H.D. J. Organomet. Chem., 1976, 116, 219.

14.   Berryhill, S.R. Ph.D. Thesis, UC Berkeley, 1978.

15.   Fischer, R.D. in "Organometallics of the f-Elements"; Marks, T.J.; Fischer, R.D., Eds.; D. Reidel Pub. Co., Boston, Mass.; 1979; p 337.

16. Eaton, D.R.; Phillips, W.D. "Advances in Magnetic Resonance";
     Waugh, J.S. Ed.; Academic Press, New York, N.Y.; 1965; vol.1,
     p 119.

17. "NMR of Paramagnetic Molecules: Principals and Applications";
     La Mar, G.N.; Horrocks, W. DeW. Jr.; Holm, R.H., Eds.;
     Academic Press, New York, N.Y.; 1973.

18. Webb, G.A. "Annual Reports on NMR Spectroscopy"; Mooney,
     E.F., Ed.; Academic Press, New York, N.Y.; 1970; vol. 3,
     p 211.

19. Webb, G.A. "Annual Reports on NMR Spectroscopy"; Mooney,
     E.F., Ed.; Academic Press, New York, N.Y.; 1975, vol. 6A,
     p 2.

20. Kurland, R.J.; McGarvey, R.B. J. Mag. Res., 1970, 2, 286.

21. Drago, R.S.; Zisk, J.I.; Richman, R.M.; Perry, W.D. J. Chem.
     Ed., 1974, 51, 371, 465.

22. McConnell, H.M.; Chestnut, D.B. J. Chem. Phys., 1958, 28,
     107.

23. McConnell, H.M.; Robertson, R.E. J. Chem. Phys., 1958, 29,
     1361.

24. Kluiber, R.W.; Horrocks, W. DeW. Jr. Inorg. Chem., 1967, 6,
     166.

25. Horrocks, W. DeW. Jr. Inorg. Chem., 1970, 9, 690.

26. Horrocks, W. DeW. Jr.; Sipe, J.P. III Science, 1972, 177,
     994.

27. von Ammon, R.; Kanellakopulos, B.; Fischer, R.D.; Laubreau,
     P. Inorg. Nucl. Chem. Let., 1969, 5, 315.

28. Sebala, A.E.; Tsutsu, M. Chem. Let., 1972, 775.

29. von Ammon, R.; Kanellakopulos, B.; Fischer, R.D. Radiochim.
     Acta., 1969, 11, 162.

30. Marks, T.J.; Kolb, J.R. J. Am. Chem. Soc., 1975, 97, 27.

31. Karraker, D.G. Inorg. Chem., 1973, 12, 1105.

32. Hayes, R.G.; Edelstein, N. J. Am. Chem. Soc., 1972, 94,
     8688.

33. Zalkin, A.; Raymond, K.N. J. Am. Chem. Soc., 1969, 91, 5667.

34. Avdeef, A.; Raymond, K.N.; Hodgson, K.O.; Zalkin, A. Inorg. Chem., 1972, 11, 1083.

35. von Ammon, R.; Fischer, R.D. Ang. Chem. Int., 1972, 11, 675.

36. Miller, M.J.; Streitwieser, A. Jr. J. Org. Chem., submitted.

37. MacDonald, C.G.; Hannon, J.S.; Sternhell, S. Aust. J. Chem., 1964, 17, 38.

38. Batiz-Hernandez, H.; Bernheim, R.A. "Prog. in NMR Spectroscopy"; Emsley, J.W.; Feeney, J.; Sutcliffe, L.H., Eds.; Pergamon Press, New York, N.Y.; 1967; p 63.

39. Streitwieser, A. Jr. "Organometallics of the f-Elements"; Marks, T.J.; Fischer, R.D., Eds.; D. Reidel Pub. Co., Boston, Mass.; 1979; p 149.

40. Streitwieser, A. Jr.; LeVanda, C. Inorg. Chem., submitted.

41. Dallinger, R.F.; Stein, P.; Spiro, T.G. J. Am. Chem. Soc., 1978, 100, 7865.

42. Karraker, D.G.; Stone, J.A.; Jones, E.R. Jr.; Edelstein, N. J. Am. Chem. Soc., 1970, 92, 4841.

43. Amberger, H.D.; Fischer, R.D.; Kanellakopolus, B. Theoret. Chim. Acta., 1975, 37, 105.

44. Streitwieser, A. Jr.; Burgherd,H.P.C.; Morrell, D.G.; Luke, W.D. Inorg. Chem., submitted.

45. Fischer, R.D.; Spiegl, A.W.; Dickert, F. personal communication.

46. Luke, W.D.; Streitwieser, A. Jr. to be published.

47. Rösch, N.; Streitwieser, A. Jr. J. Organomet. Chem., 1978, 145, 195.

48. Schilling, B. unpublished results.

49. Luke, W. unpublished results.

50. Edelstein, N.; Streitwieser, A. Jr.; Morrell, D.G.; Walker, R. Inorg. Chem., 1967, 15, 1397.

51. Harmon, C.A.; Bauer, D.P.; Berryhill, S.R.; Hagiwara, K.; Streitwieser, A. Jr. Inorg. Chem., 1977, 16, 2143.

52. Anderson, S.E.; Drago, R.S. Inorg. Chem., 1972, 11, 1564.

53. Luke, W.D. Dissertation; University of Calif., 1979.

54. Luke, W.D.; Berryhill, S.R.; Streitwieser, A. Jr. to be published.

55. Miller, M.J. unpublished results.

56. Spiegel, A.W.; Fischer, R.D. Chem. Ber., 1979, 112, 116.

57. Paquette, L.A.; Ley, W.V.; Meisinger, R.H.; Russell, R.K.; Oku, M. J. Am. Chem. Soc., 1974, 96, 5806.

58. Paquette, L.A.; James, D.R.; Birnberg, G. J. Am. Chem. Soc., 1974, 96, 7454.

59. Paquette, L.A.; Henzel, K.A. J. Am. Chem. Soc., 1975, 97, 4649.

60. Bak, B.; Led, J.J.; Nygaard, L.; Rastrup-Anderson, J.; Sorensen, G.O. J. Mol. Struct., 1969, 3, 369.

61. Levy, G.C.; Nelson, G.C. "Carbon-13 Nuclear Magnetic Resonance for Organic Chemistr"; Wiley-Interscience, New York, N.Y.; 1972.

62. Gunther, H.; Jikeli, G.; Schmicker, H.; Prestien, J. Ang. Chem. Int. Ed., 1973, 12, 762.

63. Warren, K.D. "Structure and Bonding"; Springer-Verlag, New York, N.Y.; 1977, vol. 33, p 97.

64. Kron, R.; Weiss, A.; Polzner, H.; Bischer, E. J. Am. Chem. Soc., 1975, 97, 644.

65. Gerson, F.; Moxhuk, G.; Schwyzer, M. Hel. Chim. Acta., 1971, 54, 361.

66. Scroggins, W.T.; Rettig, M.F.; Wing, R.M. Inorg. Chem., 1976, 15, 1381.

67. Heller, C.; McConnell, H.M. J. Chem. Phys., 1960, 32, 1535.

RECEIVED January 30, 1980.

COMPLEX CHEMISTRY, THERMODYNAMIC PROPERTIES, AND TRANSCURIUM CHEMISTRY

Specific Sequestering Agents for the Actinides

KENNETH N. RAYMOND,[1] WILLIAM L. SMITH, FREDERICK L. WEITL,
PATRICIA W. DURBIN, E. SARAH JONES, KAMAL ABU-DARI,
STEPHEN R. SOFEN, and STEPHEN R. COOPER

Department of Chemistry and Divisions of Materials and Molecular Research
and Biology and Medicine, Lawrence Berkeley Laboratory,
University of California, Berkeley, CA 94720

Abstract

This paper summarizes the current status of a continuing project directed toward the synthesis and characterization of chelating agents which are specific for actinide ions — especially
Pu(IV). A biomimetic approach has been used that relies on the
observation that Pu(IV) and Fe(III) have marked similarities that
include their biological transport and distribution in mammals.
Since the naturally-occurring Fe(III) sequestering agents produced
by microbes commonly contain hydroxamate or catecholate functional
groups, these groups should also complex the actinides strongly,
and several macrocyclic ligands incorporating these moieties have
been prepared. We have reported the isolation and structure analysis of an isostructural series of tetrakis(catecholato) complexes
with the general stoichiometry $Na_4[M(C_6H_4O_2)_4]\cdot21\ H_2O$ (M = Th, U,
Ce, Hf). These complexes are structural archetypes for the cavity
that must be formed if an actinide-specific sequestering agent is
to conform ideally to the coordination requirements of the central
metal ion. The $[M(cat)_4]^{4-}$ complexes have the D_2d symmetry of the
trigonal-faced dodecahedron. The complexes $Th[R'C(O)N(O)R]_4$ have
been prepared where R = isopropyl and R' = t-butyl or neopentyl.
The neopentyl derivative is also relatively close to an idealized
D_2d dodecahedron, while the sterically more hindered t-butyl compound is distorted toward a cubic geometry. A series of 2,3-di-
hydroxybenzoyl amide derivatives of linear and cyclic tetraaza-
and diazaalkanes have been prepared. Sulfonation of these compounds improves the metal complexation and excretion of plutonium
by test animals. At low dose levels, these results substantially
exceed the capabilities of compounds presently used for the thera-
peutic treatment of actinide contamination.

[1] To whom correspondence should be addressed at the Department of Chemistry,
University of California.

0–8412–0568–X/80/47–131–143$07.50/0

Introduction

With the commercial development of nuclear reactors, the
actinides have become important industrial elements. A major concern of the nuclear industry is the biological hazard associated
with nuclear fuels and their wastes ($\underline{1}$, $\underline{2}$). In addition to their
chemical toxicity, the high specific activity of alpha emission
exhibited by the common isotopes of the transuranium elements make
these elements potent carcinogens ($\underline{3}$, $\underline{4}$, $\underline{5}$, $\underline{6}$, $\underline{7}$). Unlike organic
poisons, biological systems are unable to detoxify metal ions by
metabolic degradation. Instead, unwanted metal ions are excreted
or immobilized ($\underline{8}$). Unfortunately, only a small portion of absorbed tetra- or trivalent actinide is eliminated from a mammalian
body during its lifetime. The remaining actinide is distributed
throughout the body but is especially found fixed in the liver and
in the skeleton ($\underline{5}$, $\underline{7}$, $\underline{9}$-$\underline{12}$). While the ability of some metals to
do damage is greatly reduced by immobilization, local high concen-
trations of radioactivity are produced by immobilized actinides —
thereby increasing the absorbed radiation dose and carcinogenic
potential. Removal of actinides from the body is therefore an
essential component of treatment for actinide contamination.
      Conventional chelating agents as diethylenetriaminepenta-
acetic acid, DTPA (Figure 1), remove much of the soluble actinide
present in body fluids, but are almost totally ineffective in
removing the actinide after it has left the circulation or after
hydrolysis of the metal to form colloids and polymers ($\underline{13}$, $\underline{14}$,
$\underline{15}$). The inability of DTPA to completely coordinate the tetravalent actinides is shown by the easy formation of ternary com-
plexes between Th(DTPA) and many bidentate ligands ($\underline{16}$, $\underline{17}$, $\underline{18}$).
The hydrolysis of Th(IV) and U(IV) DTPA complexes at pH near 8 is
explained by the dissociation of H^+ from a coordinated water
molecule ($\underline{19}$, $\underline{20}$, $\underline{21}$, $\underline{22}$). In addition, the polyaminocarboxylic
acids are toxic because they indiscriminately complex and remove
biologically important metals, especially zinc ($\underline{23}$, $\underline{24}$, $\underline{25}$, $\underline{26}$).
Thus there is a need to develop new and powerful chelating agents
highly specific for tetravalent actinides, particularly Pu(IV).
      While not the most toxic, plutonium is the most likely trans-
uranium element to be encountered. Plutonium commonly exists in
aqueous solution in each of the oxidation states from III to VI.
However, under biological conditions, redox potentials, complexation, and hydrolysis strongly favor Pu(IV) as the dominant
species ($\underline{27}$, $\underline{28}$). It is remarkable that there are many similarities between Pu(IV) and Fe(III) (Table I). These include the
similar charge per ionic-radius ratios for Fe(III) and Pu(IV) (4.6
and 4.2 e/Å respectively), the formation of highly insoluble
hydroxides, and similar transport properties in mammals. The
majority of soluble Pu(IV) present in body fluids is rapidly
bound by the iron transport protein transferrin at the site which
normally binds Fe(III). In liver cells, deposited plutonium is
initially bound to the iron storage protein ferritin and

Table I.   Similarities of Pu^{4+} and Fe^{3+}

	Pu^{4+}	Fe^{3+}
1) $\dfrac{\text{Charge}}{\text{Ionic radius}^a}$	$\dfrac{4}{0.96} = 4.2$	$\dfrac{3}{0.65} = 4.6$

2) $Fe(OH)_3 \rightarrow Fe^{3+} + 3OH^-$ $K \approx 10^{-38}$ (10^{-13} per OH^-)

$Fe^{3+} + H_2O \rightarrow Fe(OH)^{2+} + H^+$ $K = 0.0009$

$Pu(OH)_4 \rightarrow Pu^{4+} + 4OH^-$ $K \approx 10^{-55}$ (10^{-14} per OH^-)

$Pu^{4+} + H_2O \rightarrow Pu(OH)^{3+} + H^+$ $K = 0.031$ (in $HClO_4$)

3) Pu^{4+} is transported in the blood plasma of mammals as a complex of transferrin, the normal Fe^{3+} transport agent. The Pu^{4+} binds at the same site as Fe^{3+}.

^{a}Ref. 74.

eventually becomes associated with hemosiderin and other long term iron storage proteins (9, 29, 30). These similarities of Pu(IV) and Fe(III) suggested to us a biomimetic approach to the design of Pu(IV) sequestering agents modeled after the very efficient and highly specific iron sequestering agents, siderophores, which were developed by bacteria and other microorganisms to obtain Fe(III) from the environment (31, 32, 33).

The siderophores (Figure 2) typically contain hydroxamate or catecholate functional groups which are arranged to form an octahedral cavity the exact size of a ferric ion. Catechol, 2,3-dihydroxybenzene, and the hydroxamic acids, N-hydroxyamides, are very weak acids that ionize to form "hard" oxygen anions, which bind strongly to strong Lewis acids such as Fe(III) and Pu(IV). Complexation by these groups forms five-membered chelate rings, which substantially increases the stability compared to complexation by lone oxygen anions (34). That the hydroxamic acids strongly coordinate tetravalent actinides is supported by the formation constants presented in Table II. Due to its higher charge and strong basicity, the catecholate group forms even stronger complexes with the tetravalent actinides than the hydroxamic acids. Thus our goal has been the incorporation of hydroxamate or catecholate functional groups into multidentate chelating agents that specifically encapsulate tetravalent actinides.

The similarity between Fe(III) and the actinide(IV) ions ends with their coordination numbers. Because of the larger ionic radii of the actinide(IV) ions, their preferred coordination number found in complexes with bidentate chelating agents is eight. Occasionally higher coordination numbers are encountered with very small ligands or by the incorporation of a solvent molecule (43, 44). Theoretical calculations indicate that either the square antiprism (D_4d) or the trigonal faced dodecahedron (D_2d) is the expected geometry for an eight-coordinate complex. The coulombic energy differences between these polyhedra (Figure 3) is very small and the preferred geometry is largely determined by steric requirements and ligand field effects. Cubic coordination lies at higher energy, but may be stabilized if f-orbital interactions were important. Another important eight-coordinate polyhedron, the bicapped trigonal prism (C_2v), corresponds to an energy minimum along the transformation pathway between the square antiprism and the dodecahedron (45-50). As seen in Table III, all four of the above geometries are found in eight-coordinate complexes of tetravalent actinides with bidentate ligands. However, the mmmm isomer of the trigonal faced dodecahedron is the most prevalent in the solid state.

Actinide Catecholates

Two fundamental questions in the design of an actinide-specific sequestering agent are the coordination number and geometry actually preferred by the metal ion with a given ligand. The

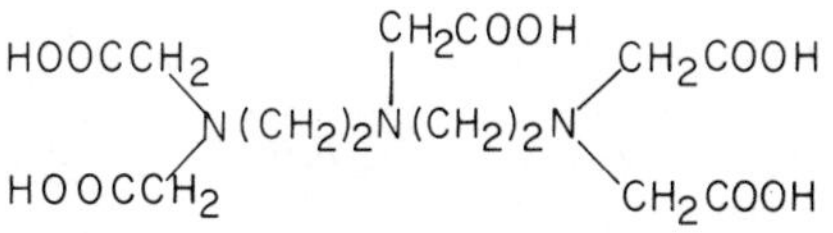

Figure 1. *Diethylenetriaminepentaacetic acid (DTPA)*

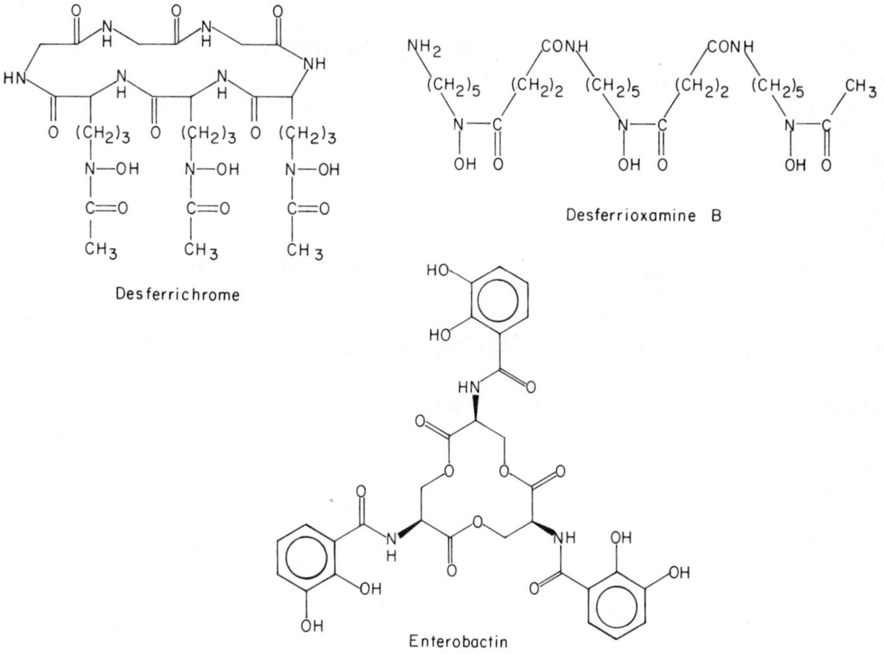

Desferrichrome

Desferrioxamine B

Enterobactin

Figure 2. *Representative siderophores*

Table II. Formation Constants for Some Actinide(IV) Hydroxamates

Metal	Temp, °C	$\log \beta_1{}^a$	$\log \beta_2$	$\log \beta_3$	$\log \beta_4$	Ref.
Benzohydroxamaic acid, Ph-C(O)-N(OH)-H						
U(IV)	25	9.89	18.00	26.32	32.94	35
Th(IV)	25	9.60	19.81	28.76		35
Pu(IV)	25	12.73				35
N-phenylbenzohydroxamaic acid, Ph-C(O)-N(OH)-Ph						
Th(IV)	20				37.70	38
Th(IV)	25				37.80	36
Th(IV)	30				37.76	37
Pu(IV)	22	11.50	21.95	31.81	41.35	39
N-phenylcinnamohydroxamic acid, Ph-C=C-C(O)-N(OH)-Ph						
Th(IV)	20	12.76	24.70	35.72	45.72	40
Catechol						
Th(IV)	30	17.72				41
4-Nitrocatechol						
Th(IV)	25	14.96	27.78	36.71	40.61	42

$^a \log \beta_n = [ML_n]/[M][L]^n$ for the reaction $M^{4+} + nL \rightarrow ML_n$ where L is the hydroxamate anion or the catecholate dianion.

Table III. Geometry of Monomeric Eight-Coordinate Actinide Complexes with Bidentate Ligands

Complex[b]	Metals	Idealized Geometry[a]	Ref.
α-M(IV)(acetylacetonate)$_4$	Th, U, Ce	$h_1h_1p_2p_2$-BTP	51,52
β-M(IV)(acetylacetonate)$_4$	Th, U, Np, Ce	ssss-SA	51,53
M(bipyridyl)$_4$	U	ssss-Cube	54
M(IV)(dibenzoylmethanate)$_4$	Th, U, Ce	mmmm-DD	63
M(IV)(N,N-diethyldithiocarbamate)$_4$	Th	mmmm-DD	52,55
[M(III)(N,N-diethyldithiocarbamate)$_4$]$^-$	Np	mmmm-DD	56
M(IV)(diisobutyrylmethanate)$_4$	U	BTP	57
M(IV)(hexafluoroacetonylpyrazolide)$_4$	Th, U	mmmm-DD	59
[M(III)(hexafluoroacetylacetonate)$_4$]$^-$	Am, Y, Eu	gggg-DD	58
M(IV)(salicylaldehydrate)$_4$	Th, U	mmmm-DD	60
M(IV)(thenoyltrifluoroacetylacetonate)$_4$	Th, U, Pu, Ce	mmmm-DD	61,62

[a] BTP = cicapped trigonal prism, DD = trigonal faced dodecahedron, SA = square antiprism. The isomer notation is taken from Ref. 45 and 48 and corresponds to the edges labelled in Figure 3.

[b] Thorium(trifluoroacetylacetonate)$_4$ was originally described as a 1111-SA (Ref. 64, but a reinvestigation established the presence of a coordinated water molecule forming a nine-coordinate complex (Ref. 65).

complexes formed by Th(IV) or U(IV) and catechol, in which the
steric restraints of a macrochelate are absent, serve as structural archetypes for designing the optimum actinide(IV) sequestering agent. Thus the structures of an isoelectronic, isomorphous
series of tetrakis-catecholato salts, $Na_4[M(C_6H_4O_2)_4] \cdot 21H_2O$; M =
Th(IV), U(IV), Ce(IV), and Hf(IV), were determined by single crystal X-ray diffraction. Suitable crystals were isolated from the
reaction of the metal chlorides or nitrates and the disodium salt
of catechol in aqueous solution under an inert atmosphere (66,
67). Measurement of magnetic susceptibility and electronic spectra of the cerium and uranium complexes verified the presence of
the +4 oxidation state.
    It was somewhat surprising that the strongly oxidizing Ce(IV)
ion (E_0 = + 1.70 V) (68) did not react with the catechol dianion,
a facile reducing agent (69). The ability of catechol to coordinate without reduction of oxidizing ions as Ce(IV), Fe(III) (70),
V(V) (71), and Mn(III) (72) is a reflection of its impressive co-
ordinating ability. The Ce(IV) complex was found by cyclic
voltammetry to undergo a quasi-reversible one-electron reduction
in strongly basic solution in the presence of excess catechol
(Figure 4). Using the Nernst equation (73) and the measured
potential of the Ce(IV)/Ce(III)(catechol)$_4$ couple of - 448 mV vs
NHE, the formation constant of the tetrakis Ce(IV) complex was
found to be greater than the corresponding Ce(III) complex by a
factor of 10^{36}. This enormous shift of the redox potential of the
Ce(IV)/Ce(III) couple is dramatic evidence of the enormous affinity of the catecholate anion for the tetravalent lanthanides and
actinides.
    The crystal structure of this isostructural series of cate-
chol complexes consists of discrete $[M(catechol)_4]^{4-}$ dodecahedra,
a hydrogen bonded network of 21 waters of crystallization and
sodium ions, each of which is bonded to two catecholate oxygens
and four water oxygens. Of the possible eight coordinate polyhedra, only the cube and the dodecahedron allow the presence of
the crystallographic $\bar{4}$ axis on which the metal ion sits. As
depicted in Figure 5 and verified by the shape parameters in
Table IV, the tetrakis(catecholato) complexes nearly display the
ideal D_{2d} molecular symmetry of the mmmm isomer of the trigonal-
faced dodecahedron.
    The symmetry of the dodecahedron, which can be regarded as
the intersection of one elongated and one compressed tetrahedron,
allows for different M-O$_A$ and M-O$_B$ bond lengths. As seen in Table
V, the experimental M-O bond lengths are equal in the thorium and
cerium complexes. However, the M-O$_B$ bond length is significantly
shorter than the M-O$_A$ bond length in the uranium and hafnium complexes. The much smaller ionic radius of the hafnium pulls the
catecholate ligands in sufficiently so that interligand contacts
become significant; the short oxygen-oxygen distance between A
sites of 2.550 Å, nearly 0.3 Å less than that for the cerium salt,
is well within the van der Waals contact distance of 2.8 Å (75).

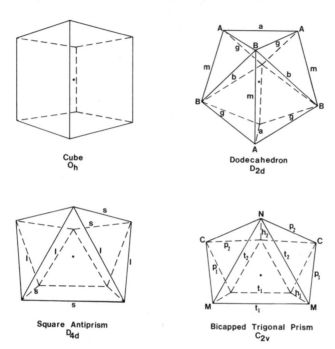

Cube
O$_h$

Dodecahedron
D$_{2d}$

Square Antiprism
D$_{4d}$

Bicapped Trigonal Prism
C$_{2v}$

Figure 3. Eight-coordinate polyhedra. The principal axes are vertical. Edge labels are taken from Refs. 45 and 48.

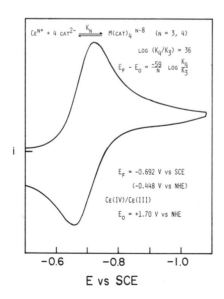

$$Ce^{N+} + 4\ CAT^{2-} \xrightleftharpoons{K_N} M(CAT)_4^{\ N-8} \quad (N = 3,\ 4)$$

$$LOG\ (K_4/K_3) = 36$$

$$E_F - E_0 = \frac{-59}{N}\ LOG\ \frac{K_4}{K_3}$$

$E_F = -0.692$ V vs SCE

$(-0.448$ V vs NHE$)$

Ce(IV)/Ce(III)

$E_0 = +1.70$ V vs NHE

$-0.6 \qquad -0.8 \qquad -1.0$

E vs SCE

Figure 4. Cyclic voltammogram of $[Ce(O_2C_6H_4)_4]^{4-}$ in 5M NaOH, 1M catechol aqueous solution, on a hanging mercury drop electrode, at 100 mV/sec scan rate

Table IV. Shape Parameters[a] (deg.) for $[M(O_2C_6H_4)_4]^{4-}$, M = Hf, Ce, U, Th, Complexes

Metal	θ_A	θ_B	ϕ	δ
Th	37.9	75.4	3.6	31.3
U	37.1	75.2	3.0	31.1
Ce	36.8	74.9	2.1	32.0
Hf	35.2	73.3	0.4	32.2
Dodecahedron[b]	36.9	69.5	0.0	29.5
Cube[b]	54.7	54.7	0.0	0.0

[a]See Ref. 45 and 49 for definitions of shape parameters.

[b]Calculated using the Hard Sphere Model.

Table V. Structural Parameters for $Na_4[M(O_2C_6H_4)_4] \cdot 21H_2O$ Complexes

Metal	Ionic Radius[a] Å	M-O_A Å	M-O_B Å	O_A-O_A Å	O_A-M-O_B deg
Th	1.05	2.421(3)	2.418(3)	2.972(6)	66.8(1)
U	1.00	2.389(4)	2.362(4)	2.883(7)	67.7(1)
Ce	0.97	2.362(4)	2.357(4)	2.831(7)	68.3(1)
Hf	0.83	2.220(3)	2.194(3)	2.554(5)	71.5(1)

[a]Ref. 74.

This lengthens the M–O_A bond of the hafnium complex relative to the others. However, since the ionic radius of uranium lies between those of cerium and thorium it is unlikely that the metal size explains the distortion in the uranium complex. As all four complexes are identical in all respects except for the metal ion, the lengthening of the M–O_A bond in the uranium complex is attributed to a ligand field effect from the f-electrons. A ligand field of D_2d symmetry will split the $^3H_4$ ground term for the $5f^2$ configuration of U(IV) into seven levels, two of which are doubly degenerate. The observed temperature-independent magnetic susceptibility of 870×10^{-6} cgs mol^{-1} is consistent with a non-degenerate ground state (76). A qualitative crystal field treatment of the D_{2d} complex predicts a nondegenerate ground state arising from either the f_{xyz} or f_{z^3} metal orbital. Thus from electron repulsion arguments, one expects the ligand oxygen that is closer to the z axis, O_A, to interact more with the filled metal orbital resulting in the observed lengthening of the M–O_A bond.

Actinide Hydroxamates

As with the actinide catecholates, we are interested in determining the optimum structure of actinide hydroxamates for use in the design of an octadentate actinide sequestering agent. Thus the structures of tetrakis(N-isopropyl-3,3-dimethylbutano- and -2,2-dimethylpropano)hydroxamatothorium(IV) have been determined by single crystal X-ray diffraction (77). Keeping the pH as low as possible, these compounds precipitate upon the addition of an aqueous solution of thorium tetrachloride to an aqueous solution of the sodium salt of the hydroxamic acid. The analogous uranium(IV) complexes were prepared similarly under an inert atmosphere using deaerated solvents. In addition to their hydrocarbon solubility, the bulky alkyl substituents impart other interesting properties to these complexes. They melt at 127–8 and 116-7°C and, under a vacuum of 10^{-3} torr, sublime at 95 and 100°C, respectively!

The alkyl substituents are also very important in determining the structures of the thorium hydroxamates. As in the tetracatecholates, the metal ion in the t-Bu complex sits on a crystallographic $\bar{4}$ axis, which limits the possible eight coordinate polyhedra to the dodecahedron and the cube (or tetragonal prism). In order to minimize steric interactions, the t-butyl groups situate themselves on the corner of a tetrahedron, resulting in the distorted cubic geometry of the complex shown in Figure 6. This steric strain also manifests itself in the C(=O)–C(t-Bu) bond length of 1.547(5) Å, which is significantly longer than 1.506(5) Å, the length normally found for an sp^2-sp^3 C–C bond (78). Because the hydroxamate anion is an unsymmetrical ligand with most of the charge localized on the nitrogen oxygen, the Th–O_N bond, 2.357(3) Å, is 0.14 Å shorter than the Th–O_C bond, 2.492(3) Å.

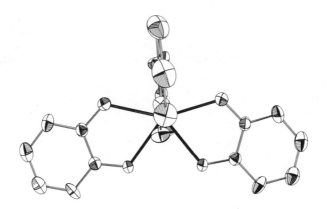

Figure 5. The [M(catechol)₄]⁴⁻ (M = Hf, Ce, Th, and U) anion viewed along the mirror plane with the 4̄ axis vertical

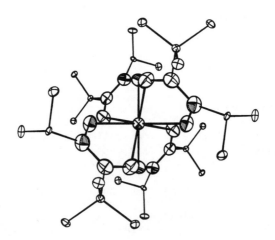

Figure 6. Th[i-Pr-N(O)-C(O)-t-Bu]₄ viewed down the 4̄ axis. In this figure and in Figure 8, the substituent carbon atoms are drawn at 1/5 scale, the hydrogen atoms are omitted for clarity, and the nitrogen and nitrogen oxygen atoms are shaded.

The average Th-O bond, 2.425 Å, is very close to the average Th-O
bond found in $[Th(catechol)_4]^{4-}$, 2.420 Å. The O_N-M-O_C (or bite)
angle observed in the t-Bu complex of 62.3(1)° is smaller than the
value required to successfully span an edge of a cube, 70.53°,
calculated using a hard-sphere model. The disparity in Th-O bond
lengths and observed bite angle cause a distortion towards the
gggg-isomer of a trigonal-faced dodecahedron, accompanied by a
10.3° twist in the BAAB trapezoid (see Figure 4 for these definitions). As expected theoretically (45, 46), the more negatively
charged nitrogen oxygens are located at the B sites of the dodecahedron, but this could also be a steric effect of the t-butyl
groups.

The relationship of the cube and the dodecahedron to the coordination polyhedron of the t-Bu complex is shown in Figure 7 and
a detailed shape parameter analysis is presented in Table VI. The
similarity of this complex to a cube is shown by the equal edge
lengths of those not spanned by the ligands, the m and g' edges,
and the dihedral angles, δ, which are close to 90° about the m and
g edges. The a and b edges are face diagonals in the cube and the
dihedral angles about these edges measure the distortion towards
the dodecahedron. Steric repulsions dominate (45, 46), since the
bulky alkyl substituents direct the geometry of the complex towards a cube. Because the ligands span alternate edges of two
parallel square faces, the complex is best designated as the ssss
isomer of a cube [after the designations for a square-antiprism
made by Hoard and Silverton (45)] with the overall symmetry of the
S_4 point group.

The influence of the alkyl substituent in determining structure is greatly reduced by the introduction of a methylene group
between the carbonyl carbon and the t-butyl group. Contrary to
the previous complex, the neopentyl derivative (shown in Figures
8 and 9) is close to the mmmm-dodecahedron found in the tetrakis-
(catecholato)thorium and the majority of other eight-coordinate
actinide complexes with bidentate ligands (Table III). While the
lack of crystallographic symmetry would allow structures other
than the dodecahedron (such as the square antiprism or bicapped
trigonal prism) the smallest dihedral angle is 35.5° and this precludes the presence of any square faces in the coordination polyhedron (for which δ = 0). As seen in Table VI the complex is,
however, distorted from an ideal dodecahedron. The bite of the
ligands, which governs the length of the m edges, is smaller than
the length of an ideal dodecahedral m edge. This results in the
flattening of the B tetrahedron as evidenced by the increased
angle between the Th-O_B vector and the pseudo $\bar{4}$ axis, θ_B, and by
the lengthened g edges. The bending of the ligands seen in Figure
8 is due to steric interactions of molecular packing. As before,
the Th-O_N bond [ave = 2.36(2) Å], is shorter than the Th-O_C bond
[ave = 2.46(4) Å]. There is no site preference for the charged
oxygen as the O_N and O_C are equally distributed over the A and B
sites of the dodecahedron, resulting in a mmmm-dodecahedron with
C_1 symmetry.

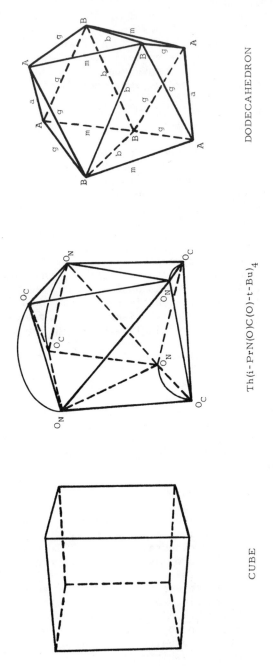

DODECAHEDRON

Th(i-PrN(O)C(O)-t-Bu)₄

CUBE

Figure 7. The coordination polyhedron of Th[i-Pr-N(O)-C(O)-t-Bu]₄ compared with a cube and a dodecahedron

Table VI. Shape Parameters[a] for Th[i-Pr-N(O)-C(O)-R]$_4$

Parameter	R = t-Butyl[b]	R = Neopentyl	Dodecahedron[c]	Cube[c]
ϕ	10.3	1.1,10.1	0.0	0.0
δ_a	33.2	67.6,70.2	51.3	0.0
δ_b	11.3	35.5,36.5,41.5,48.4	29.5	0.0
δ_g	80.5	42.3,45.4,46.9,51.0,54.0,55.6,58.9,62.7	62.5	90.0
$\delta_{g'}$	69.3		62.5	90.0
δ_m	87.9	70.2,76.2,79.7,85.5	51.3	90.0
θ_A	44.5	32.9,33.6,35.3,38.8	36.9	54.7
θ_B	60.0	78.8,82.4,84.0,84.4	69.5	54.7
M-O$_A$/r[d]	1.06	0.96,0.97,0.99,1.02	1.00	1.00
a/r	1.48	1.11,1.16	1.20	1.63
b/r	1.58	1.32,1.36,1.41,1.60	1.50	1.63
g/r	1.07	1.23,1.27,1.29,1.32,1.34,1.37,1.39,1.40	1.20	1.16
g'/r	1.26		1.20	1.16
m/r	1.27	1.03,1.04,1.04,1.04	1.20	1.16

[a] The shape parameters are defined in Ref. 45 and 49; ϕ is the twist in the BAAB trapezoid, θ is the angle between the M-O vector and the principal axis, δedge is the dihedral angle between the faces containing the edge as labeled in Figure 3.

[b] The dodecahedral g edges are divided into edges spanned by the ligands and those which are not, designated g and g' respectively.

[c] Calculated using the Hard Sphere Model.

d_r = average M-O$_B$ distance.

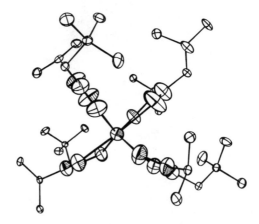

Figure 8.   Th[i-Pr-N(O)-C(O)-Neopentyl]₄ viewed down the pseudo $\overline{4}$ axis

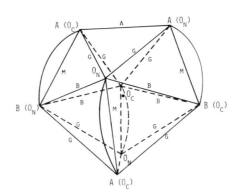

Figure 9.   Schematic structure of Th[i-Pr-N(O)-C(O)-Neopentyl]₄ emphasizing its relationship to a dodecahedron

The coordination chemistry of uranium(IV) with hydroxamic acids is complicated by the existence of the stable uranyl ion. Uranium tetrachloride is quantitatively oxidized via an oxygen transfer reaction with two equivalents of N-phenylbenzohydroxamic acid anion (PBHA) in tetrahydrofuran (THF) to form an uranyl complex and benzanilide (79). Substituents are known to be important in determining the redox behavior of hydroxamic acids (80, 81, 82, 83, 84), thus their effect on the oxidation of uranium(IV) was investigated. Reaction of UCl_4 with benzohydroxamic acid anion leads to the formation of tetrakis(benzohydroxamato)uranium(IV) as the major product. However, this compound undergoes the internal redox reaction upon heating to form an uranyl compound and benzamide. There was no evidence of oxidation-reduction in the synthesis of the tetrakis alkylhydroxamates described above, and only slight decomposition occurred on heating. The internal redox reaction displayed by the hydroxamate and uranium(IV) ions is a chemical limitation upon the use of hydroxamate groups in an actinide-specific sequestering agent. However, this may be avoided by the proper choice of substituent groups.

Actinide Sequestering Agents

The structures determined for the actinide(IV) catecholates and hydroxamates indicate that the mmmm-isomer of the dodecahedron is the preferred geometry. For maximum stability and specificity this geometry should be achieved by the ligating groups of an optimized sequestering agent that encapsulates the tetravalent actinide in a cavity with a radius near 2.4 Å. An examination of molecular models showed that this could be accomplished by the attachment of four 2,3-dihydroxybenzoic acid groups to the nitrogens of a series of cyclic tetraamines via amide linkages as shown schematically in Figure 10. The size of the cavity formed is controlled by the ring size of the tetraazacycloalkane backbone — such that a 16 membered ring appeared most promising for the actinides. Two tetra-catechol chelating agents were synthesized, as shown in Figure 11, by the reaction of 2,3-dioxomethylene- or 2,3-dimethoxybenzoyl chloride with 1,4,8,11-tetraazacyclotetradecane or 1,5,9,13-tetraazacyclohexadecane followed by the deprotection of the hydroxyl groups with BBr_3/CH_2Cl_2 (85). Subsequent biological evaluation in mice showed that these compounds reduced the accumulation of plutonium in bone and liver. However, the actinide complex apparently dissociated at low pH and the plutonium was deposited in the animals' kidneys (86). Titrimetric studies of these ligands with tetravalent actinides showed that while they strongly complex actinides, simple one-to-one complexes are not formed at or below neutral pH.

The performance of a ligand at low pH can be improved by increasing its acidity, thus reducing the competition with protons. The acidity of the catechol groups can be increased by the intro-

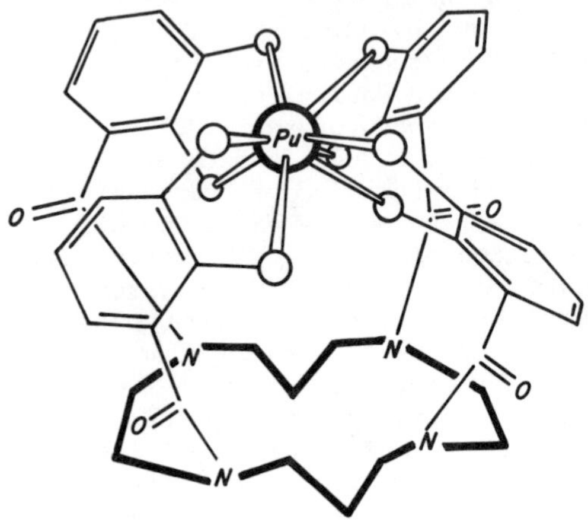

Figure 10.   Schematic structure of the tetracatechol actinide sequestering agents from a biomimetic approach based on enterobactin

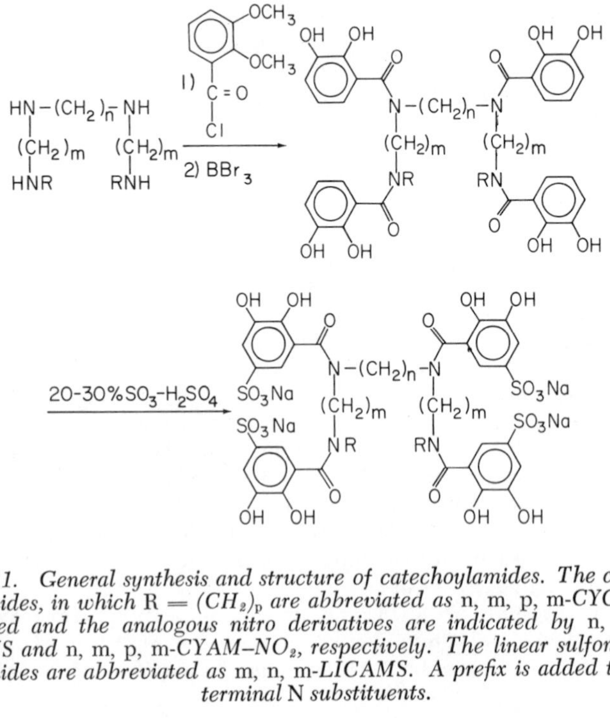

Figure 11.   General synthesis and structure of catechoylamides. The cyclic catechoylamides, in which $R = (CH_2)_p$, are abbreviated as n, m, p, m-CYCAM. The sulfonated and the analogous nitro derivatives are indicated by n, m, p, m-CYCAMS and n, m, p, m-CYAM–NO$_2$, respectively. The linear sulfonated catechoylamides are abbreviated as m, n, m-LICAMS. A prefix is added to indicate terminal N substituents.

duction of strongly electron-withdrawing groups to the aromatic
rings. A more acidic analog of the above ligands was prepared
from 2,3-dimethoxy-5-nitrobenzoic acid and 1,4,8,11-tetraazacyclo-
tetradecane. The nitro groups converted the ligand into an
acutely active poison and substantially changed its solubility
characteristics such that a large amount of plutonium was found in
the soft tissues of the injected mice (86). In sharp contrast,
sulfonation improved the water solubility, stability to air oxidation and the affinity for actinide(IV) ions at low pH. Each 2,3-
dihydroxybenzoyl group in the ligands prepared above was mono-
sulfonated regiospecifically at the 5 position by direct reaction
with 20-30% SO_3 in H_2SO_4 at room temperature (87). The increased
acidity of the sulfonated derivatives prevented the deposition of
plutonium in the kidneys of mice and promoted significant plutonium excretion without any appreciable acute toxic affects (86).

In order to examine the effect of greater stereochemical
freedom, some tetra-2,3-dihydroxy-5-sulfobenzoyl derivatives of
linear tetraamines, also shown in Figure 11, have been prepared by
similar methods (87). Maximum stability and specificity towards
the actinides, can be obtained by optimizing the length of the
methylene bridges between the amine functionalities. Butylene
bridges between the nitrogens of the linear tetraamines gave
better results in animal studies than ethylene or propylene
bridges. The linear derivatives are significantly more effective
than the cyclic catechoylamides in removing plutonium from mice
(86). In accordance with the trans configuration of amine hydrogens found in the structure of 1,5,9,13-tetraazacyclohexadecane
(88), adjacent catechoylamide groups are expected to lie on opposite sides of the macrocycle. While inversion about amides is
well known, it may not be rapid enough in these compounds to en-
able coordination of the actinide by all four catechol groups.

The catechoylamides were evaluated by their intraperitoneal
administration to mice (20 to 30 μmole/kg) 1 hour after the injection of 1.5 μCi/kg of $^{238}Pu(IV)$ citrate (86). The mice were killed by cervical dislocation 24 hours after the plutonium injection.
The effectiveness of the ligands was obtained by measuring the
plutonium in tissues and excreta using L X-rays, and the results
are presented in Table VII (86). The 4,4,4- or 3,4,3-LICAMS were
the most efficient of the catechoylamides tested; each promoted
elimination of about 65% of the injected plutonium. In addition
to sequestering the plutonium in body fluids, 3,4,3-LICAMS reduced
skeletal plutonium to 22% of the 1 hour control value (the time
of ligand injection). Monomeric dimethyl-2,3-dihydroxy-5-sulfo-
benzamide, DiMeCAMS, and 2,3-dihydroxybenzoic acid removed very
little if any plutonium. Similar results for 2,3-dihydroxy-
benzoyl-N-glycine were obtained by Bulman and co-workers (89).

This dramatic difference between the monomeric catechols and
the synthetic tetracatechol compounds confirm our original design
concept that a macrochelate would be most effective biologically
in Pu removal. Of the sulfonated catechoylamides only the

Table VII. Effect of Tetrameric Catechoylamides on the Distribution of ^{238}Pu(IV) in Mice[a]

| | Percent Absorbed Dose[b] | | | | | |
Compound	Liver	Skeleton	Soft Tissue	GI Tract	Kidneys	Whole Body
3,4,3-LICAMS	22	7	1.8	3.3	1.2	35
4,4,4-LICAMS	25	8	3.0	3.0	1.5	41
3,3,3-LICAMS	41	9	2.8	3.3	1.0	56
4,3,4-LICAMS	30	11	3.7	4.3	7.5	57
2,3,3,3-CYCAMS	30	18	7	4.7	4.0	61
2,3,2-LICAMS	26	13	12	8.1	3.9	63
3,3,3,3-CYCAMS	32	14	10	4.7	4.7	65
3,3,3,3-CYCAM	23	18	3	3.6	41	89
3,2,3,2-CYCAM-NO$_2$	37	8	26	16	6.0	93
CaNa$_3$DTPA	17	11	3	5.2	0.5	37
Desferrioxamine	23	20	4	4.2	1.6	52
DiMeCAMS	49	30	7	5.9	1.5	94
DBHA[c]	50	31	8	6.7	1.5	97
1 hr Control	30	23	32	12	3.4	~100
24 hr Control	51	31	5.1	4.6	2.6	94

[a]Ligands were administered 1 hour and the mice were killed 24 hours after injection of ^{238}Pu(IV) citrate, from Ref. 86.

[b]Five mice per group except for 4,4,4-LICAMS, 10 mice; 1 hour control, 7 mice; and 24 hour control, 34 mice.

[c]2,3-Dihydroxybenzoic acid.

4,4,4-LICAMS showed significant acute toxic effects in mice. For comparison, DTPA, the most effective conventional chelating agent, was examined and found to remove 63% of the injected plutonium. However, the dose-response curve, Figure 12, shows that 3,4,3-LICAMS is much more effective than DTPA at lower doses — up to two orders of magnitude difference (90). This is a good indication that endogenous metals are not strongly bound by 3,4,3-LICAMS, while metals such as calcium and zinc bind strongly to DTPA, reducing the effective concentration of the ligand. Thus a much larger amount of DTPA is required to achieve the same effective concentration of a smaller quantity of 3,4,3-LICAMS because of both a lower intrinsic affinity for actinide(IV) ions as well as a lower specificity.

The greater efficacy of plutonium decorporation by 3,4,3-LICAMS compared to DTPA has also been observed in beagles (91). Beagles were treated with a single intravenous injection of 30 μmole/kg of either Ca-DTPA or 3,4,3-LICAMS or 30 μmole/kg of both 30 minutes after an intravenous injection of 0.233 μCi ^{239}Pu(IV), 0.087 μCi ^{237}Pu(IV) and 0.575 μCi ^{241}Am(III) in a citrate buffer. Retention of the radionuclides was determined seven days after their injection. Serious toxic effects were seen in the kidneys of all dogs treated with 3,4,3-LICAMS. The dose response curve of Figure 12 would suggest that smaller doses should be nearly as effective, and avoid such toxic effects. As seen in Table VIII, 3,4,3-LICAMS removed about 86% of the injected plutonium, much better than the 70% removed by DTPA. In contrast, DTPA was much more effective in americium decorporation. This was expected since the affinity of catechol ligands for the larger and less acidic Ln(III) or An(III) ions is quite low. The measured ratio of the tetrakis(catecholato)Ce(IV)/Ce(III) formation constants of 10^{36} is an indication of the decreased affinity of 3,4,3-LICAMS for the trivalent actinides (67).

Summary

For the first time a class of sequestering agents has been designed and synthesized for the specific role of complexing plutonium and other actinide(IV) ions. This has resulted from the combination of two observations: (1) The chemical properties of Pu(IV) and Fe(III) are similar in many respects and this similarity extends to the biological transport and distribution properties of Pu(IV), which accounts for much of the biological hazard of this element. (2) The design of specific sequestering agents for Fe(III) was solved by microbes a few billion years ago with the production of low molecular-weight chelating agents (siderophores) that incorporate chelating groups such as hydroxamic acids and catechol.

Synthetic macrochelates have been designed such that the chelating groups can form a cavity that gives eight-coordination about the metal and the dodecahedral geometry observed in the

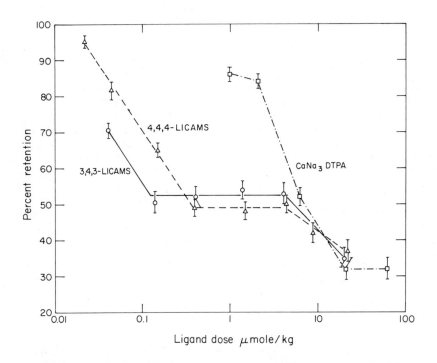

Figure 12. Dose response comparison between LICAMS and CaNa₃DTPA for
²³⁸Pu removal from mice. Percent retention = 100 − percent removed (90).

Table VIII.   Effect of 3,4,3-LICAMS on Plutonium Retention in Beagles[a]

| | Percent Injected Dose[b] | | |
	Whole Body	Liver	Non-liver
Control	90	35	55
LICAMS	14	6.3	7.4
CaNa$_3$ DTPA	30	9.9	20
CaNa$_3$ DTPA + LICAMS	12	4.9	6.7

[a]Ligands were administered 30 minutes and the dogs were killed 7 days after injection of the radionuclides, from Ref. 91.

[b]Two dogs per group, except that the control values were taken from References 92, 93, 94.

Table IX.   Summary of Actinide Sequestering Properties of Tetra-meric Catechoylamides

Cyclic

3,3,3,3-CYCAM	Mobilizes Pu but deposits it in kidneys.
3,2,3,2-CYCAM-NO$_2$	Acutely toxic.
3,3,3,3-CYCAMS	Sulfonation increases acidity and solubility, prevents Pu deposition in kidneys.
2,3,3,3-CYCAMS	Effective Pu removal.

Linear

2,3,2-LICAMS	Least effective in linear compounds.
3,3,3-LICAMS	Longer chain length, slight improvement
4,3,3-LICAMS	in Pu removal, still not very effective.
4,4,4-LICAMS	Slightly toxic
3,4,3-LICAMS	Derivative of spermine (a natural product)

Longer central bridge gives optimum geometry and maximum Pu removal

Less constrained linear structures are superior to corresponding cyclic compounds.

unconstrained actinide complexes composed of monomeric ligands. The most promising actinide squestering agents yet prepared (Table IX) are the sulfonated catechoylamide derivatives of linear tetra-amines. These compounds appear to strongly bind tetravalent actinides, while only weak complexation has been observed for trivalent and divalent metals. A derivative of the natural product spermine, 3,4,3-LICAMS, is more effective in plutonium removal at low dosages than any other sequestering agent tested to date.

Acknowledgment

This work was supported by the Division of Nuclear Sciences, Office of Basic Energy Sciences, U.S. Department of Energy under Contract No. W-7405-Eng-48, and the Office of the Director, Lawrence Berkeley Laboratory.

Literature Cited

1. Blomeka, J. O.; Nichols, J. P.; Mclain, W. C. Physics Today, August 1973, 26, 36-42.

2. Kube, A. S.; Rose, D. J. Science, 1973, 182, 1205-11.

3. Bienvenu, P.; Nofre, C.; Cier, A. C. R. Acad. Sci., 1963, 256, 1043-4.

4. Stannard, J. N. "The Health Effects of Plutonium and Radium";Jee, W. S. S., Ed.; J. W. Press: Salt Lake City, 1976; pp 362-72.

5. Durbin, P. W. "Handbook of Experimental Phramacology, Vol. 36: Uranium, Plutonium, Transplutonic Elements"; Hodge, H. C.; Stannard, J. N.; Hursh, J. B., Eds.; Springer-Verlag: New York, 1973; pp 739-896.

6. Denham, D. H. Health Physics, 1969, 16, 475-87.

7. Bair, J. C.; Thompson, R. C. Science, 1974, 183, 715-22.

8. Jones, M. M.; Pratt, T. H. J. Chem. Ed., 1976, 53, 342-7.

9. Durbin, P. W. Health Physics, 1975, 29, 495-510.

10. International Commission on Radiological Protection, Publication 19: "The Metabolism of Compounds of Plutonium and Other Actinides"; Pergamon Press: New York, 1972.

11. Rundo, J.; Starzyk, P. M.; Sedlet, J.; Larsen, R. P.; Oldham, R. D.; Robinson, J. J. "Diagnosis and Treatment of Incorporated Radionuclides"; International Atomic Energy Agency: Vienna, 1976; pp 15-23.

12. Vaughan, J.; Bleany, B.; Taylor, D. M. "Handbook of Experimental Pharmacology, Vol. 36: Uranium, Plutonium, Transplutonic Elements"; Hodge, H. C.; Stanndard, J. N.; Hursh, J. B., Eds.; Springer-Verlag: New York, 1973; pp 349-502.

13. Catsch, A. "Diagnosis and Treatment of Incorporated Radionuclides"; Interantional Atomic Energy Agency: Vienna, 1976; pp 295-305.

14. Smith, V. H. Health Physics, 1972, 22, 765-78.

15. Catsch, A.; Harmuth-Hoene, A-E. Biochemical Pharmacology, 1975, 24, 1557-62.

16. Pachauri, O. P.; Tandon, J. P. J. Inorg. Nucl. Chem., 1975, 37, 2321-3.

17. Pachauri, O. P.; Tandon, J. P. Indian J. Chem., 1977, 15A, 57-8.

18. Pachauri, O. P.; Tandon, J. P. J. Gen. Chem. USSR (Engl. Transl.), 1977, 47, 398-401; Zh. Obshch. Khim., 1977, 47, 433-6.

19. Carey, G. H.; Martell, A. E. J. Am. Chem. Soc., 1968, 90, 32-8.

20. Bogucki, R. F.; Martell, A. E. J. Am. Chem. Soc., 1958, 80, 4170-4.

21. Fried, A. R.; Martell, A. E. J. Am. Chem. Soc., 1971, 93, 4695-700.

22. Grimes, J. H. "Diagnosis and Treatment of Incorporated Radionuclides"; International Atomic Energy Agency: Vienna, 1976; pp 419-60.

23. Cohen, N.; Guilmette, R. Bioinorganic Chem., 1975, 5, 203-10.

24. Seven, M. J. "Metal-Binding in Medicine"; Seven, M. J.; Johnson, L. A., Eds.; J. B. Lippincott: Philadelphia, 1960; pp 95-103.

25. Foreman, H.; Nigrovic, V. "Diagnosis and Treatment of De-
     posited Radionuclides"; Kornberg, H. A.; Norwood, W. D.,
     Eds.; Excerpta Media Foundation: Amsterdam, 1968; pp 419-23.

26. Planas-Bohne, F.; Lohbreier, J. "Diagnosis and Treatment of
     Incorporated Radionuclides"; International Atomic Energy
     Agency; Vienna, 1976; pp 505-15.

27. Taylor, D. M. "Handbook of Experimental Pharmacology, Vol.
     36: Uranium, Plutonium, Transplutonic Elements"; Hodge, H.
     C.; Stannard, J. N.; Hursh, J. B., Eds.; Springer-Verlag:
     New York, 1973; pp 323-47.

28. Bulman, R. A. Structure and Bonding, 1978, 34, 39-77.

29. Taylor, D. M. Health Physics, 1972, 22, 575-81.

30. Popplewell, D. S. "Diagnosis and Treatment of Incorporated
     Radionuclides"; International Atomic Energy Agency: Vienna,
     1976; pp 25-34.

31. "Microbial Iron Metabolism"; Neilands, J. B., Ed.; Academic
     Press: New York, 1974.

32. Raymond, K. N. "Advances in Chemistry Series, No. 162:
     Bioinorganic Chemistry — II"; Raymond, K. N., Ed.; American
     Chemical Society: Washington, D.C., 1977; pp 33-54.

33. Raymond, K. N.; Carrano, C. J. Acc. Chem. Res., 1979, 12,
     183-90.

34. Huhey, J. E. "Inorganic Chemistry: Principles of Structure
     and Reactivity"; Harper and Row: New York, 1972; pp 418-22.

35. Barocas, A.; Baroncelli, F.; Biondi, G. B.; Grossi, G.
     J. Inorg. Nucl. Chem., 1966, 28, 2961-7.

36. Dyrssen, D. Acta Chem. Scand., 1956, 10, 353-9.

37. Reidel, A. J. Radioanal. Chem., 1973, 13, 125-34.

38. Zharovskii, F. G.; Ostrovskaya, M. S.; Sukhomlin, R. I.
     Izv. Vyssh. Ucheb. Zaved., Khim. Khim. Teknol., 1967, 10,
     989-93; Chem. Abstr., 1968, 69, 30696y.

39. Chimutova, M. K.; Zolotov, Yu A. Soviet Radiochem. (Engl.
     Transl.), 1964, 6, 625-30; Radiokhimiya, 1964, 6, 640-5.

40. Zharovskii, F. G.; Sukhomlin, R. I.; Ostrovskaya, M. S. Russ. J. Inorg. Chem. (Engl. Transl.), 1967, 12, 1306-9; Zh. Neorg. Khim., 1967, 12, 2476-80.

41. Agrawal, R. P.; Mehrotra, R. C. J. Inorg. Nucl. Chem., 1962, 24, 821-7.

42. Avdeef, A.; Bregante, T. L.; Raymond, K. N. Manuscript in preparation.

43. Casellato, U.; Vidali, M.; Vigato, P. A. Inorg. Chim. Acta, 1976, 18, 77-112.

44. Moseley, P. T. "MTP Int. Rev. Sci.: Inorg. Chem., Ser. Two"; Bagnall, K. W., Ed.; Butterworth: London, 1975; Vol. 7, pp 65-110.

45. Hoard, J. L.; Silverton, J. V. Inorg. Chem., 1963, 2, 235-43.

46. Burdett, J. K.; Hoffmann, R.; Fay, R. C. Inorg. Chem., 1978, 17, 2553-68.

47. Blight, D. G.; Kepert, D. L. Inorg. Chem., 1972, 11, 1556-61.

48. Porai-Koshits, M. A.; Aslanov, L. A. J. Struct. Chem. USSR (Engl. Transl.), 1972, 13, 244-53; Zh. Strukt. Khim., 1972, 13, 266-76.

49. Muetterties, E. L.; Guggenberger, L. J. J. Am. Chem. Soc., 1974, 96, 1748-56.

50. Kepert, D. L. Prog. Inorg. Chem., 1978, 24, 179-249.

51. Allard, B. J. Inorg. Nucl. Chem., 1976, 38, 2109-15.

52. Steffen, W. L.; Fay, R. C. Inorg. Chem., 1978, 17, 779-82.

53. Lenner, M. Acta Crystallogr., Sect. B, 1978, 34, 3770-2.

54. Piero, G. D.; Perego, G.; Zazzetta, A.; Brandi, G. Cryst. Struct. Comm., 1975, 4, 521-6.

55. Brown, D.; Holah, D. G.; Rickard, C. E. F. J. Chem. Soc., Sect. A., 1970, 423-5.

56. Brown, D.; Holah, D. G.; Rickard, G. E. F. J. Chem. Soc., Sect. A, 1970, 786-90.

57. Day, V. W.; Fay, R. C. "Abstracts of Papers"; American Crystallographic Association, Summer Meeting, 1976, p 78.

58. Burns, J. H.; Danford, M. D. Inorg. Chem., 1969, 8, 1780-4.

59. Volz, K.; Zalkin, A.; Templeton, D. H. Inorg. Chem., 1976, 15, 1827-31.

60. Hill, R. J.; Rickard, C. E. F. J. Inorg. Nucl. Chem., 1977, 39, 1593-6.

61. Lenner, M.; Lindquist, O. Acta Crystallogr., Sect. B, 1979, 35, 600-3.

62. Baskin, Y.; Prasad, N. S. K. J. Inorg. Nucl. Chem., 1963, 25, 1011-19.

63. Wolf, L.; Bärnighausen, H. Acta Crystallogr., 1960, 13, 778-85.

64. Wessels, G. F. S.; Leipoldt, J. G.; Bok, L. D. C. Z. Anorg. Allg. Chem., 1972, 393, 284-94.

65. Hambley, T. W.; Kepert, D. L.; Raston, C. L.; White, A. H. Aust. J. Chem., 1978, 31, 2635-40.

66. Sofen, S. R.; Abu-Dari, K.; Freyberg, D. P.; Raymond, K. N. J. Am. Chem. Soc., 1978, 100, 7882-7.

67. Sofen, S. R.; Cooper, S. R.; Raymond, K. N. Inorg. Chem., 1979, 18, 1611-16.

68. Latimer, W. M. "Oxidation States of the Elements and Their Potentials in Aqueous Solution", 2nd ed.; Prentice-Hall: Englewood Cliffs, N.J., 1952; p 294.

69. Ho, T.-L.; Hall, T. W.; Wong, C. M. Chem. Ind. (London), 1972, 729-30.

70. Raymond, K. N.; Isied, S. S.; Brown, L. D.; Fronczek, F. F.; Nibert, J. H. J. Am. Chem. Soc., 1976, 98, 1767-74.

71. Cooper, S. R.; Freyberg, D. P.; Raymond, K. N. Manuscript in preparation.

72. Magers, K. D.; Smith, C. G.; Sawyer, D. T. Inorg. Chem., 1978, 17, 515-23.

73. Meites, L. "Polarographic Techniques"; Wiley: New York, 1965; p 279.

74. Shannon, R. D. Acta Crystallogr., Sect. A, 1976, 32, 751-67.

75. Pauling, L. "The Nature of the Chemical Bond", 3rd ed.;
     Cornell University Press: Ithica, N.Y., 1960; p 260.

76. Figgis, B. N. "Introduction to Ligand Fields"; Interscience:
     New York, 1966.

77. Smith, W. L.; Raymond, K. N. Manuscript in preparation.

78. "Tables of Interatomic Distances and Configurations in
     Molecules and Ions, Chem. Soc., Spec. Publ., No. 18, Suppl.
     1956-1959"; The Chemical Society: London, 1965.

79. Smith, W. L.; Raymond, K. N. J. Inorg. Nucl. Chem., 1979,
     41, 1431-6.

80. Minor, D. F.; Waters, W. A.; Ramsbottom, J. V. J. Chem. Soc.,
     Sect. B, 1967, 180-4.

81. Mackor, A.; Wajer, Th., A. J. W.; de Borer, Th. J. Tetra-
     hedron, 1968, 24, 1623-31.

82. Rawson, G.; Engberts, J. B. F. N. Tetrahedron, 1970, 26,
     5653-64.

83. Ozaki, S.; Masui, M. Chem. Pharm. Bull., 1977, 25, 1179-85.

84. Balaban, A. T.; Pascaru, I.; Cuiban, F. J. Mag. Res., 1972,
     7, 241-6.

85. Weitl, F. L.; Raymond, K. N.; Smith, W. L.; Howard, T. R.
     J. Am. Chem. Soc., 1978, 100, 1170-72.

86. Durbin, P. W.; Jones, E. S.; Raymond, K. N.; Weitl, F. L.
     Radiat. Res., in press.

87. Weitl, F. L.; Raymond, K. N. Submitted for publiciation in
     J. Am. Chem. Soc.

88. Smith, W. L.; Ekstrand, J. D.; Raymond, K. N. J. Am. Chem.
     Soc., 1978, 100, 3539-44.

89. Bulman, R. A.; Griffin, R. J.; Russel, A. T. NRPB Report
     RANDD-1, 1977, 87-9.

90. Durbin, P. W.; Jones, E. S.; Raymond, K. N.; Weitl, F. L.
     Unpublished data.

91. Bruenger, F. W.; Atherton, D. R.; Jones, C. W.; Weitl, F. L.; Durbin, P. W.; Raymond, K. N; Taylor, G. N.; Stevens, W.; Mays, C. W. "Research in Radiobiology"; University of Utah Radiobiology Laboratory, Annual Report COO-119-254, 1979.

92. Stover, B. J.; Atherton, D. R.; Bruenger, F. W.; Buster, D. S. Health Physics, 1968, 14, 193-7.

93. Stover, B. J.; Atherton, D. R.; Buster, D. S. Health Physics, 1971, 20, 369-74.

94. Stover, B. J.; Atherton, D. R.; Buster, D. S. "The Radiobiology of Plutonium"; Stover, B. J.; Jee, W. S. S., Eds.; J. W. Press: Salt Lake City, 1972; pp 149-69.

RECEIVED December 26, 1979.

Inner- vs. Outer-Sphere Complexation of Lanthanide (III) and Actinide (III) Ions

GREGORY R. CHOPPIN

Department of Chemistry, Florida State University, Tallahassee, FL 32306

Abstract

   The thermodynamic data for complexation of trivalent lanthanide and actinide cations with halate and haloacetate anions are
reported. These data are analyzed for estimates of the relative
amounts of inner (contact) and outer (solvent separated) sphere
complexation. The halate data reflected increasing inner sphere
character as the halic acid pK_a increased. Use of a Born-type
equation with the haloacetic acid pK_a values allowed estimation
of the effective charge of the carboxylate group. These values
were, in turn, used to calculate the inner sphere stability constants with the M(III) ions. This analysis indicates increasing
the inner sphere complexation with increasing pK_a but relatively
constant outer sphere complexation.

   Although the concept of outer sphere complexation was introduced by Werner (1) in 1913 and the theory first given a mathema-
tical base by Bjerrum (2) in 1926, progress in understanding the
factors involved in the competition between inner and outer sphere
complexation has been very slow.
   We use the term "outer sphere complex" to refer to species in
which the ligand does not enter the primary coordination sphere of
the cation but remains separated by at least one solvent molecule.
Such species are known also as "solvent separated" ion pairs to
distinguish them from inner sphere complexes in which the bonding
is ionic ("contact" ion pairs). Mironov (3) offered some empirical
rules for outer sphere complexation but these provide no insight
into the basis of the inner-outer sphere competition. Beck (4)
and Gutmann (5) have reviewed outer sphere complexation and attribute a significant role to hydrogen bonding. This would agree
with a correlation between the pK_a of ligand acid (i.e., the ligand
basicity) and the competition of inner vs outer sphere complexation for Ln(III) cations (6).

0–8412–0568–X/80/47–131–173$05.00/0
© 1980 American Chemical Society

For labile complexes, it is often quite difficult to distinguish between inner and outer sphere complexes. To add to this confusion is the fact that formation constants for such labile complexes when determined by optical spectrometry are often lower than those of the same system determined by other means such as potentiometry, solvent extraction, etc. This has led some authors to identify the former as "inner sphere" constants and the latter as "total" constants. However, others have shown that this cannot be correct even if the optical spectrum of the solvated cation and the outer sphere complex is the same ($\underline{4}$, $\underline{7}$). Nevertheless, the characterization and knowledge of the formation constants of outer sphere complexes are important as such complexes play a significant role in the Eigen mechanism of the formation of labile complexes ($\underline{8}$). This model describes the formation of complexes as following a sequence:

$$M_{(aq)} + X_{(aq)} \rightarrow [M(H_2O)_n X]_{aq} = [M(H_2O)X]_{aq} = MX_{aq}$$

The first step is diffusion controlled while the second represents the fast formation of the outer sphere complex. The final step involves the conversion of the outer to the inner sphere complex. This is the rate determining step and is dependent on the equilibrium concentration of the outer sphere complex. Consequently, calculations of rate constants by the Eigen model involves estimation of the formation constant of the outer sphere species.

Trivalent lanthanide and actinide cations form labile, ionic complexes of both inner and outer sphere character. Consequently, they are useful probes to study inner-outer sphere complexation competition due to ligand properties. Two earlier papers have reported complexation of these cations by two series of related anions, the halates ($\underline{9}$) and the chloroacetates ($\underline{10}$). In this paper we offer a more extensive analysis of the inner-outer sphere competition in these complexes.

Halates

The data for formation of $EuXO_3^{+2}{}_{(aq)}$ complexes of the halates are given in Table I. A number of authors ($\underline{11}$, $\underline{12}$, $\underline{13}$) have proposed that enthalpy and entropy changes should be more positive for inner sphere complexation and, in fact, outer sphere formation may be reflected by small enthalpy values and even negative entropies. Based on this concept, we interpreted the halates data as reflecting essentially complete outer sphere complexation for $EuClO_3^{+2}$ ($HClO_3$, $pK_a = -2.7$) and similarly complete inner sphere complexation for $EuIO_3^{+2}$ (HIO_3, $pK_a = 0.7$). We can attempt a crude estimate for $EuBrO_3^{+2}$ by using $\Delta S_o = -20$ and $\Delta S_i = +60$ J/m/K (the subscript o refers to outer sphere (e.g., β_o, ΔH_o, etc.) and the subscript i to inner sphere (e.g., β_i, ΔH_i, etc.)) as the entropy changes for outer and inner sphere complexation respectively with:

$$a \; \Delta S_o + (1-a) \; \Delta S_i = \Delta S_{exp}$$

For $\Delta S_{exp} = 3 \pm 5$, we obtain an estimate of 65-75% of outer sphere for $EuBrO_3^{+2}$ ($HBrO_3$, $pK_a = -2.3$). However, such a calculation of simple additive entropies is probably too naive to be of much value. Morris, et al (14) have used similar reasoning to assign predominant outer sphere character to $ScClO_3^{+2}$ and $ScBrO_3^{+2}$.

Fuoss (15) proposed an equation which has been used frequently to calculate outer sphere formation constants. The equation has the form:

$$K_o = \frac{4\pi N a^3}{3000} \; e^{\mu/kT}$$

with $a = 5 \overset{\circ}{A}$ and $\mu =$ the electrostatic energy of attraction between cation and anion. This equation, when correction is included for the ionic strength, gives a calculated stability constant for $EuClO_3^{+2}$ in good agreement with the experimental value. A value for the dielectric constant of 70 was used in this calculation.

In summary, the halate data reaffirm the tendency of increased inner sphere character in Ln(III) complexes as the ligand pK_a increases.

Chloroacetates

As chlorine substitutes for hydrogen in the methyl group of the acetate anion, the carboxylate basicity decreases (i.e., pK_a decreases). The thermodynamic data for the Eu(III) and Am(III) complexes and the ligand pK_a values are listed in Table II. Data for β-chloropropionate (16) are also included.

Analysis of the entropy changes, indicates essentially 100% inner sphere formation for the Ac, β-ClPr and ClAc complexes, 50% inner sphere for the Cl_2Ac complexes. However, a study of [139]La nmr shifts (18) was interpreted to show only 50% inner sphere character for $LaClAc^{+2}$ and 20-25% for $LaCl_2Ac^{+2}$. In light of this lack of agreement, we have analysed the complexation by another approach which would seem to be more justified than the entropy based estimations.

Munze (19) has used a Born-type equation to calculate stability constants of Ln(III) and An(III) complexes of carboxylates as well as other ligands which agreed well with experimental values. His approach was modified by allowing the dielectric constant to be a parameter (the "effective" dielectric constant, D_e) in an analysis of fluoride complexation by M(II), M(III) and M(IV) cations (20). A value of $D_e = 57$ was found satisfactory to calculate trivalent metal fluoride stability constants which agreed with experimental values for Ln(III), An(III) and group IIIB cations (except Al(III)). Subsequently, the equation was used

Table I

Thermodynamic Parameters for Halate Complexation[9]

$T = 25.0°C$; $I = 0.1$ M(NaClO$_4$)

Complex	$-\Delta G$ (kJ mol^{-1})	ΔH (kJ mol^{-1})	ΔS (JK^{-1}mol^{-1})
EuClO$_3^{+2}$	0.25±0.42	− 6.3±1.7	−20±7
EuBrO$_3^{+2}$	3.39±0.25	− 2.5±1.3	+ 3±5
EuIO$_3^{+2}$	6.53±0.42	+11.0±0.8	+59±6

Table II

Thermodynamic Parameters for Monocarboxylate Complexation

$T = 25.0°C$; $I = 2.0$ M(NaClO$_4$)

Ligand	$-\Delta G$ (kJ mol^{-1})	ΔH (kJ mol^{-1})	ΔS (JK^{-1}mol^{-1})	pK$_a$	Ref.
		a) Eu(III)			
Ac	10.92 0.04	5.9 0.4	62	4.80	16
β-ClPr	9.17 0.04	9.97 0.21	63	4.13	17
ClAc	6.15 0.20	12.35 0.20	62	2.73	10
Cl$_2$Ac	4.32 0.20	7.54 0.17	40	1.1	10
Cl$_3$Ac	1.84 0.22	0.25 0.02	7	−0.5	10
		b) Am(III)			
Ac	11.22 0.12	18.0 1.2	98	4.80	16
ClAc	6.49 0.12	7.70 0.42	51	2.73	10
Cl$_2$Ac	4.48 0.20	3.31 0.33	26	1.1	10
Cl$_3$Ac	2.84 0.22	8.84 0.08	9	−0.5	10

successfully with $D_e = 57$ to calculate stability constants for complexation of Ln(III) by oxocarbon ligands (21). The equation has the form:

$$\Delta G = \frac{Ne^2 Z_1 Z_2}{(4.187 \times 10^2)D_e d_{12}} - RT\nu\ln 55.5 + RT\Sigma\ln f \qquad (1)$$

where N = Avogadro's number
    e = unit charge, 4.80×10^{-10} esu
Z_1, Z_2 = ionic charge of cation and anion
    $\nu = -1$
    d_{12} = internuclear distance in the ion pair M-X

$$\Sigma\ln f = \frac{-\Delta Z^2\ 0.511\ I^{\frac{1}{2}}}{1 + Ba^{\frac{1}{2}}} - CI^{\frac{1}{2}} - DI$$

$$\Delta Z^2 = [Z^2_{MX} - (Z^2_M + Z^2_X)]$$

    B = 0.33, C = 0.75, D = -0.015, a = 4.3

Our approach is the following. We assume that variation in ligand pK_a values as chlorine is substituted for hydrogen is due to differences in the effective charge on the carboxylate oxygens. We redefine Z_2 in (1) as Z'_2, the effective anion charge, in contrast to Z_2, the formal anion charge of -1. In principle; use of equation (1) with the proper values of ΔG, D_e, d_{12}, etc. allows calculation of Z'_2 the effective anionic charge. For the system of ligands of Table II, we found it necessary to use values of $D_e = 15.5$ and $d_{12} = 2.33$ Å to obtain physically reasonable values of Z'_2 for all 5 ligands (i.e., Z'_2 between -1 and 0) with the experimental ΔG_{HA} (although both D_e and d_{12} could vary 10-20% with little net effect). It is not possible to comment on possible physical meanings for these values as we have no simple physical model for the protonation of carboxylate groups in aqueous solution. The values of Z'_2 obtained with these values are listed in Table III. One bit of support for these results is found in a plot of pK_a vs Z'_2 which indicates $Z'_2 = -1$ at $pK_a \sim -1.5$ which corresponds with the range of reported values of monocarboxylic acids (22).

We assume that the effective charge remains the same upon complexation by Ln(III) and An(III). Based on the success of the fluoride and oxycarbon calculations, we use $D_e = 57$. We also use $d_{12} = 2.38$ Å ($r_{+3} = 1.0$ Å, $r_{o-} \simeq 1.83$ Å) and with equation (1) calculate inner sphere stability constants. The numbers we obtained are listed in Table III along with the values of β_o (based on Eu(III) complexing) and the per cent inner sphere complexation. The latter data is obtained from the relation: $\beta_T = \beta_o + \beta_i$.

In Figure 1, we compare the estimated per cent $LnXi^{+2}$ from the nmr results and from the calculations based on ΔS and on equation (1). For the ΔS calculations we used $\Delta S_i \simeq 60$, $\Delta S_o \simeq$

Table III

Values Calculated by Equation (1)

Ligand	Z_2'	$\beta_i(Eu)$*	$\beta_o(Eu)$	% Inner Sphere
Ac	-0.93	80	--	100
β-ClPr	-0.85	40	--	100
ClAc	-0.60	7.3	4.7	60
Cl_2Ac	-0.37	1.4	4.3	25
Cl_3Ac	-0.19	0.40	1.7	18

*Uncertainty estimated as 10-20%.

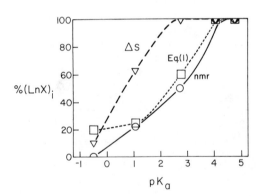

Figure 1.   Dependency of the percentage of inner-sphere complexation on ligand pK$_a$ as estimated by ΔS, Equation 1, and NMR

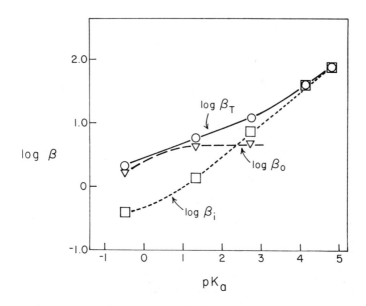

Figure 2.   Variation of log β_T (experimental), log β_o, and log β_i (estimated with Equation 1 with ligand pK$_a$)

0 J/m/K. The agreement between the nmr estimates and those from equation (1) add weight to the estimates in Table III. In Figure 2 the variation of log β_i and log β_o as functions of pK_a reflect the vital role of ligand basicity in the inner-outer sphere competition. These curves indicate that the cross-over from predominantly outer sphere to predominantly inner sphere occurs near pK_a values of 2. However, since the enthalpy and entropy changes for inner sphere complexation are larger than for outer sphere formation, both ΔH and ΔS would still be endothermic (characteristic of inner sphere reaction).

Summary

Both the halate and chlorocarboxylate systems show a relation between ligand pK_a and inner vs outer sphere complexation. For trivalent actinide and lanthanide cations it seems that carboxylate ligands form predominantly inner sphere complexes if their pK_a value exceeds 2 although the relative concentration of outer sphere complex is still significant for $pK_a \sim 3$. Moreover, since inner-outer sphere competition also seems to be a function of cation charge density, (e.g., MSO_4^o are predominantly outer sphere complexes while MSO_4^{+1} complexes are predominantly inner sphere (23)) inner sphere formation should remain dominant for M(IV) cations and ligands of lower values of pK_a than 2. Whereas, Pu(III) would form roughly equal amounts of $(PuClAc^{+2})_i$ and $(PuClAc^{+2})_o$, Pu(IV) would be expected to form predominantly inner sphere complexes with $ClAc^{-1}$ and, perhaps, even with Cl_2Ac^{-1}.

This research was supported through a contract with the Office of Basic Energy Sciences, U.S.D.O.E.

Literature Cited

1. Werner, A.; Neue Anschauungen auf dem Gebiet der anorganischen Chemie, 1913, 3rd ed., Vieweg Sohn, Braunschweig.
2. Bjerrum, N., Kgl. Danske Vidensk. math. fysike medd., 1926, 9, 7.
3. Mironov, V.E., Russian Chem. Rev., 1966, 35, 455; UDC 541.49: 541.571.53 (Trans.).
4. Beck, M.T., Coord. Chem. Rev., 1968, 3, 91.
5. Gutmann, V., Chimia, 1977, 31, 1.
6. Choppin, G.R. and Bertha, E.L., J. Inorg. Nucl. Chem., 1973, 35, 1309.
7. Johansson, L., Acta Chem. Scand., 1971, 25, 3579.
8. Eigen, M. and Wilkins, R., Adv. Chem. Ser., 1965, No. 49, 55.
9. Choppin, G.R. and Ensor, D.D., J. Inorg. Nucl. Chem., 1977, 1226.
10. Ensor, D.D. and Choppin, G.R., J. Inorg. Nucl. Chem., in press.
11. Rossotti, F.J.C., Modern Coordination Chemistry, J. Lewis and R.G. Wilkins (ed.) Interscience, New York, 1960.

12. Choppin, G.R. and Strazik, W.F., Inorg. Chem., 1965, 4, 1254.
13. Ahrland, S., Coord. Chem. Rev., 1972, 8, 21.
14. Morris, D.F.C.; Haynes, F.B.; Lewis, P.A. and Short, E.L., Electrochim. Acta, 1972, 17, 2017.
15. Fuoss, R.M., J. Am. Chem. Soc., 1958, 80, 5059.
16. Choppin, G.R. and Schneider, J.K., J. Inorg. Nucl. Chem., 1970, 32, 3283.
17. Orebaugh, E.G.; Degischer, G. and Choppin, G.R., unpublished results.
18. Rinaldi, P.L.; Khan, S.A.; Choppin G.R. and Levy, G.C., J. Am. Chem. Soc., 1979, 101, 1350.
19. Munze, R., Phys. Chemie (Leipzig), 1972, 249, 329; ibid, 1973, 252, 145; J. Inorg. Nucl. Chem., 1972, 34, 661; ibid, 1972, 34, 973.
20. Choppin, G.R. and Unrein, P.J., "Thermodynamic Study of Actinide Fluoride Complexation", Transplutonium Elements, W. Muller and R. Lindner (ed.), North-Holland, Amsterdam, 1976.
21. Choppin, G.R. and Orebaugh, E.G., Inorg. Chem., 1978, 17, 2300.
22. Brown, H.C.; McDaniel, D.H. and Hafliger, O., "Dissociation Constants", Determination of Organic Structures by Physical Methods, Braude, E.A., and Nachod, F.C. (ed.), Academic Press Inc., N.Y., 1955.
23. Ashurst, K.G. and Hancock, R.D., J. Am. Chem. Soc., 1977, 1701.

RECEIVED December 26, 1979.

Actinides: *d*-or *f*-Transition Metals?

WERNER MÜLLER

Commission of the Europen Communities Joint Research Centre,
Karlsruhe Establishment, European Institute for Transuranium Elements,
Postfach 2266, D-7500 Karlsruhe, Federal Republic of Germany

The chemical and physical properties of elements depend on their electron configuration and, in particular, on the number and the nature of the bonds formed by the outer ("valence") electrons. Crystal structure and lattice parameters, as well as thermodynamic and electronic properties reflect the nature and strength of these chemical bonds. For instance, the crystal radii of elements are a function of their atomic number and oxidation number, hence of their electron configuration in the solid state. Thermodynamics give access to the strength of the bond: melting temperatures, vapour pressures and enthalpies of dissolution are measures of the tendency of the elemental crystal to pass into the liquid, gaseous or solution phase. Specific heat and electrical resistivity as well as magnetic susceptibility furnish additional information on the electrons, their energy and their participation in bonding. Finally, spectroscopy (optical, photoemission) can provide information on the mixing of different quantum characters in the bond (hybridization).

Actinide elements (Z = 89 - 103) include the heaviest natural and most of the synthetic transuranium elements. They form a series of transition elements, characterized by the filling of an inner - the 5f-electron shell. The elements from Ac (Z = 89) to Es (Z = 99) are available in quantities sufficient for solid state studies. Elemental actinides are metallic. The methods of metal preparation and characterization have been improved to yield samples of known purity and crystal structure, sometimes in the form of single crystals. Recent measurements of structural, thermodynamic and electronic properties have emphasized elements in the beginning and in the centre of the actinide series.

In the following, methods for preparation, purification and characterization of actinide metals are reviewed. Properties are presented, the theoretical interpretation of which underlines the special nature of the actinides in comparison with d or 4f (lanthanide) transition metals.

0–8412–0568–X/80/47–131–183$05.00/0

Metal Preparation Chemistry

Availability and Handling.   Actinides are natural or synthetic radioelements, in many instances with limited element or isotope availability (Table I). Half-life of the radioactive decay is not always a sufficient criterion for the choice of a nuclide for solid state studies (1): For example, the handling risk of $^{248}Cm(T_{1/2} = 3.6 \times 10^5 y)$ is determined by neutron emission due to spontaneous fission, that of ^{243}Am $(T_{1/2} = 7.4 \times 10^3 y)$ by the gamma emission of the ^{239}Np daughter nuclide. Safe handling of actinides, therefore, demands adequate containment and shielding; handling of the chemically reactive metals requires, in addition, an inert atmosphere. Preparation and investigation techniques have to be adapted to the mass and the form of the sample available; the removal of the decay heat of short-lived isotopes may pose a problem.

Table I. Actinides for solid state studies

Nuclide	Half-life (years)	Availability	Problems
^{227}Ac	2.2×10^1	mg	γ (daughters), (α)
^{231}Pa	3.2×10^4	g	α, γ (^{231}Ac)
^{237}Np	2.1×10^6	g-kg	(α)
^{239}Pu	2.4×10^4	g-kg	α, heat
^{242}Pu	3.9×10^5	g	n(sf)
^{241}Am	4.3×10^2	g-kg	α, γ(60keV), heat
^{243}Am	7.4×10^3	g	γ(^{239}Np), α
^{244}Cm	1.8×10^1	g	α, γ, n(α,n), heat
^{248}Cm	3.6×10^5	mg	n(sf)
^{249}Bk	0.9×10^0	mg	(β)
^{249}Cf	3.5×10^2	mg	(α)
^{252}Cf	2.6×10^0	mg-g	n(sf), γ, α, heat
^{253}Es	5.5×10^{-2}	$\ll$mg	α, heat

Preparation Methods. Actinide metal preparation is based
on methods known or developed to yield high purity material by
metallothermic reduction or thermal dissociation of prepurified
compounds. Electrolytic reduction is possible from molten salts,
but not from aqueous solutions. Further purification of the metals
can be achieved by electrorefining, selective evaporation or
chemical vapour transport.

1. Metallothermic reduction of compounds. Metallothermic
reduction of halides (fluorides), a method used for lanthanide
metal preparation, was among the first methods to be successfully
applied to actinides:

$$AnF_x + xLi = An + xLiF \nearrow \quad (An = Ac, Am, Cm, Bk, Cf)$$

Due to its drawbacks (difficult preparation of water-free starting
material, neutron emission from (α,n) reactions, presence of
non-volatile impurities in the product), methods involving vapor-
isation of the actinide metal after reduction of a compound
(oxide, carbide) are preferred. If the vapour pressure of the
reductant and that of the actinide compound are markedly lower
than that of the metal formed, the latter can be removed from the
reaction mixture via the vapour phase and condensed in high
purity:

$$An_2O_3 + 2La = 2An \nearrow + LaO_3 \quad (An = Am, Cf, Es)$$

Starting from prepurified oxides and using lanthanum metal as
reductant, volatile Am, Cf or Es have been isolated and purified
by repeated distillation or sublimation. Thorium was used
successfully as reductant for actinium and curium oxides (2):

$$AnO_2 + Th = An \nearrow + ThO_2 \quad (An = Ac, Cm)$$

Actinide metals with very low vapour pressures (U,Np,Pu) have
been obtained by metallothermic reduction of their carbides by
non-volatile reductants (Ta, W) (3):

$$AnC + Ta = An \nearrow + TaC \quad (An = U, Np, Pu)$$

The tantalothermic reduction of PuC and UC requires reaction
temperatures of 1700 and 1900°C, respectively. The starting
carbides are prepared by carboreduction of their oxides; too low
a CO partial pressure during carboreduction may lead to actinide
losses by evaporation.

2. Thermal dissociation of compounds. Similar to the removal
of volatile actinide metals from a condensed reactant mixture is
the thermal dissociation of compounds that have components of
very different vapour pressures.

By thermal dissociation of intermetallic compounds with noble
metals (Pt, Ir), the volatile metals americium, curium and californium have been obtained in high purity (4,5):

$$AnPt = An + Pt \quad (An = Am, Cm, Cf)$$

The intermetallic compounds are synthesized by heating mixtures
of actinide oxides or halides with finely divided noble metal
powders in pure hydrogen. Protactinium metal was prepared in a
modified version of the van Arkel-de Boer procedure; protactinium
iodide, formed by reaction between iodine and protactinium carbide,
was thermally dissociated on a resistance heated tungsten wire
(6,7):

$$AnI_x = An + x/2I_2 \quad (An = Th, Pa)$$

Gramme quantities of protactinium could be deposited when the
dissociation wire was replaced by an induction heated tungsten
or protactinium sphere (3). Table II lists selected methods of
actinide preparation via the vapour phase.

Table II. Examples of actinide preparation via the
vapour phase

An	Starting compound	Method	Temperature
Ac	Ac_2O_3	reduction by Th	$1750^{\circ}C$
Pa	PaC	dissociation of PaI_x	$1200-1400^{\circ}C$
Pu	PuC	reduction by Ta	$1700^{\circ}C$
Am	$AmPt_5$	dissociation	$1300^{\circ}C$
Cf	Cf_2O_3	reduction by La	$1000^{\circ}C$

Characterization. Actinide metal samples for the determination of properties related to bonding have to be characterized
for chemical purity and phase homogeneity. Purity is checked by
chemical or physical analysis, crystal structure is determined
by X-ray or neutron diffraction techniques; phase heterogeneities
can be observed by metallography.
1. Purity. The use of evaporation methods for the preparation of
actinide metals reduces the number and quantities of impurities.
Nevertheless, possible chemical contaminations from reactions
with reducing agents, container vessel or crucible material or with
constituents of the atmosphere as well as the accumulation of
products of radioactive decay have to be taken into account.

Spectral methods (spark source mass spectrometry SSMS, secondary ion mass spectrometry SIMS, inductively coupled argon plasma for emission spectroscopy ICAP-ES) which avoid separation steps are increasingly applied for multi-element analysis. Hot extraction is used for O, N, H determinations. Oxygen is also determined by activation analysis, nitrogen after adaptation of classical methods (micro-Kjeldahl). Combination and comparison of different, independent methods are desirable, but hampered by the often limited availability of samples of actinides.

Very low impurity contents have been detected by the measurements of impurity sensitive properties like residual resistivity. Examples of impurity contents in less-common actinide metals are published for Pa (7), Am (8) and Cm (9). Isotope dilution mass spectrometry is expected to be increasingly applied to the accurate determination of selected elements, or to standardisation of routine methods or reference samples.

2. Crystal structure and phase homogeneity. X-ray diffraction is routinely employed for the determination of the crystal structure of the metal samples. For the structure analysis of polycrystalline (powder) material, film techniques involving the use of thin (breakable) glass or quartz capillaries are increasingly replaced by diffractometer techniques. Single crystals are investigated by Weissenberg or Gandolfi techniques. Due to their low penetration depth in heavy element samples, X-rays are unable to probe the bulk of actinide samples. Neutron diffraction, however, has enabled the checking of the crystal structure of a curium sample at cryogenic temperatures despite its encapsulation in a double metal container (10). Classical metallography can be applied to detect phase transitions or phase heterogeneities, even with rare actinide metals (11).

Properties

Crystal Structure and Phase Stability. Crystal structure and (formal) valence of metals depend on the configuration of the outer ("valence") electrons (12). For simple metals of the main groups of the periodic table, e.g. Na (I), Mg (II), or Al (III), the electron configurations 3s, 3s3p, or $3sp^2$ correspond to the highly symmetrical structures bcc, hcp and fcc respectively. It is more difficult to assign the crystal structures of d transition elements to definite electron configurations. The polymorphism of the lanthanide and actinide metals reflects transitions between configurations of the outer electrons, and, hence, is determined also by a possible participation of f electrons in the chemical bonding. The bonding forces are evidenced also in the temperatures and heats of melting and of evaporation.

1. Structure and metal radii. Most of the trivalent lanthanide metals display a close packed structure at room temperature. The light lanthanides crystallize in a dhcp, the heavier lanthanides in a hcp form. The high temperature form is, in general,

bcc. In contrast to the lanthanides with their simple metal
lattices, the light actinides crystallize in numerous, in part
unusual modifications of highly directional bonding, but with
bcc as the normal high temperature form. Transplutonium metals,
however, have at room temperature the dhcp structures known from
the lanthanides, while their high temperature forms are fcc.
Table III lists structures of lanthanides and actinides.

Table III. Crystal structures of lanthanides and actinides

Lanthanides				Actinides						
La dhcp	fcc	bcc		Ac	fcc					
Ce dhcp	fcc	bcc		Th	fcc					bcc
Pr dhcp		bcc		Pa	bct					fcc
Nd dhcp		bcc		U	orth		t			bcc
Pm dhcp		bcc(?)		Np	orth		t			bcc
Sm α-Sm				Pu	m	bcm	fcrh	fcc	fct	bcc
Eu(II) bcc				Am	dhcp	β-Am←(fcc)→γ-Am				
Gd hcp		bcc		Cm	dhcp					fcc
Tb hcp		bcc		Bk	dhcp					fcc
Dy hcp		bcc		Cf	dhcp				fcc	fcc(16)
Ho hcp		bcc		Es					fcc	

Recent work (7) has confirmed a previous observation of a
fcc phase (13) as the high temperature form of Pa; the bcc form
predicted (14) by extrapolating the variation of the expansion
coefficients in the different lattice directions was never detected. Dilatometry and differential thermal analysis were used in
an attempt to clear up controversy in the literature on the
polymorphism of Am (15). There seem to be at least 3 different
phases, the dhcp (" α "-) phase stable up to about 650°C, a
" β "-phase" existing until 1050°C, followed by the high temperature
form between 1050°C and the melting point. It is unknown which of
the latter phases corresponds to the fcc form observed. At variance
with previous work, but according to recent results (16), the
room temperature form of Cf is dhcp (not hcp), and there are two
different fcc forms above 600 and 725°C, respectively, with different lattice parameters.
    Caution is indicated with regard to interpretations (or
speculations) on the basis of fcc high temperature modifications
of the rare transplutonium metals; fcc phases of similar lattice

parameters form also with constituents of the air (nitrides,
"monoxides" (2)), and their unambiguous identification as a metal
phase requires the observation of a reversible phase transformation in a different, but well-known metal modification and/or
elemental analysis of the sample.

The polymorphism of the lighter actinides reflects the
existence of numerous bonding (including 5f) electron states of
almost identical energies. The observation of dhcp structures for
the transplutonium metals indicates only a slight participation
of the predominantly localized 5f electrons in the bonding.

Johansson (17) expects the transplutonium metals to transform
to a bcc phase immediately before melting; in this bcc phase, the
5f wave function overlap is reduced, hence, also the 5f electron
contribution to the bond.

The metal radii, calculated for the same coordination number,
demonstrate the special position of the actinides: the radii of
the lighter actinides (like those of the d transition metals) pass
through a minimum. However, when the middle of the series is
approached, the actinide radii decrease again with increasing
atomic number, as is the case for the lanthanide series because
of the 4f electron contraction (Figure 1).

The metal radii (table IV) have been correlated (18, 19)
to formal valences, which, for the lighter actinides, are markedly
higher than 3, and show some similarity with those of d transition
metals. The dhcp transplutonides are trivalent (like most of the
lanthanides); the larger fcc radii indicate a tendency for divalency which seems to be attained with the high temperature forms
of Cf and Es (r ~ 2.0 Å).

Table IV. Metal radii of lanthanides and actinides (in Å)

Lanthanides		Actinides			
La	1.88	Ac	1.88		
Ce	1.71	Th	1.79		
Pr	1.83	Pa	1.63		
Nd	1.82	U	1.56		
Pm	1.81	Np	1.54		
Sm	1.80	Pu	1.58		
Eu	2.04	Am	1.72	1.73	
Gd	1.80	Cm	1.74	1.78	
Tb	1.78	Bk	1.70	1.77	
Dy	1.77	Cf	1.69	1.75	2.03
Ho	1.76	Es			~2.0

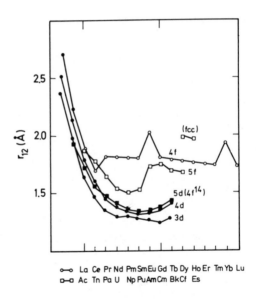

Figure 1. Metal radii (coordination number 12)

The metal radii of the trivalent actinides are equal (La/Ac) or smaller (Gd/Cm) than those of the corresponding lanthanides. The similarity of the metal radii of La and Ac had been expected by H. Hill [20] because of the possibility of heavy element electrons attaining relativistic velocities. As a consequence of the experimental confirmation [2] of H. Hill's expectation, Zachariasen [21] reduced his predicted radii of trivalent actinide metals by 0.08 Å. Recent calculations of the atomic volumes of the actinide metals [22] support the experimental values of the radii of the lighter actinides and account for the localisation of the 5f electrons taking place in Am.

2. Thermodynamics of phase transitions. The conditions of phase transitions like melting or sublimation reflect the crystal stability (Table V).

Table V. Melting temperatures and sublimation enthalpies of lanthanides and actinides

Lanthanides			Actinides		
	T_f(°C)	ΔH_s [24] (kJ/mole)		T_f(°C)	ΔH_s [25] (kJ/mole)
La	920	430	Ac	1050[x]	420
Ce(>III)	795	420	Th	1750	600
Pr	935	355	Pa	1230	595
Nd	1024	325	U	1130	530
Pm	1027	330	Np	640	465
Sm	1072	205	Pu	640	350
Eu(II)	826	175	Am	1170	285
Gd	1312	400	Cm	1350	385
Tb	1356	390	Bk	990	295
Dy	1407	290	Cf	900[xx]	195
Ho	1461	300	Es	860[xx]	140

x estimated from temperature dependence of metal preparation yield [26]

xx electron microscope observation of temperature at which vapour deposited metal particles coalesce [27, 28]

The melting temperatures of the trivalent lanthanides increase steadily with increasing atomic number, hence with decreasing metal radius. Melting temperatures are below the curve, when metal valences are higher (Ce) or lower (Eu, Yb) than three. The melting temperatures of the actinides vary in an irregular way. The surprisingly low melting temperatures of some of the light actinides are ascribed to the fact (23) that in the liquid phase the atoms are free to arrange themselves in such a way as to take maximum advantage of the 5f electron overlap – in contrast to the supposed configuration of the bcc high temperature solid phase.

Vapour pressures and sublimation enthalpies are correlated to the strength of the bonding, hence to the number and nature of bonding electrons. During evaporation, all the bonds of the crystal are broken; the heat of sublimation is identical to the bond strength, when the electron configurations in the solid and in the vapour phase are identical. For main group or d transition metals, sublimation enthalpies increase with the metal "valence" in a regular way; the sublimation heats are about 85, 170, 380 590 kJ/mole for mono-, di-, tri- and tetravalent metals, respectively. For 4f metals, as well, the vapour pressures decrease with increasing valence; divalent Eu and Yb ($\Delta H_s \approx$ 170 kJ/mole) are more volatile than the typical trivalent lanthanides. In general, the vapour pressures of actinides show this expected behaviour.

Selected thermodynamic properties (enthalpies of sublimation, of dissolution, and –if required– the energy difference between the lowest level of the ground configuration of the gaseous atom and the lowest level of the excited configuration $f^n s^2 d$ of the gaseous atom) of all actinides have been used by Nugent et al.(25) to establish a correlation function P(M). This function P(M) was re-examined utilizing new experimental data (26). Correlation between physical properties and electronic structure of the actinides have been discussed in a comprehensive approach by Fournier (29).

Early predictions of sublimation enthalpies made by analogy to trends in the lanthanides (25) have not always been confirmed. The evaporation enthalpies of Am (30) and of Cf (31) were higher than anticipated. The experimental value for Am(ΔH_s= 285 kJ/mole) could be interpreted taking into account a magnetic contribution during the transition from the condensed to the vapour phase; despite its high vapour pressure (ΔH_s= 200 kJ/mole) Cf must be considered a trivalent metal.

Mortimer (32) has reviewed data on specific heats of actinide metals.

The specific heat of Am was determined with vapour deposited samples of [241]Am and [243]Am using an adiabatic technique (33). An anomaly in Cp of Am is centred around 50-60 K, similar to the

anomaly of the specific heat of Pu. The entropy of Am at room
temperature is close to the value obtained from vapour pressure
measurements (30). Recently (34), the electronic specific heat
coefficient of Pa has been determined as $\gamma = 5mJ/mole\ K^2$; that
of Am seems to be smaller than 4 mJ/mole K^2 (35). The values and
variations of the electronic specific heat coefficients of the
lighter actinides (Table VI) are ascribed to the participation
of the 5f electrons in the bonding of these elements.

Table VI. Specific heats of actinides

	$C_{p(298)}$ (J/mole K)	γ (mJ/mole K^2)
Th	27.3	4
Pa	$33^{\pm}1$	5 (34)
U	27.6	10
Np	29.6	14
Pu	31.2	16
Am	28	$\leqslant 4$ (35)

Electronic Structure.

1. Electrical Resistivity. Electrical resistivity is composed of contributions from imperfections and impurities, both
temperature dependent and temperature independent (residual resistivity), lattice scattering, magnetic interactions and electron-
electron interactions (36).

Resistivity measurements in actinide systems are very sensitive to crystal imperfections from radioactive decay (self-
irradiation damage), self-heat and impurities. Recent measurements
on Pa (37), Am (33) and Cm (38), were, therefore, made with
freshly prepared, vapour deposited samples.

The electrical resistivities of most of the lighter actinide
metals - due to their f electron participation in bonding -
differ remarkably from those of "normal" metals. The resistivities
start to increase along the actinide series after Pa, and reach
a maximum at Pu, before localization of 5f electrons sets in.
(Figure 2).

A new measurement of the resistivity of Pa (37) confirms
the general features reported in a previous, preliminary publication (39), but puts the resistivity of Pa just below that of
Th. The temperature dependence of Am resistivity (31) has been

explained by assuming a s-d scattering mechanism; the occupied
5f levels in Am are probably too low to influence the electrical
properties. The room temperature electrical resistivity is around
70 $\mu \Omega$ cm for ^{241}Am and ^{243}Am, both bulk and film samples. The
low temperature power dependence of the resistivity suggests a
reduced f electron participation in the conduction process, as
expected from the electronic specific heat coefficient. The electrical resistivity of Am reflects the increasing localization of
5f electrons when the centre of the actinide series is approached.
This tendency is confirmed in the resistivity of ^{244}Cm metal (38).

Superconductivity of actinides seems to be well understood
(40) on the basis of recent progress in the calculation of their
band structures and their vibrational spectra. Thorium and
uranium are known to be superconductors. By using ac susceptibility
techniques, superconductivity could be detected in both Am (41)
and Pa (42). Vapour deposited Am was used as sample material;
Pa had been prepared by the (modified) van Arkel - de Boer
procedure (3). The transition temperatures are 0.79 K for Am, and
0.42 K for Pa.

2. Magnetism. Brodsky (36) has reviewed the magnetic properties of actinides. The magnetic susceptibilities of Th, Pa, U, Np,
Pu and Am, are almost temperature independent (Figure 3), whereas
Cm, Bk, Cf show CURIE-WEISS temperature dependencies (43, 44, 45),
with effective moments of the order of 8, 8.5 and 10 μB respectively. The reciprocal susceptibility of ^{244}Cm metal shows a
minimum around 52 K which was confirmed by neutron diffraction
studies (10) to be due to antiferromagnetic ordering. The magnetism of actinide metals confirms the band-like character of 5f
electrons in the first half of the series (with participation in
the bond), and their lanthanide like localization in the heavier
metals.

3. Spectroscopy. The application of optical and photoelectron spectroscopy to elucidate electron energy states of pure
actinide metals is still in the initial stages (46). Reflectivity
measurements on Th samples (mechanically polished, electropolished,
or as grown from the vapour phase) demonstrate the importance of
sample and surface preparation (47), and explain reasons for
discrepancies in published results (48, 49). Preliminary measurements of the optical reflectivity of Am films evaporated on
different window materials (50) seem to indicate that the 5f levels
are lying more than about 6 eV below the FERMI level, thus
supporting the interpretation of the electrical resistivity results.

4. Theory. The difference between the lighter and heavier
actinides is supported by electronic structure calculations (51)
permitting a comparison of the electron masses μ (Table VII) which
are inversely proportional to the bandwidth and a measure of the
electron localization: μ_f of the early actinides is similar to
μ_d of the 3d transition metals Fe, Co, Ni; μ_f in the second half
of the heavy actinides is similar to μ_f of the light 4f lanthanides.

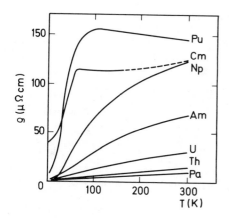

Figure 2.   *Electrical resistivity of acti-*
nides

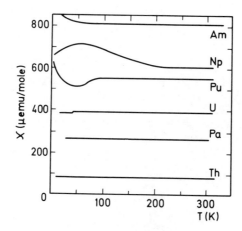

Figure 3.   *Magnetic susceptibility of actinides*

Table VII. Relative masses of d/f electrons (51)

$3s^2 3p^6$ μ_d	Sc 5.60	Ti 5.96	V 6.28	Cr 7.14	Mn 8.27	Fe 9.87	Co 11.3	Ni 13.2
$4s^2 3d^{10} 4p^6$ μ_d	Y 3.62	Zr 3.60	Nb 3.68	Mo 3.96	Tc 4.44	Ru 5.13	Rh 6.21	Pd 7.57
$5s^2 4d^{10} 5p^6$ μ_d	La 3.29	Hf 3.24	Ta 3.22	W 3.40	Re 3.71	Os 4.10	Ir 4.82	Pt 5.74
$5s^2 4d^{10} 5p^6$ μ_f	Ce 43.0	Pr 56.4					Gd 152.6	Tb 205.6
$6s^2 5d^{10} 6p^6$ μ_f	Th 12.0	Pa 15.98	U 21.00	Np 19.02	Pu 23.78	Am 29.80	Cm 36.76	Bk 45.52

Conclusion

The physicochemical properties of actinide metals confirm
the presence of band-like 5f electrons up to Pu. The participation
of these 5f electrons in the metallic bond is assumed to begin
with Pa. In the first half of the actinide series, 5f electrons
are similar to d electrons in typical transition metals: the 5f
electron orbitals are more extended than 4f orbitals; for the
light actinides, 5f electrons are "delocalized" and hybridized
in a rather large band with 6d and/or 7s electrons. Starting with
Am, the 5f electrons are localized again, like 4f electrons in
the lanthanides.

Acknowledgements

Thanks are due to Drs. R. Bett, G.R. Choppin, J.M. Fournier
and L. Manes for helpful discussions during the preparation of
this review.

Literature Cited

1. Peterson, J.R. J. Phys. (Paris), Colloque C4, 1979, 40, 152.
2. Baybarz, R.D.; Bohet, J.; Buijs, K.; Colson L.; Müller W.;
   Reul, J.; Spirlet, J.C.; Toussaint J.C. in Müller W., and
   Lindner R., Eds., "Transplutonium 1975"; North Holland :
   Amsterdam, 1976; p.61.
3. Spirlet, J.C. J. Phys. (Paris), Colloque C4, 1979, 40, 87.
4. Keller C.; Erndmann, B. J. Inorg. Nucl. Chem., Suppl. 1976,
   274.
5. Müller W.; Reul, J.; Spirlet, J.C. Atomwirtschaft, 1972, 17,
   415.
6. Brown, D.; Tso, T.C.; Whittacker, B. J. Chem. Soc. Dalton,
   1977, 22, 2291.
7. Bohet, J.; Müller, W. J. Less Common Metals, 1978, 57, 185.
8. Ramthun, H.; Müller W. Intern. J. of Appl. Rad., 1975, 26, 589.
9. Müller, W.; Reul, J.; Spirlet, J.C. Rev. Chim. Min. 1977, 14,
   212.
10. Fournier, J.M.; Blaise, A.; Müller, W.; Spirlet, J.C.
    Physica B, 1977, c86, 30.
11. Sari, C; Müller, W.; Benedict, U. J. Nucl. Mat. 1972/1973, 45
    73.
12. Brewer, L. Science, 1968, 161, 115.
13. Asprey, L.B.; Fowler, R.D.; Lindsay, J.D.G.; White, R.W.;
    Cunningham, B.B. Inorg. Nucl. Chem. Lett., 1971, 7, 977.
14. Marples, J.A.C. Acta Cryst.,1965, 18, 815.
15. Rose, R.L.; Kelley, R.E.; Lesuer, D.R. J. Nucl. Mat., 1979,
    79, 414.
16. Noé, M.; Peterson, J.R. in Müller, W.; Lindner, R., Eds.
    "Transplutonium 1975";North Holland: Amsterdam, 1976; p.61.
17. Johansson, B. private communication, 1978.
18. Zachariasen, W.H. in Coffinberry, A.S.; Miner, W.N., Eds.
    "The Metal Plutonium", Chicago Univ. Press, 1961; p.99.
19. Cunningham, B.B.; Wallmann, J.C. J. Inorg. Nucl. Chem., 1964,
    26, 271.
20. Hill, H. Chem. Phys. Lett., 1972, 114.
21. Zachariasen, W.H., comment to (2).
22. Skriver, H.; Andersen, O.K.; Johansson, B. J. Phys. (Paris),
    Colloque C4, 1979, 40, 130.
23. Kmetko, E.A.; Hill, H. J. Phys., 1976, F6, 1025.
24. Nugent, L.J.; Burnett, J.L.; Morss, L.R. J. Chem. Thermod.,
    1973, 5, 663.
25. David, F.; Samhoun, K.; Guillaumont, R.; Edelstein, N.
    J. Inorg. Nucl. Chem., 1978, 49, 69.
26. Stites, J.G.; Salutsky, M.L.; Stone, B.D. J. Am. Chem. Soc.,
    1955, 77, 237.
27. Haire, R.G.; Baybarz, R.D. J. Inorg. Nucl. Chem., 1974, 36,
    1295.
28. Haire, R.G.; Baybarz, R.D. J. Phys. (Paris), Colloque C4,
    1975, 40, 101.

29. Fournier, J.M., Thesis, University of Grenoble (1975)
30. Ward, J.W.; Müller, W.; Kramer, G.F. in Müller, W.; Lindner, R.
    Eds. "Transplutonium 1975"; North Holland: Amsterdam, 1976,
    p.161.
31. Ward, J.W.; Kleinschmidt, D.D.; Haire, R.G. J. Phys. (Paris),
    Colloque C4, 1979, 40, 233.
32. Mortimer, M.J. J. Phys. (Paris),Colloque C4, 1979, 40, 124.
33. Müller, W.; Schenkel, R.; Schmidt, H.E.; Spirlet, J.C.;
    McElroy, D.L; Hall, R.D.A.; Mortimer, M.J. J. Low Temp. Phys.,
    1978, 30, 561.
34. Stewart, G.R.; Smith, J.L.; Spirlet, J.C.; Müller, W.,
    3rd Conf. on Supercond. in d- and f- Band Metals, La Jolla,1979.
35. Smith, J.L.; Stewart, G.R.; Huang, C.Y.; Haire, R.G. J. Phys.
    (Paris),C4, 1979, 40, 138.
36. Brodsky, M.B. Rep. Prog. Phys., 1978, 41, 1548.
37. Bett, R., to be published.
38. Schenkel, R. Solid State Comm., 1977, 23, 389.
39. Hall, R.O.A.; Lee, J.A.; Mortimer, M.J. J. Low Temp. Phys.,
    1977, 27, 305.
40. Smith, J.L., 3rd Conf. on Supercond. in d- and f- Band Metals,
    La Jolla, 1979.
41. Smith, J.L.; Haire, R.G. Science, 1978, 200, 535.
42. Smith, J.L.; Spirlet, J.C.; Müller, W. Science, 1979,
    205, 188.
43. Kanellakopulos, B.; Blaise, A.; Fournier, J.M.; Müller, W.
    Solid State Comm., 1975, 17, 713.
44. Peterson, D.R.; Fahey, J.A.; Baybarz, R.D. in Miner, W.N. Ed.
    "Plutonium 1970 and Other Actinides", Metallurgical Society
    AIME: New York, 1970, p. 20.
45. Fujita, D.K.; Parsons, T.C.; Edelstein, N.; Noé, M.;
    Peterson, J.R. in Müller, W.; Lindner, R., Eds. "Transplutonium
    1975", North Holland: Amsterdam, 1976, p. 173.
46. Veal, B.W. J. Phys. (Paris),Colloque C4, 1979, 40, 163.
47. Alvani, C; Naegele, J. J. Phys. (Paris),Colloque C4, 1979, 40,
    131.
48. Weaver, J.H.; Olson, C.G. Phys. Rev. 15B, 1977, 4602.
49. Veal, B.W.; Koelling, D.D.; Freeman, A.J. Phys. Rev. Lett.,
    1973, 30, 1061.
50. Naegele, J.; Spirlet, J.C.; Winkelmann, H. in Mulak, J.;
    Suski, W.; Troc, R. Eds. "Proceedings of the 2nd Int. Conf.
    on the Elect. Structure of the Actinides";Polish Acad. of
    Sciences: Wroclaw, 1977, p.275.
51. Brooks, M.S.S. J. Phys. (Paris), Colloque C4, 1979, 40, 155.

RECEIVED December 26, 1979.

Vapor Pressure and Thermodynamics of Actinide Metals

JOHN W. WARD and PHILLIP D. KLEINSCHMIDT—Los Alamos
Scientific Laboratory, Los Alamos, NM 87545

RICHARD G. HAIRE—Oak Ridge National Laboratory, Oak Ridge, TN 37830

DAVID BROWN—AERE Harwell, Oxfordshire, England OX11 ORA

The chemistry of the actinides is easily the most complex of
any grouping in the periodic table. Whereas reasonably self-consistent models (1,2,3) have been developed to describe the general
properties of other elemental groups (e.g., transition metals, rare
earths), such attempts have had only limited success for the actinide metals. The series begins apparently as another transition
metal group, Ac being quite similar to La (Y, etc.), Th following
as the homologue of Hf, but here the effects of the broad f-bands
lurking just above the Fermi level are seen already, as will be
demonstrated later. With Pa we see the effect of the broad ($\sim$3 eV)
f-band beginning to hybridize with the valence electrons, somewhat
analogously to the beginning of the 3d transition metal series, but
becoming enormously more complex. In the transition metals the d-
electrons also undergo progressive band-narrowing while remaining
near the Fermi level as the shells are filled, but here the symmetries are regular, the orbitals not polarized. In the 5f series
the multiple f-electron orbital symmetries are, in the broad-band
situation, superimposed upon the metal valence orbitals, with the
added complication of the polarization (like p-electrons) which
produces added repulsive and attractive binding forces.

The low-temperature phases of U, Np, and Pu exhibit tortuous
crystal structures as the asymmetrical, polarized f-orbitals begin
to fill, and each system seeks the lowest energy possible for a
metallic bonding configuration. Relief is provided by entropy effects at higher temperatures, resulting in simpler (less f-bonded)
structures, but the effect of additional f-electrons is still
greater than that of progressive band narrowing, at least until Pu.
The multi-phase behavior of Pu is famous; however, even here, band-
narrowing results in the high-temperature fcc δ-phase becoming
almost a normal trivalent rare-earth-like metal (4). Nevertheless,
the bonding forces eventually predominate, and the metal surrenders
by melting at a low temperature (as is also the case for Np, and to
a lesser extent, U). The resultant liquid has a high cohesive
energy, abnormal viscosity, and exhibits a very high boiling point,
as the f-electrons can now freely adjust to bond in optimum fashion.

0–8412–0568–X/80/47–131–199$05.50/0

Band-narrowing at Am (to ~0.6 eV (5)) results in an abrupt shift to rare-earth-like behavior, though this is still hidden by the "magic" f^6 non-magnetic ground-state configuration. The dhcp low-temperature phase and higher-temperature phase transition of Am, Cm, Bk, and Cf are reminiscent of the early lanthanides, where the progressive localization of (albeit far fewer) f-electrons starting near the Fermi level is seen also for La, Ce, Pr, Nd.... A very useful correlation has been noted by Johansson (6), in which the localizations of the two series are compared as chemical and structural homologues, rather than in the usual periodic fashion:

Ce	Pr	Nd	Pm	Sm	Eu
Pu	Am	Cm	Bk	Cf	Es....

Some of the close similarities in the physico-chemical properties of these pairs will be noted below. This pairing is not intended to be all-inclusive; the half-filled shell effect at Cm introduces a major perturbation, as does the facet that the f-electron energy levels retreat much more rapidly from the Fermi level in the rare earths. Johansson has also tried to assign some values to the part of the cohesive energy due to f-electron bonding in the early actinides (7), as an extension of the model developed by Nugent (3) and extended by David et al (8). The Engel-Brewer treatment, which relates spectroscopically determined energy levels to spd bonding configurations and various types of structures, has no way of coping with f-bonding in the early actinides. Nugent (3) and David (8) have attempted to improve and extend these techniques by including additional information, particularly heats of solution and entropy estimates, but with only limited success, the prediced values for Th through Am scattering in random fashion (the schemes work quite well for compounds, e.g., the oxides). Johansson's (7) estimates shed light on the participation of f-electrons in metallic bonding, but do not provide any improvement in the modelling schemes for cohesive energies. We can only say that the early actinides are unique, non-magnetic because of f-electron bonding, and admit to no metallic valence that can be defined in normal terms.

The situation is noticeably improved beyond Am, but the progressive stabilization of the divalent state soon introduces yet another complexity, beginning possibly already with Cf. In any case, an accurate number for the cohesive energy remains the single most important parameter that can be provided for both the experimentalist and theoretician. From this value the complete thermodynamics can be eventually derived, giving a firm data base from which to draw meaningful conclusions about the physics and chemistry observed.

Nature of the Thermodynamic Analysis

From an accurate vapor pressure determination we derive a heat of vaporization, which is then essentially the cohesive energy if the phase transition is not accompanied by a change in electronic structure or magnetism. In addition to this heat of vaporization (the experimental value is always for the mid-point of the temperature range of study), an entropy of vaporization is produced, and since good spectroscopic data are fortunately now available for the actinide gases through Es, a condensed-phase entropy (at temperature) can be obtained from the data. There remains then the extrapolation and calculations of the data to 298 K, and how this can be reliably done is described below.

The Entropy Correlation--Metallic Radius and Magnetism. Ward and Hill (9) have established a correlation relating the crystal entropy S_{298}^o to metallic radius, atomic weight, magnetic properties and electronic structure. This correlation permits calculating entropy values for unmeasured metals based on comparison with a closely similar metal that has been measured; the technique has been shown to be accurate to within a few-tenths of an entropy unit for most metals in the periodic table. A comparison of metallic radius with crystal entropy is shown in Fig. 1. The general effect of valence and the smooth nature of both sets of curves is apparent; both the radius effect and magnetic effects are far larger than the correction for atomic weight. The general formula is

$$S_{298}^o = [S_{298}^k] [r^u/r^k] + [3/2]R\ln[A^u/A^k] + S\mu \qquad (1)$$

where u refers to the unknown metal, k to the known (they are not exponents), A is the atomic weight, and Sμ is the magnetic entropy term. An interesting result is found if the magnetic term is deliberately subtracted; the "non-magnetic" rare-earth entropies decrease smoothly as a function of radius.

For the actinides the crystal entropies follow approximately the decreasing average radius produced by f-electron participation in metallic bonding. They are also clearly shown to be non-magnetic, as the f's are itinerant. However, the entropy correlation itself cannot predict these values, since there is no model in terms of a like metal that can be used to compare these totally unique early actinides. There are also of course perturbations due to the high electronic specific heats, caused by high densities of states at the Fermi level.

At Am there is an abrupt change as discussed above. The trivalency of Am is clearly shown in its chemistry, heat of vaporization (10) and superconductivity (11).

Part of the complexity of Am thermodynamics is due to the very large magnetic entropy change (4.1 cal/mol-deg.) upon vaporization to the free divalent atom, which is also seen in the early actinides, the free atoms being fully magnetic with localized f-electrons. This is shown in Fig. 2, where the gaseous entropies S_{298}^o of both the lanthanides and actinides reflect the changing magnetic entropies (lower plot). Note that here again, subtraction of these magnetic entropies results in a relatively smooth non-magnetic base line for both gaseous series; perturbations can in each case be shown to result from low-lying electronic levels. Even at 1400 K (upper plot) the behavior is not remarkably changed, through many

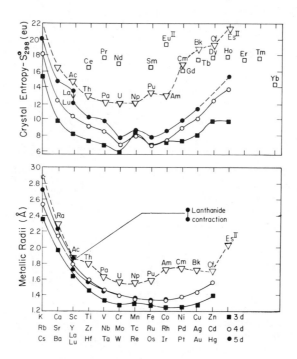

Figure 1. *Plots of metallic radii* (lower) *and crystal entropies* S°$_{298}$ (upper) *for the metallic elements*

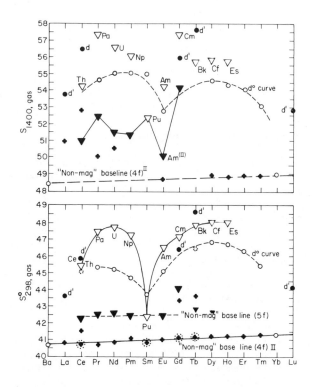

Figure 2.   Comparison of gaseous entropies for the lanthanides and actinides at 298 K and 1400 K: (▽), Magnetic actinides; (▼), nonmagnetic actinides; (○), magnetic lanthanides; (+), nonmagnetic lanthanides; (◌), low-lying levels removed.

new energy states have been introduced for the earlier metals.

The importance of these observations lies in the fact that the S_{298}° value is the basis point for the free-energy functions for both the solid and the gas. With a reasonably accurate estimate of the solid crystal entropy, the gaseous spectroscopic data and precise vapor pressure measurements, it is possible to calculate all the thermodynamic values for the metal, up to the highest temperatures of measurement. Also, a self-consistent heat capacity curve starting at 298 K is produced; for the intensely-radioactive and scarce trans-curium metals, normal calorimetry may never be possible, and these techniques become extremely important tools. A detailed example of a typical calculation will be given under Californium below.

Experimental Techniques

Of the several vapor pressure methods used, the Knudsen effusion method is the simplest and most accurate for high-temperature work with limited quantities of material. Although the principles of the method were based for many years on the tenets of the Kinetic Theory of gases, Ward and coworkers (12-17) established that Optical Theory actually describes the effusion process in the Knudsen (no gaseous collision) region. This knowledge was turned to advantage in the studies with Am (10) and Cm (18) and for several of the studies to be reported here.

The method takes advantage of the optical line-of-sight from the sample to the target, and the geometry is chosen so the solid angle subtended by the collimated molecular beam impinging upon the target sees only the sample surface within the cell. Thus, particularly for a metal where the evaporation/condensation coefficient is considered to be unity, precise data can be obtained from a thin disk of sample weighing only a few milligrams, up to almost the moment of depletion.

At Los Alamos a new Knudsen effusion apparatus has been constructed to take specific advantage of the principles described. The equipment incorporates a 1200 ℓ/sec Ultek ion pump, stainless-steel chamber with cryopump and ancillary turbomolecular pumpdown capability. This system is mounted above an argon-inerted glovebox; sample loading is accomplished from below. Details are shown in Fig. 3. In the center is the effusion cell, machined by the spark-erosion process and usually made of W. Specific details are given in Ref. (18). The sample is contained in an inner cup, variously made of single-crystal Ta, W, or W-coated Y_2O_3. Heating is provided by radiation from a Ta strip furnace. The effusion cell and furnace are surrounded by Ta radiation shields and a water-cooled copper shroud and collimator.

The effusing beam is condensed on copper targets in a water-cooled target cassette mechanism. The cassette target holder will contain up to 50 targets between loadings; exposed targets can be removed for analysis without disturbing the vacuum in the main chamber. Radioactive target deposit analysis produces data dependent only on the counting statistics of the deposit, and thus only a function of exposure time, and nearly independent of pressure. Combination of these data with mass spectrometric identification of the gaseous species permits an exact evaluation of the vaporization process. In the mass spectrometer mode, the effusing beam passes through the cross-beam ionizer of the Extranuclear quadrupole mass spectrometer, allowing interchangeable target or mass spectrometric measurements to be made. This instrument has good resolution and sensitivity to better than 10^{-9} atmospheres pressure in the cell.

Temperatures are measured with a Pyro optical pyrometer, or, for temperatures below 750°C, a calibrated thermocouple. The entire cell assembly may be moved freely in the x, y, and z directions during operations, for exact alignment.

Results and Discussion

The intent of the following section is to consider the relationship of the thermodynamic heats and entropies to the metallic state(s) of each actinide, particularly in terms of electronic structure and bonding. Detailed experimental content will not be included, except to indicate the latest references and to describe new results, insofar as they are available. The reader is especially referred to the compilation by Oetting, Rand, and Ackermann (19) for a thorough compilation of thermodynamic information on the actinide metals, up to 1976.

Before considering each metal, let us reexamine the entropy-radius relationship, this time entirely in terms of non-magnetic entropies. This is shown in Fig. 4, and clearly delineates the progression of divalent, trivalent, tetravalent, etc. metals (dashed lines); the monovalent row is far off to the right of the plot. The 6d row is shown in detail (solid curve) through Au, and the latest data on the actinides are plotted as triangles. These relationships will be considered in the following discussions.

Thorium. Though Th seems to be in many respects a "normal" tetravalent transition metal, the broad f-bands just above the Fermi level already cause perturbations, as shown by Koelling and Freeman (20) and by Glötzel (21), who also considered the similar situation for La and the descent of the f-band below the Fermi level for Ce (which occurs also, but without narrowing, for Pa). The heat of vaporization for Th has been definitively measured by Ackermann and Rauh (22) as $\Delta H_{298}^{o} = 142.7 \pm 0.3$ kcal/mol. This is, however, a change in the sequence of rising heats of vaporization for Ti (112.3), Zr (143.4), and Hf (148.4). The melting temperature is also anomalously low.

Even more striking is the anomalous position of Th in the entropy-radius relationship of Fig. 4. In following the IVA elements, the shift to smaller radius at Hf corresponds to the gross effect of the lanthanide contraction in the previous row. Note that Th is far over into the trivalent metal area, corresponding to a very large radius (i.e., lower valence for a supposedly tetravalent metal).

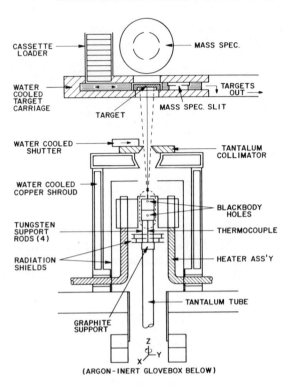

Figure 3.   Cross-sectional view of UHV target–mass spectrometer Knudsen effusion apparatus

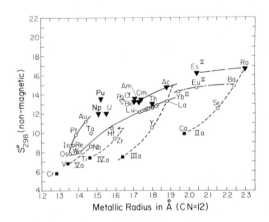

Figure 4.   Plot of the nonmagnetic crystal entropy against metallic radius

The vaporization of Th shares the problems of several of the early actinides, where the great stability of the monoxide gas drives the reaction

$$M_{(s)} + MO_{2(s)} \rightarrow 2\ MO_{(g)} \tag{2}$$

when there are only a few ppm of oxygen present. For Th the ratio ThO/Th can approach 200 when sufficient oxygen is present (23); fortunately for this system the dioxide vapor pressure is still very small.

Protactinium. In terms of equilibria with various oxygen-bearing species, Pa represents the worst possible situation. The PaO/Pa is predicted to be of the order of 50 (23) with any oxygen present, and the PaO_2 pressure is very appreciable. In this respect the system is similar to the uranium metal plus oxygen system, which took more than 20 years of careful work to solve properly.

Preliminary results from work presently under way with a 50 mg sample of Pa-231, shows three gaseous species, PaO_2, PaO, and Pa, in the apparent ratio 10/1/0.02, the Pa signal being very hard to measure. Data have been taken in the temperature range 1800-2400 K at an electron energy of 17 eV; appearance potential curves taken for the three species clearly delineate various fragmentation processes above this energy. The scarcity and expense of Pa metal make waiting for the disappearance of the oxygen-bearing species a serious problem, in contrast to the case for U, where sample availability is no problem. A number of possible equations can be proposed:

$$Pa_{(\ell)} + PaO_{2(s)} \longrightarrow 2PaO_{(g)} \tag{3}$$

$$PaO_{2(s)} \dashrightarrow PaO_{2(g)} \tag{4}$$

$$PaO_{2(s)} \dashrightarrow PaO_{(g)} + O_{(g)} \tag{5}$$

$$Pa_{(\ell)} \dashrightarrow Pa_{(g)} \tag{6}$$

$$Pa_{(\ell)} \xrightarrow{\ [0]\ } PaO_{(g)} \tag{7}$$

In the present work, Eq. 4 is no doubt the major contributor to the vapor in view of the smoothness and reproducibility of the first data points; this implies a large amount of PaO_2 present, perhaps floating on the surface of the molten metal. Possible contenders for the PaO signal are Eqns. 3 and 7. Again, in view of the reproducibility of the data we have, it is hard to imagine Eqn. 7 as a serious contender. We believe that Eqn. 3 is occurring at the metal-oxide interface, the PaO vapor leaving the surface under the influence of a diffusion mechanism through the dioxide. Eqn. 7 would imply large quanties of oxygen being supplied near the cell area.

Analysis of our preliminary data in the manner of Ackermann, Rauh, and Chandrasekharaiah (24) gives a net total pressure of all species of 7.0 x 10^{-5} atm at 2412 K, about an order of magnitude lower than for the UO$_2$ system.

The thermodynamic analysis for Pa in Ref. 19 represents an educated guess with no measured values, except for spectroscopic data for the gas. The choice of 12.4 for the solid entropy is an average between Th and U; however, a recent though still somewhat preliminary measurement in the UK (25) on the same metal we are using indicates the value may be of the order of 13.2. This could well be the result of electronic perturbations from the first (or at least partial) 5f electron below the Fermi level. This entropy value is also shown in Fig. 4, and has moved toward the unique position of the other f-bonded actinides.

Uranium. The vaporization behavior of uranium and the uranium-oxygen system has been carefully analyzed, and well documented in Ref. 19. The earlier work before better instrumentation and good mass spectrometry was applied to the problem reflects the trouble with the UO$_{(g)}$ contribution, which is now properly understood; the situation with U is similar to but somewhat less severe than that alluded to above with Pa. The entropy of 12.00 for the solid shows the small-radius, highly f-bonded, nonmagnetic state of the room-temperature metal; this value is plotted in Fig. 4.

Neptunium. Here again the earlier work was compromised by the presence of NpO$_{(g)}$, which at the onset produces an NpO/Np ratio in the gas of about 7. The more recent work of Ackermann and Rauh (26, 27) gives excellent precision and second/third law agreement. The measured entropy for the solid of 12.06 is also plotted in Fig. 4, and reflects the even smaller radius of the orthorhombic room-temperature phase.

Plutonium. The famous multi-phase behavior of this trouble-some metal is a result of many f-states combined with rapid band-narrowing at the higher temperatures. Since the thermodynamic contribution of the monoxide gas is now quite small (23), even the earliest studies were little different than later ones with much higher-purity metal. Troublesome second/third law problems were partly due to incomplete or inaccurate thermodynamic functions and questions about the real crystal entropy. The presently accepted value of 13.42 shows the effect of an abnormally large electronic specific heat, which is a measure of the many complexities in energy relationships near the Fermi level for this metal.

The recent study by Bradbury and Ohse (28) extends measurements to 2219 K and 9.5 x 10^{-4} atm, connecting smoothly to earlier, lower-temperature data. The second-law and third-law heats of vaporization of 82.25 and 81.66 kcal/mol, respectively, are in quite good agreement with the summary assessment in Ref. 19.

Americium. The physico-chemical properties of Am are clearly rare-earth-like and the gas is quite simply divalent. Therefore the continuing discrepancy over the heat of sublimation between theoretical correlations and the experimental data is still somewhat puzzling. Ward and Hill (9) correctly predicted the non-magnetic crystal entropy, later confirmed by experiment. Ward, Müller and Kramer (10) measured the vapor pressure with high precision on very pure 241-metal, and calculated the complete thermodynamics of the element, from 298 K to the boiling point. The heat of vaporization of 67.9 ± 0.5 kcal/mol and entropy data were combined to show the effects of the large magnetic entropy transition upon vaporization, upon the boiling point, and also the change of the heat of vaporization with temperature, through the F=H-TS term. Knowledge of these effects helps explain some (but not all) of the discrepancies between the measured and predicted values.

A recent study by Ward, Kleinschmidt, and Haire (29) has extended measurements with the new apparatus at Los Alamos into the liquid range, using the 243-isotope. The results of this work are given in Fig. 5, which shows a smooth extension of the Am-241 data (heavy line, limits indicated by small arrows) with the new results. An additional series of experiments were performed with various concentrations of Am dissolved in La. The system showed nearly ideal behavior, giving the correct heat of vaporization, Raoult's law calculation of vapor pressure, and activity coefficients which show the effects of a small heat of mixing.

The data from these experiments are summarized in Table I. On the basis of these comparative studies, the thermodynamics of Am appears to be well-established, the heat of vaporization reported in Ref. 10 being confirmed. It should be noted that the crystal entropy estimate for Am was based on La as a model. In terms of the rare-earth/actinide pairing of Johansson discussed earlier, the entropy for Pr with the magnetic contribution removed would have given a similar result.

The f-electrons of Am, though quite localized, are still very close to the Fermi level (like Ce) compared to most rare-earths, and the energy bandwidth is comparable to that for Ce. Therefore, we should still expect some complexities in bonding and valence-level interactions. The entropy position of Am in Fig. 4 is representative of a rather normal trivalent metal exhibiting an "actinide-contraction" beginning at Ac.

Curium. The vapor pressure of high-purity Cm-244 metal was measured by Ward, Ohse and Reul (18), in both the solid and liquid phases. This work is reported as definitive in Ref. 19; however, the Tables A1.14 are improperly calculated and the ΔH°_{298} value given is incorrect. Their value quoted in the discussion is 92.6 ± 1.0 kcal/mol, and the thermodynamic data for Cm should be recalculated on the basis of the correct heats and entropies.

The crystal entropy value for Cm used in Ref. 19 is 17.2 ± 0.2 and was calculated by Ward and Hill (9) using Gd as a model, citing

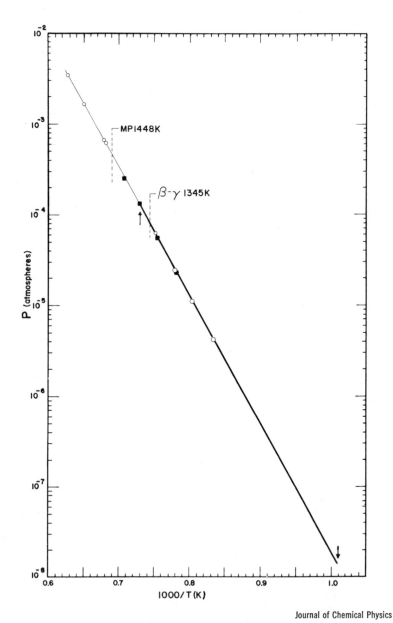

Figure 5.   Vapor pressure data for americium: (○, ■), americium-243 data; (—), americium-241 data, upper and lower limits indicated by arrows (29).

Table I. Thermodynamic Data for Americium

1. Am-243 liquid

Log P (atm) = $6.279 \pm 0.119 - (13935 \pm 180)/T$ [1469-1575 K]

$\Delta H^\circ_T = 63776 \pm 824$ cal/mol

$\Delta S^\circ_T = 28.73 \pm 0.54$ cal/mol-deg.

$\Delta H^\circ_{298} = 70423 \pm 741$ cal/mol (2nd law)

$\Delta H^\circ_{298} = 68073 \pm 50$ cal/mol (3rd law)

2. Am-243 solid

Log P (atm) = $6.296 \pm 0.0088 - (13983 \pm 114)/T$ [1200-1415 K]

$\Delta H^\circ_T = 63986 \pm 522$ cal/mol

$\Delta S^\circ_T = 28.81 \pm 0.40$ cal/mol-deg.

$\Delta H^\circ_{298} = 67241 \pm 501$ cal/mol (2nd law)

$\Delta H^\circ_{298} = 68205 \pm 30$ cal/mol (3rd law)

3. Am-241 solid (Data from Ref. 10)

Log P (atm) = $6.578 \pm 0.046 - (14315 \pm 55)/T$ [990-1358 K]

$\Delta H^\circ_T = 65503 \pm 250$ cal/mol

$\Delta S^\circ_T = 30.07 \pm 0.21$ cal/mol-deg.

$\Delta H^\circ_{298} = 67871 \pm 273$ cal/mol (2nd law)

$\Delta H^\circ_{298} = 68051 \pm 25$ cal/mol (3rd law)

many similar physico-chemical processes as the basis for comparison. Again, to emphasize the smoothly-changing nature of the entropy-radius relationship, a metal like Ce with a trivalent gas (plus the proper magnetic contribution) could have been used successfully if the dhcp ground-state crystal structure is considered important. However, the energies associated with phase changes are usually small, and the half-filled shell effect at Cm (and Gd) is major.

Berkelium. Theoretical correlations (3,7,8) have predicted a cohesive energy for Bk in the range of 65-75 Kcal/mol. The metal is supposedly a homologue of Tb, but would be placed under Nd or Pm in the pairing of Johansson noted earlier.

Bk-249 metal is a very soft beta-emitter with a half-life of only 320 days and a fairly high spontaneous fission cross-section, so that the target counting of effusion samples is quite difficult.

The soft betas are in themselves difficult to measure, and there
are in addition various other betas, gammas and alphas from daughter
decay and fission products. The data reported here are necessarily
preliminary; after a number of months sufficient Cf-249 (an alpha-
emitter) will grow in to permit a much more accurate assay of our
target results.

Approximately 1.5 mg of Bk-249 metal was loaded into a single-
crystal Ta cup within a Mo effusion cell. First time/temperature
estimates based on a heat of vaporization in the neighborhood of
70 kcal/mol gave no detectable signals. Temperatures were finally
raised to more than 1300 K, and the data reported here cover the
range 1326-1582 K. Even at these temperatures the signal to the
mass spectrometer was very small, while the deposit reaching the
targets in terms of counts/min was very large. These preliminary
data for Bk are from the direct beta-counts taken with a window-
less counter, and are plotted in Fig. 6. Because of the complexities noted above the precision is not very high; the targets
will be re-counted when sufficient Cf-249 has grown in.

Nevertheless, it is quite clear the Bk is much more like Tb
than was expected. The heat of vaporization from these data is
$\Delta H_{298}^{\circ} = 91.2 \pm 4.4$ kcal/mol (1326-1582 K). Tb gives by contrast
a value of 90.6 (for the same temperature range) and Nd 73.5. The
provisional vaporization equation is

$$\text{Log P (atm)} = 5.04 \pm 0.67 - (19926 \pm 969)/T . \qquad (8)$$

We estimate the crystal entropy of Bk to be 18.4 ± 0.3 cal/
mol-deg. based on either Tb or Nd, the proper atomic weight, radius
and magnetic corrections being made according to Eqn. 1. The non-
magnetic part of this value is included in Fig. 4, and falls in with
the other rare-earth-like actinides. Final thermodynamic tabulations must await proper counting of the target data.

Californium. Ward, Kleinschmidt and Haire (30) have measured
the vapor pressure of Cf-249 metal from 771-1026 K, using a 2 mg
sample. The data are described by the equation

$$\text{Log P (atm)} = 5.675 \pm 0.039 (9895 \pm 34)/T. \qquad (9)$$

Second-law and third-law heats were in close agreement, to give
$\Delta H_{298}^{\circ} = 46900 \pm 300$ cal/mol as the heat of vaporization. The
thermodynamic data clearly establish Cf as a trivalent metal, though
with a very high vapor pressure, actually midway between Sm and Eu,
as shown in Fig. 7.

Californium is the first element in the sequence of actinides
to show strong divalent tendencies. The progressive stabilization
of the divalent ground-state (7), which is apparently complete at
Fm and perhaps may occur already for Es (the estimated crystal entropy for Es is plotted in Figs. 1, 2, and 4 as being divalent).
The properties of Cf are closely similar to those of Sm, and we

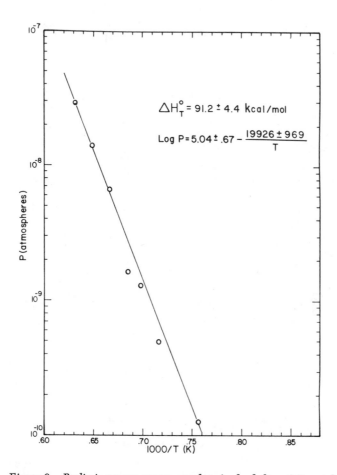

$$\triangle H_T^\circ = 91.2 \pm 4.4 \text{ kcal/mol}$$

$$\text{Log } P = 5.04 \pm .67 - \frac{19926 \pm 969}{T}$$

Figure 6.   Preliminary vapor pressure data for berkelium-249 metal

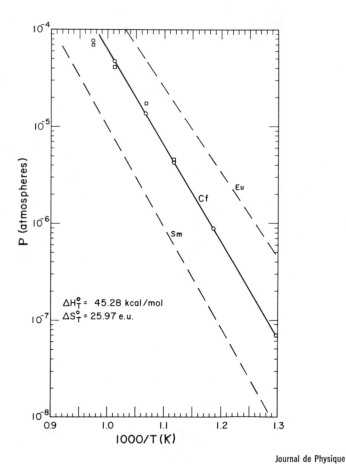

Figure 7.   Vapor pressure of solid californium: (○), target data; (□), MS data (30).

have used $\underline{Sm}$ as our model to establish the crystal entropy value
to be 19.25 cal/mol-deg.

Several experiments at Oak Ridge and Los Alamos have indicated
the possible presence of a divalent form of the metal, at least in
thin films (31,32). Divalent surface states have recently been
demonstrated for $\underline{Sm}$ (33). A tentatively-observed temperature of
transition of about 800°C implied that vapor pressure studies above
this temperature might reveal divalent behavior (because of the
scarcity of material and extremely high vapor pressure, direct studies above 800°C are precluded). We have recently performed
Raoult's-Law vaporization studies for a 1.5 a/o solution of $\underline{Cf}$ in
liquid $\underline{Eu}$ over the temperature range 958-1140 K. The first data
points fell directly upon the trivalent line of Fig. 7, slowly
rising above the line as the more volatile $\underline{Eu}$ solvent evaporated
faster than the $\underline{Cf}$. We concluded from this experiment that vaporization above 800°C of $\underline{Cf}$ from a divalent solvent does not induce
the divalent state in $\underline{Cf}$. This phase may only be stable (as in $\underline{Sm}$)
as a thin film.

Thermodynamic Calculations

We will use data from the $\underline{Cf}$ studies as an example of the meth-
odology of calculations of thermodynamic values from entropy and
vapor pressure data. As indicated in the introduction, we have immediately established two major tie-points with the ΔS_T° of vaporization from the vapor pressure work, and from the estimate of
crystal entropy at 298 K for the solid. This gives us then (from
the spectroscopic data) the entropy of vaporization at 298 K, and
the condensed-phase entropy at TK; this is illustrated in Fig. 8.
Experience with measured values from other metal systems shows the
ΔS_V line to be nearly straight. Similarly, the free-energy function values for the condensed phase change smoothly with temperature, starting with S_{298}°. The gaseous values come directly from
the spectroscopic data. If one were to compare the data for $\underline{Cf}$,
for example, with the published data for $\underline{Sm}$, the curves would
shift downward to new $\underline{Sm}$ tie points, and the general curvature
would change slightly also, depending partly on the melting temperature for each metal (indicated on Fig. 8 as 1173 K for $\underline{Cf}$).

We then generate our thermodynamic heats and entropies self-
consistently using the general formula for the free-energy function:

$$\frac{F_T^\circ - H_{298}^\circ}{T} = \frac{H_T^\circ - H_{298}^\circ}{T} - S_T^\circ \qquad (9)$$

The complete data compilation for Cf is shown in Tables II and III.
Note, as a final point, that from the analysis a heat capacity for
the condensed phase is derived, from 298 K to the highest temperatures of measurements. These data are shown in Table II and plotted
in Fig. 9. We believe this to be a highly-reliable curve, and, for
these exotic materials, not likely to be measured directly for some
time.

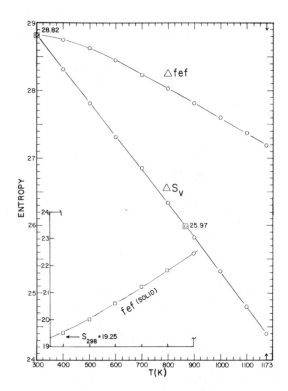

Figure 8.   Entropy and free-energy function construction plots for the analysis of californium thermodynamic data

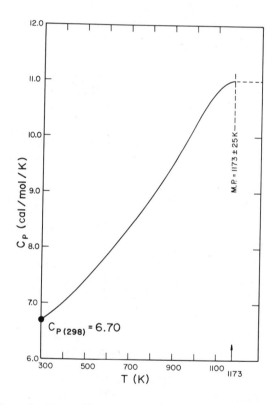

Figure 9. Derived heat capacity for californium, 298–1173 K

Table II. Thermodynamic Data for Californium

	Californium Gas			Californium Solid			
$T(K)$	$S^\circ_{(g)}$	$-fef_{(g)}$	$C_{p(s)}$	$S^\circ_{(s)}$	$-fef_{(s)}$	$H^\circ_T - H^\circ_{298}$	ΔH°_T
298	48.07	48.07	6.70	19.25	19.25	---	46 900
400	49.53	48.26	7.03	21.23	19.51	686	46 720
500	50.64	48.63	7.42	22.83	20.00	1410	46 493
600	51.54	49.04	7.86	24.29	20.59	2170	46 180
700	52.31	49.46	8.40	25.48	21.21	2990	45 906
800	52.97	49.85	8.94	26.65	21.83	3853	45 580
900	53.56	50.23	9.53	27.74	22.43	4772	45 118
1000	54.08	50.59	10.20	28.76	23.00	5757	44 640
1100	54.55	50.93	10.86	29.76	23.57	6814	44 070
1173	54.87	51.16	11.00	30.50	23.98	7617	43 630

Table III. Pressure-Temperature Data for Solid Californium

$T(K)$	$P(atm)$	3rd-law ΔH°_{298}
771	6.83×10^{-8}	46 926
843	8.80×10^{-7}	46 899
895	4.22×10^{-6}	46 902
938	1.32×10^{-5}	46 944
989	4.64×10^{-5}	46 921
1026	7.64×10^{-5}	------*

*Sample depleting

In summary, we have tried to describe the most recent state of understanding for the cohesive energies of the actinide metals. New results show there are still some surprises, even beyond the f-bonded early actinides. Vapor pressure measurements on Ra, Ac, and Es are planned, along with completion of the Bk studies. The Pa and Pa-oxygen systems will obviously require extensive work. Es vapor pressures will probably be studied as Raoult's-Law evaporations from liquid metal solvent. This will complete the vapor pressure measurements possible on the actinide metals, as there are no isotopes stable enough beyond Es.

Present best values of the crystal entropy and heats of vaporization are given in Table IV. Unknown values are left blank, and tentative data are given in parenthesis.

Table IV. Entropies and Heats for the Seventh-Row Metals

	S^0_{298}, cal/mol-deg.	ΔH^0_{298}, kcal/mol
Radium	*16.4 ± 0.2	----
Actinium	*14.8 ± 0.2	----
Thorium	12.76	142.7
Protactinium	(13.2)	----
Uranium	12.00	127.0
Neptunium	12.06	111.1
Plutonium	13.42	82.0
Americium	*13.2 ± 0.2	67.9
Curium	*17.2 ± 0.2	92.6
Berkelium	*18.4 ± 0.3	(91.2) [at temp.]
Californium	*19.2 ± 0.2	46.9

*Estimate from Eqn. 1.

Summary

Precise vapor pressure measurements by target collection/mass spectrometric Knudsen effusion techniques were combined with crystal entropy estimates to produce self-consistent free-energy functions, permitting calculation of heats, entropies and free energies from 298 K to the highest temperatures of measurement. The vapor pressures and thermodynamics of vaporization of americium, curium, berkelium, and californium are compared in terms of electronic structure and bonding trends in the trans-plutonium elements. These results are contrasted with the behavior of the early actinides, with attention to energy states and possible effects of f-electron bonding.

Acknowledgements

This work was funded by the USDOE Office of Basic Energy Sciences, Nuclear Sciences Division. Many thanks are due Dr. G. M. Matlack for radioanalytical analysis and Dr. R. C. Feber for spectroscopic analysis. We are especially indebted to Drs. J. Blaise, J. Verges, J. F. Wyart, J. Conway, and E. F. Worden for supplying their recent spectroscopic data.

Literature Cited

1. Brewer, L.; Science, 1968, 161, 115.
2. Brewer, L.; J. Opt. Soc. Am., 1971, 61, 1101.
3. Nugent, L.; Burnett, J. L.; Morss, L. A.; J. Chem. Thermo., 1973, 5, 665.
4. Bonnelle, C.; Structure and Bonding, 1976, 31, 23.

5. Hill, H. H.; Kmetko, E. A.; "Proc. Plutonium 1970 and Other Actinides"; Nucl Met. Series, Vol 17, AIME, 1970, p 233.
6. Johansson, B.; Phys. Rev. B, 1975, 11, 2836.
7. Johansson, B.; Phys. Rev. B, 1975, 11, 1367.
8. David, F.; Samhoun, K.; Guillaumont, R.; Edelstein, N.; J. Nucl. Chem., 1978, 40, 69.
9. Ward, J. W.; Hill, H. H.; "Heavy Element Properties"; Müller, W. and Blank, H., Eds.; North Holland Publ. Co.: Amsterdam, 1976; p 161.
10. Ward, J. W.; Müller, W.; Kramer, G. F.; "Transplutonium Elements"; Müller, W; Lindner, R., Eds., North Holland Publ. Co.:Amsterdam, 1976, p 161.
11. Smith, J. L.; Haire, R. G.; Science, 1978, 200, 535.
12. Ward, J. W.; Mulford, R. N. R.; Kahn, M.; J. Chem Phys., 1967, 47, 1710.
13. Ward, J. W.; Mulford, R. N. R.; Bivins, R. L.; J. Chem. Phys., 1967, 47, 1718.
14. Ward, J. W.; J. Chem. Phys., 1967, 47, 4030.
15. Ward, J. W.; Fraser, M. V.; J. Chem. Phys.,1968, 49, 3743.
16. Ward, J. W.; J. Chem. Phys., 1968, 49, 5129.
17. Ward, J. W.; Fraser, M. V.; J. Chem. Phys., 1969, 50, 1877.
18. Ward, J. W.; Ohse, R. W.; Reul, R.: J. Chem. Phys., 1975, 62, 2366.
19. Oetting, F.; Rand, M.; Ackermann, R., Eds. "The Chemical Thermodyanics of Actinide Elements and Compounds, Part I: The Actinide Elements"; IAEA; Vienna, 1976.
20. Koelling, D. D.; Freeman, A. J.; Phys. Rev. B, 1975, 12, 5622.
21. Glötzel, D.; J. Phys. F, 1978, 8, L163.
22. Ackermann, R. J.; Rauh, E. G.; J. Chem. Thermo., 1972, 4, 521.
23. Ackermann, R. J.; Rauh, E. G.; Rev. Hautes. Temp. Refract. Fr., 1978, 15, 259.
24. Ackermann, R. J.; Rauh, E. G.; Chandrasekharaiah, M. S.; J. Phys. Chem., 1974, 73, 762.
25. Hall, R. O. A.; private communication.
26. Ackermann, R. J.; Rauh, E. G.; J. Chem. Thermo., 1975, 7, 211.
27. Ackermann, R. J.; Rauh, E. G.; J. Chem. Phys., 1978, 62, 102.
28. Bradbury, M. H.; Ohse, R. W.; J. Chem. Phys., 1979, 70, 2310.
29. Ward, J. W.; Kleinschmidt, P. D.; Haire, R. G.; J. Chem. Phys., 1979, 71, in press.
30. Ward, J. W.; Kleinschmidt, P. D.; Haire, R. G.; J. de Physique, 1979, Colloque C4, 233.
31. Haire, R. G.; Asprey, L. B.; Inorg. Nucl. Chem. Lett., 1976, 12, 73.
32. Burns, J. H.; Peterson, J. R.; "Proc. Int. Conf. Rare Earths and Actinides", Durham, England, 1977, July 4-6, p 52-54.
33. Allen, J. W.; Johansson, L. I.; Bauer, R. S.; Lindau, I.; Hagstrom, S. B. A., Phys. Rev. Let., 1978, 41, 1499.

RECEIVED March 4, 1980.

Techniques of Microchemistry and Their Applications to Some Transcurium Elements at Berkeley and Oak Ridge

J. R. PETERSON

Department of Chemistry, University of Tennessee, Knoxville, TN 37916 and
Transuranium Research Laboratory, Oak Ridge National Laboratory,
Oak Ridge, TN 37830

Research on the transcurium elements requires specialized
techniques. The inherent radioactivity of these elements often
precludes otherwise routine manipulations, and the small amounts
available require the development of novel techniques to facilitate the study of their basic chemical and physical properties.
Much of our present knowledge of the inorganic and physical
chemistry of the transuranium elements was first obtained from
the application of microchemical techniques to submicrogram
quantities of material. Indeed, the primary justification for
the techniques of microchemistry is found in their application
to the investigation of rare materials. Prior to 1942 these
applications were chiefly in the fields of organic and biochem-
istry. With the production of the first few micrograms of
plutonium in June 1942, it became necessary to develop a broad
array of microchemical methods suited to submilligram quantities
of material. It is not the purpose here to review all these
techniques, but instead to focus on those which could be or have
been used for the study of some properties of Bk-249, Cf-249, and
Es-253 on the microgram to milligram scale.

Excluded here are those techniques relating to tracer-level
work (below weighable quantity of sample; measurement by radio-
assay only), for the concern here will be with the determination
of bulk properties of these elements. Tracer-scale studies
usually reveal directly only one property of the element under
investigation, that is, its relative preference for one environment over another, or more simply, its phase distribution. Nevertheless, as each new transuranium element was discovered and was
available only in trace quantities, a great deal of chemistry was
learned by inference from tracer-scale studies, including the
identity of oxidation states, approximate values of oxidation or
reduction potentials, the composition and stability of complex
ions, and relative volatilities.

In an effort to provide both some historical perspective of
the development and some current usage of microchemical techniques, as well as to provide some results of their respective

0–8412–0568–X/80/47–131–221$05.50/0
© 1980 American Chemical Society

applications, the present discussion will be limited to two main
areas of research. The first, synthesis, is the more important,
since any program to study the bulk properties of the transcurium
elements requires the synthesis of the particular metal or compound of interest. Treated here as examples will be the preparations of the metallic state and several binary compounds like
oxides, halides, chalcogenides, and pnictides. The second area,
investigative methods, will deal primarily with absorption spectrophotometry but with some mention of the X-ray and electron
diffraction methods which have contributed much to the elucidation of the structural properties of Bk, Cf, and Es.

Criteria for Selection of Microchemical Techniques

Although space is not available for a complete discussion,
the reader should be aware of the following factors which influence the choice of a particular technique for use in transcurium element research:
  1. Applicability – the technique must be able to accom-
      plish the desired goal on the scale of operation
      mandated by the available sample size. One proves
      out any new technique by using it on a substitute
      sample (most often these are lanthanide materials)
      where confirmation of an already known property is
      possible.
  2. Safety of experimenter and equipment – here containment
      of the radioactive sample is the key feature; also
      important is the ease of manipulation in order to
      minimize the chances of radioactive contamination.
  3. Maintenance of sample purity – following the difficult
      task to synthesize samples of high purity, it is
      necessary to avoid their chemical contamination by
      the very application of some particular technique.
Indeed, one of the most formidable problems encountered in working with very small samples is that of maintaining a high degree
of sample purity through a series of chemical and mechanical
manipulations. Because of the great increase of the surface-to-
volume ratio [$\propto$ (sample radius)$^{-1}$] on the microscale, as compared
to that on the macroscale, "chance" contamination is much more
probable with small samples. A general guideline to use in
microscale research work is to keep the actinide sample in a concentrated form and in a small volume container. This limits the
source of radioactivity and minimizes the effects of chemical
contamination of the sample by "chance" contact of the sample
with some impurity.

Single Ion-Exchange Resin Bead Technique

One of the pioneering microchemical techniques developed in
the laboratory of the late Professor Burris B. Cunningham at the

(now) Lawrence Berkeley Laboratory was the single ion-exchange resin bead technique for the concentration and manipulation of purified actinide ions. Individual resin beads are loaded to saturation by equilibration with a dilute acid solution of the actinide ion. Excess actinide and surface contaminants are easily removed by washing the loaded bead in water or dilute acid. The amount of actinide sorbed is controlled by the size of the resin bead chosen; for example, 1 µg of a typical trivalent actinide ion is sorbed by a Dowex 50 x 4 resin bead whose air-dried (from H_2O) diameter is 0.15 mm. An actinide-loaded bead is easily manipulated on a quartz fiber, represents an actinide concentration of about 2 M, and, being spherical, has only a single point of contact with its container, thus minimizing surface or "contact" chemical contamination. Examples of the use of the single bead technique for the preparation of actinide metal and compounds and for the study of spectroscopic and magnetic properties of trivalent actinide ions are found in the literature (1,2,3,4,5). Here the preparations of binary compounds and the pure metals are discussed first. Then the development of microtechniques for obtaining absorption spectra is traced from the use of single beads of ion-exchange resin to our present-day microscope spectrophotometer facility at the Oak Ridge Transuranium Research Laboratory (TRL).

Compound Preparation on the Microscale

Starting with an air-dried, actinide-loaded, single resin bead, an oxide is produced by calcining the bead in air or oxygen at 1200 °C. At the TRL the apparatus shown diagrammatically in Figure 1 is used; the bead is placed in a Pt crucible which is heated by radiation from an encircling Pt induction shield. The resulting oxide sample might be transferred to a silica capillary tube for attachment to a general preparation/vacuum system (Figure 2) for subsequent chemical treatment. Alternatively, it might be used directly for study by an applicable physical property measurement technique, like X-ray powder diffraction, magnetic susceptibility, solution calorimetry, etc.

The chemistry required to convert the oxide to other binary compounds is independent of the scale of operation. However, with microscale synthetic methods applied to radioactive materials, successful preparations are achieved more readily by carrying out the chemistry in situ, that is, in such a manner that eliminates, or at least minimizes, the necessity of having to "handle" the sample during or following its synthesis. Thus, actinide compounds are usually prepared in silica capillary tubes which can be flame sealed at the conclusion of a synthesis to provide the desired sample for study in a small volume, quartz container. A special feature of the preparation/vacuum system in the TRL is the capability to interrupt a synthesis, isolate (by means of a stopcock) and remove the sample, examine it in

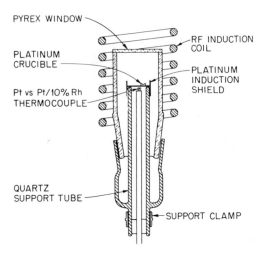

Figure 1.   Schematic of resin-bead calcination apparatus

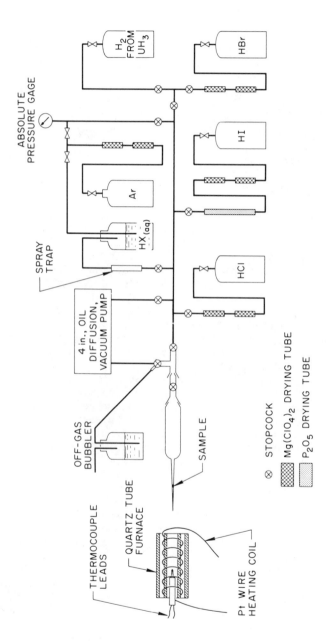

Figure 2. Schematic of preparation–vacuum system used for synthesizing transcurium element compounds

situ via absorption spectrophotometry and/or X-ray diffraction, and then return it to the preparation/vacuum system should additional chemistry be required.

In Table I are summarized some generally useful chemical reactions for the preparation of transplutonium element metal and some compounds. For simplicity, and because of variable oxidation states, the equations are not necessarily balanced. The oxalate precipitation and subsequent calcination to the oxide is reserved for multimicrogram and greater quantities of actinides and for Es-253, whose intense radiation precludes the use of the resin bead technique. The preparation of fluoride compounds is not carried out in quartz but in Monel (6). Details of the conditions of temperature, pressure, etc. to effect these chemical reactions are available in the literature (1,3,6-11).

Metal Preparation on the Microscale

Two methods for producing transcurium element metal are listed in Table I. In both routes the actinide compound, which might have been prepared using the single bead technique, and reductant metal are placed in a metal crucible (usually Ta), which is heated to promote the reduction reaction. The method of choice depends upon the quantity of actinide available and the physical form of the product metal required. For bulk product metal with only a limited amount of material (μg quantities), the fluoride reduction method is better. The essential difference in the two synthetic routes is that in the fluoride reduction, the product metal remains inside the crucible system (byproducts, excess reductant, and volatile impurities leave), whereas in the oxide reduction, the product metal leaves the heated crucible (along with any volatile impurities) and deposits on a cooler surface, completely separated from the byproduct oxide, nonvolatile impurities, and excess reductant. Unless at least several hundred micrograms of metal are being produced, the metal product obtained by oxide reduction is in the form of a thin foil. Advantage has been taken of this form of metal product for structural studies by electron diffraction (12).

The apparatus used for the production of metal on the few microgram scale via fluoride reduction has been improved considerably between its use for the first preparation of Bk metal (2) and the more recent preparations of Bk metal on the half-milligram scale (13,14) and Cf metal on the $\leq$ 10 μg scale (8). The interested reader is referred to the literature cited for further details on this metal-making technique.

Metal samples produced on the microscale have been studied mainly by X-ray powder and electron diffraction methods, both as a function of temperature (8,12,15,16) and pressure (17). Recently a microsusceptometer incorporating a superconducting quantum interference device (SQUID) has been constructed at Oak Ridge that has sufficient sensitivity to determine the magnetic

Table I.   Preparative Chemistry for Transplutonium Element (An)
Metal and Some Compounds

<u>Metal</u>

An fluoride + Li → An metal + LiF ↑

An oxide + La(Th) → An metal ↑ + La(Th) oxide

<u>Oxides</u>

An(III) (sorbed in resin bead) + O_2 → An oxide

An(III) + $H_2C_2O_4$ → $An_2(C_2O_4)_3$ ↓ $\xrightarrow{\Delta}$ An oxide

<u>Halides</u>

An oxide + HX → AnX_3 + H_2O (X = F,Cl,Br)

An + HX → AnX_3 + H_2 (An + X_2 → AnX_3)

AnX_3 + HI → AnI_3 + HX

An oxide (AnF_3) + F_2 → AnF_4

AnX_3 + H_2 → AnX_2 + HX

<u>Oxyhalides</u>

An oxide (AnX_3) + HX/H_2O $\xrightarrow{\text{carrier gas}}$ AnOX

<u>Semimetallics</u>

An metal + H_2 → An hydride

An metal + VA element → An pnictide

An metal + VIA element → An chalcogenide

properties of these small samples (18). Also the solution
microcalorimeter (19) at Oak Ridge has sufficient sensitivity to
measure the heats of reaction of these metal samples with aqueous
acid. The precision of the results of the calorimetric and
magnetic susceptibility measurements is severely limited on the
microscale by the precision to which the samples can be weighed.
Both investigative devices were specifically designed for the
capability of obtaining data from samples of transcurium elements
and compounds.

Absorption Spectrophotometry on the Microscale

Techniques for obtaining absorption spectra from small
samples of transcurium element species have advanced considerably over the last twenty years. In a first attempt to observe
the solution absorption spectrum of Bk(III) in a "risk free"
manner, Cunningham and colleagues (5) in the late 1950s attached
a capillary absorption cell ($\sim$ 7 µL) to the lower end of the ion-
exchange column used in the final purification step, so that the
purified Bk(III) solution passed through the cell on its way to
its final container. A bench spectrometer served as the light
analyzer in the visible wavelength region of the spectrum. Although no Bk(III) absorption bands were detected, these workers
were able to set an upper limit on the molar extinction coef-
ficient($\leq$ 20) of any Bk(III) absorption band in this wavelength
region from preliminary experiments with Am(III) and Nd(III)
using the same cell.

Utilizing the capability of a single ion-exchange resin
bead to concentrate the sorbed actinide ion and to provide good
optical transparency in the visible and near infrared wavelength
regions, Cunningham and Wallmann attempted to obtain the absorption spectra of Bk(III) and Cf(III). Their apparatus (Figure
3) was tested using Am(III) and they observed the 503 nm Am(III)
absorption band ($\varepsilon \sim 350$) through a hand spectroscope with only
1 ng of Am(III) sorbed in the bead. Later improvements of this
same basic technique included better masking of stray light
(Figure 4), provisions for increasing the effective pathlength
by stacking several actinide-loaded beads, inclusion of a quartz
"light pipe" to gather more effectively the transmitted light,
and automated recording of the spectrum via film techniques.
With this multibead-stack apparatus Green and Cunningham (4)
recorded the Cf(III) absorption spectrum and demonstrated, using
Pr(III), that the "bead" spectrum was very similar to an absorption spectrum obtained in acid solution.

The next development took place during the course of the
Ph.D. research of the author at Berkeley. With a total of only
2 µg of Bk-249 with which to work, the first spectroscopic measurements were made using the single bead technique, but in an apparatus (Figure 5) designed for use with a Cary Model 14 Recording Spectrophotometer. The Bk-loaded bead was placed in the

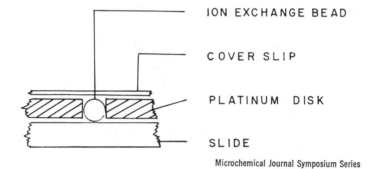

ION EXCHANGE BEAD

COVER SLIP

PLATINUM   DISK

SLIDE

Microchemical Journal Symposium Series

Figure 3.   Schematic of first single-bead microabsorption cell (1)

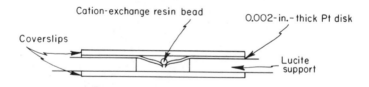

Figure 4.   Schematic of improved, single-bead microabsorption cell

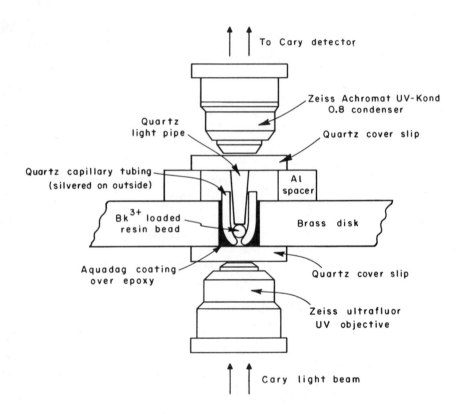

Figure 5. Schematic of single-bead microabsorption cell for use in Cary spectro-photometer

cell and secured by the quartz light pipe making optical contact
with the bead. The difficulty with this apparatus was its inability to transmit more than a percent or two of the Cary light
beam. Despite poor spectral resolution, repeated wavelength
scans confirmed the observation of about six absorption peaks.
Another distinct disadvantage in this particular case was that
the bead was less transparent in the near ultraviolet wavelength
region, where Bk(III) seemed to have significant absorption.
These limitations on the single bead absorption cell for the
study of the spectral properties of berkelium stimulated the development of a new experimental technique – one which had the
capability of higher resolution and also allowed investigation
into the ultraviolet wavelength region. The chosen method
centered around the suspension of a drop of Bk(III) solution
between two tapered quartz rods. Prototype cells loaded with
drops of Nd(III) solution were found to yield significantly
improved spectra. The first "suspended drop" or "light-pipe"
cell constructed for berkelium work is shown schematically in
Figure 6. A 1-mm diameter quartz rod was drawn down in a flame
to about 100 μm in diameter and mounted in a brass disk. This
"entrance light pipe" was aligned with and spaced about 200 μm
from the "catcher light pipe", a short ($\sim$ 1 mm) section of quartz
rod about 100 μm in diameter. The Bk(III) solution was transferred to the light-pipe cell as a 50–60 nL drop suspended on
the end of a micropipette. The cell also included water-soaked
paper to maintain a wet atmosphere in an effort to prevent
vaporization of the droplet to the point where solid formation
would restrict light transmission through the cell system. The
light source and analyzer was again the Cary spectrophotometer.
The Bk(III) spectrum obtained with this light-pipe cell was
similar to that obtained from the bead cell but provided new
evidence for Bk(III) absorption in the near ultraviolet wavelength region. Unfortunately, however, only a small percentage
of the Cary light beam was transmitted through the cell system.
     Improvement of the cell optics was the prime motivation for
further developmental work on the light-pipe cell design, but
other considerations were drop stability, control of drop size,
and ease of handling. The cell optics were eventually improved
to the point where $\sim$ 30% of the Cary light beam condensed by the
objective lens was transmitted through the cell. These improvements were realized by shortening the length of the entrance
light pipe, by coating the entrance light pipe with silver or
aluminum, by polishing the ends of the light pipes, and by including a nearly spherical bulge in the entrance light pipe to
catch more of the incident light and to reverse the direction of
back-reflected light.
     The size of the drop of solution in the cell, which de-
termined the concentration of the absorbing species, was controlled by the heating effect of the infrared source lamp of the
spectrophotometer and by the addition of dilute, high purity acid

solution. This addition was accomplished by a pump, which consisted of a closed reservoir system operated by solution expansion induced by Pt wire resistance heating. A photomicrograph of the pump nozzle, positioned between the two light pipes, is shown in Figure 7.

An improved cell design allowed greater ease in the loading, drop observation, and optical alignment procedures. A photograph of this cell with its sliding cover is shown in Figure 8. With the receipt of an additional 28 μg of Bk-249 at Berkeley, this cell system was used with droplets containing ∿ 4 μg of Bk(III). Considering the usual operating volume of solution across the light-pipe gap, the emerald green droplet represented a Bk(III) concentration of about 4 M. A typical Bk(III) solution absorption spectrum recorded over the wavelength range 320 to 680 nm is shown in Figure 9. The high background absorption in the lower wavelength region is probably caused by the presence of H_2O_2 and/or Cl_2, generated radiolytically from the aqueous HCl solution. Later experiments with Es-253 in an essentially identical microabsorption cell have been reported in the literature (20,21).

The microscope spectrophotometer system in routine use at the TRL is described in reference (7), so no details of the apparatus and its use are given here. Instead a brief description of the reason for developing and continually refining the microscope spectrophotometer facility will be presented. Historically the way to characterize a solid-state sample of a transplutonium element has been by standard X-ray powder diffraction analysis. When a systematic study of element 99, einsteinium, was undertaken, it was found that obtaining useful diffraction data from Es-containing materials was a very difficult, if not an impossible, task (22). The intensely radioactive Es-253 not only caused rapid blackening of the film used to record the diffraction pattern, but more importantly, it degraded the crystallinity of the sample.

In contrast to the necessity of having repetitive long-range order for obtaining an X-ray powder diffraction pattern, an absorption spectrum results from the summation of all the local actinide ion environments in the analyzing light path through the sample. Absorption spectrophotometric analysis is also faster and has greater sensitivity for the detection of minor components than does X-ray analysis. Because of this sensitivity, spectral studies of Es compounds can be, and have been, undertaken to investigate progeny growth in Es compounds. This research also requires the knowledge of the absorption spectra of Bk and Cf compounds in their respective pure states, because of the genetic relationship

$$^{253}_{99}Es \xrightarrow[\quad t_{1/2} = 20.5 \text{ d} \quad]{\alpha} {}^{249}_{97}Bk \xrightarrow[\quad t_{1/2} = 314 \text{ d} \quad]{\beta^-} {}^{249}_{98}Cf.$$

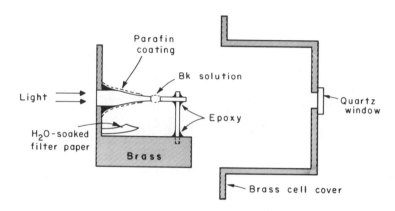

Figure 6.   *Schematic of first light-pipe microabsorption cell for berkelium solution*

Figure 7.   *Light-pipe area of microabsorption cell showing position of pump nozzle*

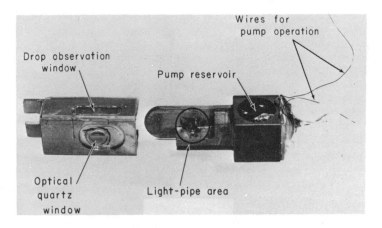

Figure 8.   Overview of light-pipe microabsorption cell with sliding cover off

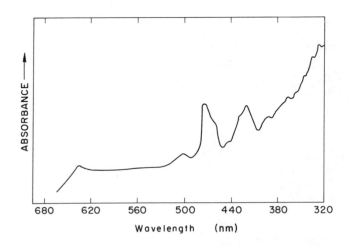

Figure 9.   Absorption spectrum of berkelium(III) in aqueous HCl solution

With such spectral data identification of the progeny species
in an Es compound can be made by assignment of the peaks in its
absorption spectrum. The change in percentage composition of an
initially pure Es compound as a function of time is shown in
Figure 10. Following an initial interest in just characterizing
trivalent (23) and divalent (24,25) Es in the solid state, the
more recent emphasis has been directed toward elucidation of the
chemical consequences of radioactive decay via studies of some
halide compounds of Es (and Bk) over long time periods. Although
bulk Bk(II) is unknown in the solid state, does nature produce
it via the alpha decay of Es(II) compounds? The decay of the
Es dihalides has been monitored spectrophotometrically, and the
granddaughter Cf(II) products have been identified on the basis
of the knowledge of Cf(II) spectra obtained from direct synthesis
of Cf dihalides (26,27,28,29). No absorption peaks attributable
to Bk(II) have been observed. Can Cf(II) result from the decay
of Es(II) without going through Bk(II)? Are the characteristic
absorption peaks of Bk(II) outside the useful wavelength range
(300–1100 nm) of the microscope spectrophotometer, or are they
masked by the absorption peaks of Es(II) and/or Cf(II)?

Current Attempts to Synthesize and Characterize Bk(II)

At the present time this problem is being attacked by attempting to synthesize bulk Bk(II) directly (30). Because the
microchemical techniques employed combine most of the ones
discussed here, a brief description of the experimental approach
will be presented. Although H_2 is a sufficiently strong reductant to reduce the Cf and Es trihalides to the corresponding
dihalides, it will not reduce $BkBr_3$ to $BkBr_2$. Therefore Bk metal
was chosen to be the reducing agent for the reaction $Bk + 2BkBr_3$
$\rightarrow 3BkBr_2$. The $BkBr_3$ was prepared in a quartz capillary by treatment of BkF_3 with anhydrous HBr. The Bk metal was prepared by Li
metal reduction of BkF_4. In an inert atmosphere enclosure a
piece of Bk metal was placed into the capillary containing the
sample of $BkBr_3$, which was subsequently evacuated and flame
sealed to an overall length suitable for mounting in an X-ray
powder camera. Following positioning of the piece of Bk metal
on top of the sample of $BkBr_3$ (performed by vibrating the sealed
capillary), the bromide was melted and quenched. Absorption
spectral analysis confirmed the presence of Cf(II), the beta decay
daughter of Bk, and X-ray diffraction analysis produced a poor
powder pattern different from those known for Bk metal and $BkBr_3$.
Detailed analysis of these data is currently in progress.
There are both advantages and disadvantages of these complementary analysis methods. One advantage is the possible confirmation of the X-ray results by the results of the spectral
analysis and vice versa, lending support for the conclusions
drawn on the basis of the results of either analysis alone. A
disadvantage is that for obtaining high-quality powder diffraction

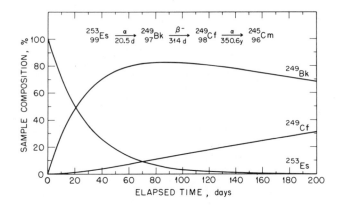

Figure 10. Ingrowth of berkelium-249 and californium-249 from initially pure einsteinium-253 as a function of time

data the sample should be microcrystalline in nature, whereas for obtaining high-quality absorption spectral data the sample should be macrocrystalline in nature (this is usually achieved by melting the sample).

Additional experiments to synthesize and characterize Bk(II) in bulk will be carried out by reduction of a Bk trihalide with Bk metal and analysis of the products by X-ray powder diffraction and absorption spectrophotometry.

Acknowledgments

A number of colleagues, graduate students, and postdoctoral research associates have contributed much toward the development and application of microchemical techniques to research with the transplutonium elements. The author gratefully acknowledges their assistance and patience. In view of space only their names are listed here. At the Berkeley laboratory (LBL): Burris Cunningham, Jim Wallmann, Tom Parsons, Jere Green, Dennis Fujita, and Judy Copeland. At the Oak Ridge laboratory (TRL): Rus Baybarz, Dick Haire, George Werner, Jack Young, Paul Huray, Jim Fahey, Jim Stevenson, Bob Fellows, Maxy Noé, Mickey Raschella, Daniel Damien, and Stanley Nave.

This research was sponsored by the Division of Nuclear Sciences, U.S. Department of Energy under contracts DE-AS05-76ER04447 with the University of Tennessee (Knoxville) and W-7405-eng-26 with the Union Carbide Corporation.

Literature Cited

1. Cunningham, B. B. Microchem. J. Symp. Ser., 1961, 1, 69.
2. Peterson, J. R.; Fahey, J. A.; Baybarz, R. D. J. Inorg. Nucl. Chem., 1971, 33, 3345.
3. Peterson, J. R.; Cunningham, B. B. Inorg. Nucl. Chem. Letters, 1967, 3, 327.
4. Green, J. L.; Cunningham, B. B. Inorg. Nucl. Chem. Letters, 1966, 2, 365.
5. Cunningham, B. B. J. Chem. Ed., 1959, 36, 32.
6. Stevenson, J. N.; Peterson, J. R. J. Inorg. Nucl. Chem., 1973, 35, 3481.
7. Young, J. P.; Haire, R. G.; Fellows, R. L.; Peterson, J. R. J. Radioanal. Chem., 1978, 43, 479.
8. Noé, M.; Peterson, J. R. "Transplutonium Elements 1975"; Müller, W. and Lindner, R., Eds.; North-Holland Publishing Co., Amsterdam, 1976; p. 69.
9. Damien, D. A.; Haire, R. G.; Peterson, J. R. J. de Physique, 1979, 40, C4-95.
10. Peterson, J. R. J. Inorg. Nucl. Chem., 1972, 34, 1603.
11. Keller, C. "The Chemistry of the Transuranium Elements"; Verlag Chemie, Weinheim, Germany, 1971.

12. Haire, R. G.; Baybarz, R. D. J. Inorg. Nucl. Chem., 1974,
     36, 1295.
13. Stevenson, J. N.; Peterson, J. R. Microchem. J., 1975, 20,
     213.
14. Fuger, J.; Peterson, J. R.; Stevenson, J. N.; Noé, M.;
     Haire, R. G. J. Inorg. Nucl. Chem., 1975, 37, 1725.
15. Haire, R. G.; Asprey, L. B. Inorg. Nucl. Chem. Letters,
     1976, 12, 73.
16. Haire, R. G.; Baybarz, R. D. J. de Physique, 1979, 40,
     C4-101.
17. Burns, J. H.; Peterson, J. R. "Rare Earths and Actinides
     1977"; Corner, W. D. and Tanner, B. K., Eds.; Institute of
     Physics, London, 1978; p. 52.
18. Nave, S. E.; Huray, P. G. J. de Physique, 1979, 40, C4-114.
19. Raschella, D. L. "Solution Microcalorimeter for Measuring
     Heats of Solution of Radioactive Elements and Compounds";
     Ph.D. Dissertation, The University of Tennessee, Knoxville,
     December 1978; U.S. Department of Energy Document No. ORO-
     4447-081, 1978.
20. Cunningham, B. B.; Peterson, J. R.; Baybarz, R. D.;
     Parsons, T. C. Inorg. Nucl. Chem. Letters, 1967, 3, 519.
21. Fujita, D. K.; Cunningham, B. B.; Parsons, T. C.; Peterson,
     J. R. Inorg. Nucl. Chem. Letters, 1969, 5, 245.
22. Haire, R. G.; Peterson, J. R. "Advances in X-Ray Analysis",
     Volume 22; McCarthy, G. J.; Barrett, C. S.; Leyden, D. E.;
     Newkirk, J. B. and Ruud, C. O., Eds.; Plenum Publishing
     Corp., New York, 1979; p. 101.
23. Fellows, R. L.; Peterson, J. R.; Noé, M.; Young, J. P.;
     Haire, R. G. Inorg. Nucl. Chem. Letters, 1979, 11, 737.
24. Fellows, R. L.; Peterson, J. R.; Young, J. P.; Haire, R. G.
     "The Rare Earths in Modern Science and Technology"; McCarthy,
     G. J. and Rhyne, J. J., Eds.; Plenum Publishing Corp., New
     York, 1978; p. 493.
25. Peterson, J. R.; Ensor, D. D.; Fellows, R. L.; Haire, R. G.;
     Young, J. P. J. de Physique, 1979, 40, C4-111.
26. Young, J. P.; Vander Sluis, K. L.; Werner, G. K.; Peterson,
     J. R.; Noé, M. J. Inorg. Nucl. Chem., 1975, 37, 2497.
27. Young, J. P.; Haire, R. G.; Fellows, R. L.; Noé, M.;
     Peterson, J. R. "Transplutonium Elements 1975"; Müller, W.
     and Lindner, R., Eds.; North-Holland Publishing Co.,
     Amsterdam, 1976; p. 227.
28. Peterson, J. R.; Fellows, R. L.; Young, J. P.; Haire,
     R. G. Radiochem. Radioanal. Letters, 1977, 31, 277.
29. Wild, J. F.; Hulet, E. K.; Lougheed, R. W.; Hayes, W. N.;
     Peterson, J. R.; Fellows, R. L.; Young, J. P. J. Inorg.
     Nucl. Chem., 1978, 40, 811.
30. Peterson, J. R.; Ensor, D. D.; Haire, R. G.; Young, J. P.
     University of Tennessee and Oak Ridge National Laboratory,
     unpublished results, 1979.

RECEIVED December 26, 1979.

Chemistry of the Heaviest Actinides: Fermium, Mendelevium, Nobelium, and Lawrencium

E. K. HULET

Lawrence Livermore Laboratory, University of California, P.O. Box 808, Livermore, CA 94550

From data gathered in a rather small number of experiments and limited by working with scarcely more than a few atoms, we can now discern that the chemical properties of the heavy actinides systematically deviate from those of their lanthanide counterparts. The differences between the later elements of the 4f and 5f series can be generally interpreted on the basis of subtle changes in electronic structure. The most important change is a lowering of the 5f energy levels with respect to the Fermi level and a wider separation between the 5f ground states and the first excited states in the 6d or 7p levels. Thus, in comparison with analogous 4f electrons, the later 5f electrons appear more tightly bound to the atom. Our conclusions regarding these shifts toward greater stabilization of 5f orbitals with increasing atomic number are mainly supported by the appearance of the divalent oxidation state well before the end of the actinide series and the predominance of the divalent state in the next to last element in the series. It is these conclusions and the underlying experimental evidence that will be the main subject of this review.

Because of the uniqueness of divalency within a series of elements that are commonly trivalent, most of the chemical research concerning the heaviest actinides has been concentrated on studies of lower oxidation states. The chemical properties of the trivalent ions of the lathanides and actinides are virtually the same throughout both series and, for this reason, there has been little incentive to specifically study this oxidation state in Md, No, and Lr. This close relationship between the scientific significance and the research completed up to now is strongly correlated with the extraordinary effort required to produce experimental information concerning these elements. In a scientific sense, the principle of cost-effectiveness has governed the selection of research topics. Beside the scientific effort required, there are additional restrictions to obtaining extensive experimental data.

0–8412–0568–X/80/47–131–239$06.25/0

Chemical studies of these elements must be performed with isotopes having not only a fleeting existence but producible only in atom quantities. In Table 1 we list the most frequently made isotopes, their half lives, and the atoms that have been synthesized for each data point. Except for ^{255}Fm, the nuclides listed can be created only by nuclear reactions between accelerated charged particles and transplutonium target nuclei. For this reason and the short lifetimes of the isotopes, all chemical studies are carried out at large heavy-ion accelerators. Such research calls upon nuclear physics for the methods of element synthesis and detection while the research goals are aimed toward atomic and chemical properties. Therefore, this field of research most easily falls into the domain of the nuclear chemist.

Element	Half Life	Average Atoms/Experiment
^{255}Fm	20.1 h	10^{12}
^{256}Md	77 min	10^{6}
^{255}No	3.1 min	10^{3}
^{256}Lr	31 s	10

Table 1. The isotopes commonly produced for chemical studies of Fm, Md, No, and Lr. Their half lives and numbers of atoms available seriously limit the information obtainable by experiment.

To insure that a statistical average behavior is observed in the chemical experiments with No and Lr, it has been necessary to make repeated measurements for each data point. Indeed, the determination of the distribution coefficients for Lr in a solvent extraction experiment required over 200 experiments to define the behavior of about 150 atoms of Lr (1). Experiments of this kind are exceptionally difficult and computer-controlled equipment has been devised to perform either a portion or all of operations needed for the chemical tests and the analysis of samples. Computer automation, although requiring a larger effort to implement, permits an experiment to be repeated many times in rapid sequence with the added advantage of doing each quickly before the complete decay of the radioactive atoms of a short-lived isotope.

It is clear that many fundamental and important physical constants, electronic and molecular structures, and magnetic and thermodynamic properties cannot be determined when only a few atoms of these elements are available. As an example, the energies of low-lying electronic levels, obtainable from optical emission spectroscopy, would provide information essential to understanding ionic and bonding properties and would allow the calculation of some thermodynamic constants. Yet, with a small number of atoms, we are presently unable to obtain this kind of basic knowledge. Nevertheless, many other measurements are feasible and these can provide qualitative and a detailed knowledge from the behavior of only a few atoms. Among the demonstrated possibilities are the study of ionic properties in aqueous and nonaqueous solutions, measurement of the magnetic moment of free atoms, and the volatility of the halide compounds. The list of feasible experiments will undoubtedly expand in time as advancements are made in technology and as we stretch our ingenuity.

Undoubtedly, there is some skepticism with regard to deductions and conclusions about the true chemical properties of an element when they are based upon observing the behavior of less than one hundred atoms. This question has never been fully addressed by any underlying theoretical treatment using thermodynamic and kinetic arguments. In some instances, a serious case could be made for a cautious view, and one that we can imagine, is the vapor pressure of a metal. Since the volatility of a metal is dependent on the strength of bonds between like atoms, it seems likely that vapor pressures would be perturbed when there are too few atoms present to constitute a majority that are interbonded. However, there are a vast number of cases where the tracer chemistry of an element is identical to its bulk behavior. New actinide elements have been identified on the basis of the elution position of 17 atoms from an ion-exchange column (2). At least for actinide ions in aqueous solutions, we would not anticipate any unusual behavior dependent on their concentration until they become one of the major constituents. The principal justification for this view is that in any given solution, every element on earth is likely to be present at the level of one to a million atoms together with major concentrations of added reagents. The few atoms of heavy actinides introduced into the solution are not likely to be singled out for extraneous side reactions because of the presence of larger numbers of other metal cations with similar chemical properties. Thus, the behavior of a single actinide ion should be close to average because of the dilution with greater numbers of chemically-similar ones. These intriguing aspects of "one-atom chemistry" are now being explored from a theoretical viewpoint and should be on firmer ground in the future (3).

Fermium

Several Fm isotopes with half lives of nearly a day to 100 days are available in amounts of at least 10^9 atoms. The nuclides ^{255}Fm and ^{257}Fm are conveniently used for chemical investigation of Fm and they are obtainable as products from long neutron irradiations of ^{242}Pu and ^{244}Cm. The 20-h ^{255}Fm is

generated by the beta decay of 40-d ^{255}Es produced in the neutron irradiations. By chemically isolating the Es and periodi-
cally reseparating Fm from its parent, one can secure a fairly
long-term source of ^{255}Fm adequate for all tracer experiments.

The ground-state electronic configuration of Fm is $5f^{12}7s^2$
or an $^3H_6$ level (4). This was established by an atomic-beam
measurement of the magnetic moment g_j of 3.24-h ^{254}Fm. In this
elegant measurement, FmF_3 was reduced with ZrC_2 in an atomic-beam
apparatus to produce a beam of neutral Fm atoms. Three magnetic
resonances were detected and the best value for g_j was calculated.
To obtain the level term, it was necessary to extrapolate the
mixing due to intermediate coupling in the electron spin-orbit
interactions (j-j and L-S). These extrapolations were made from
lower actinides and supplemented by Hartree-Fock calculations
for free atoms. From similar calculations, the next higher
level is predicted to be $^5G_7$ starting about 20,000 cm^{-1} above
the ground state and having the configuration $5f^{11}6d7s^2$. However, the $f^{12}sp$ and $f^{11}s^2p$ configurations are very close in
energy (5) to the $f^{11}ds^2$ so that it is impossible to unambigu-
ously estimate the next level above the ground state.

The electron binding energies of Fm have been measured for
the K, L_{1-3}, N_{1-5}, $N_{6,7}$, O_{1-3}, $O_{4,5}$, and $P_{2,3}$ shells (6).
These were determined to an accuracy of ~10 eV by conversion-
electron spectroscopy in the beta decay of ^{254m}Es to ^{254}Fm. A
surprisingly low binding energy for the $P_{2,3}$ ($6p_{1/2,3/2}$) shell
of 24 $\pm$ 9 eV was found. Predicted values derived either from
extrapolations of those measured in lower actinides or calculated
by Hartree-Fock methods are about 20 to 60 eV higher in energy.
As the authors suggested, a binding energy of 24 eV might provide
a possibility for 6p involvement in chemical and spectroscopic
interactions.

The properties of Fm metal and of its solid compounds are
for the most part unknown because there are insufficient quantities to prepare even microsamples. In the numerous thermo-
chromatographic studies by Zvara and coworkers, the evaporation
of Fm and Md tracer from molten La at 1150°C was compared with
the behavior of other selected lanthanides and actinides (7).
The volatility of Md and Fm was found to be greater than that of
Cf and Cf was about equivalent to Yb and Eu, and all were much
more volatile than Am. The volatilities are correlated by the
number and energy of the valence bonds minus the energy needed to
promote electrons to the valence bands in the metals. Therefore,
within the normally trivalent lanthanides and actinides, the more
volatile elements are associated with the divalent metals. The
unusual volatility of Fm and Md was then construed by Zvara as
evidence for divalency in the metallic state.

The separation methods for Fm are the same as those used for
separating other trivalent lanthanides and actinides. For separating the adjacent elements, Es and Md, a high-resolution chromatographic method is necessary. Either ion exchange, using strongly

acidic resins (9), or extraction chromatography employing alkylphosphoric acids (8) is strongly preferred. A complexing agent
(α-hydroxyisobutyric acid) is required to selectively elute the
actinides from cation-exchange resins. The separation factors,
defined as the ratio of the distribution coefficients of two
metal ions, are small for both cation exchange and extraction
chromatography. These factors range from 1.7 to 2.04 for Es-Fm
separations using a Dowex-50 cation exchanger (9) or extraction
chromatography with HCl as the eluant and bis(2-ethylhexyl)-
phosphoric acid diluted with heptane as the extractant (10). The
Fm-Md separation factors obtained by these two methods were 1.4
and 4.0, respectively (9,10). The major difference between these
methods of chromatographic separation lies in the elution sequence. With alkylphosphoric acid extractants, the elements are
eluted in order of atomic number while in cation exchange, the
order is reversed.

The solution chemistry of Fm deals largely with the highly-
stable tripositive oxidation state although the dipositive state
is also known. Formation constants for citrate complexes (11)
and the first hydrolysis constant have been accurately determined
for Fm^{3+} (12,13). Since the formation and hydrolysis constants
for Am, Cm, Cf, and Es were measured simultaneously with those
for Fm, the complex strengths of many of the trivalent actinides
can be compared (13). All constants were determined at an ionic
strength of $\mu = 0.1$ in a perchlorate medium by measuring the
partitioning of the radioactive tracers between a thionyltri-
fluoracetonate-benzene phase and the aqueous phase. The results
for Fm may be expressed as follows:

$$Fm^{3+} + H_2O \rightleftharpoons FmOH^{2+} + H^+; \qquad\qquad \log K = -3.80 \pm 0.2$$

$$Fm^{3+} + 2H_3Cit \rightleftharpoons Fm(HCit_2)^{2-} + 5H^+; \qquad \log \beta_1 = 11.17$$

$$Fm^{3+} + 2H_3Cit \rightleftharpoons FmCit_2^{3-} + 6H^+; \qquad \log \beta_1' = 12.40$$

Compared to the other actinide ions investigated, Fm formed
stronger complexes with citrate and hydroxyl ions because of its
smaller ionic radius. The smaller radius is a direct consequence
of the increased nuclear charge and partial shielding of the
outermost 6p electrons by the inner f electrons.

The reduction of Fm^{3+} to Fm^{2+} was first reported in 1972 by
N. B. Mikheev and coworkers (14). The reduction was accomplished
with Mg metal in the presence of Sm^{3+} which was coreduced in an
aqueous-ethanol solution. Identification of the divalent state
of Fm was established by determining the extent of its cocrystallization with $SmCl_2$ and this was compared to the amount of tracer
Sr^{2+} also carried with $SmCl_2$. A milder reductant, Eu^{2+}, failed
to reduce Fm^{3+}, which placed the standard reduction potential of
Fm^{3+} between Eu^{2+} and Sm^{2+} or -0.43 to -1.55 V relative to the
standard $Pt,H_2|H^+$ electrode. Later work (15) by these scientists

narrowed the range to betweem -0.64 and -1.15 V and most recently, they were able to estimate the potential was the same as the $Yb^{3+} \rightarrow Yb^{2+}$ couple within 0.02 V, or -1.15 V (16). The reduction of Fm to a divalent ion with $SmCl_2$ has also been observed recently by Hulet et al. (17).

In further work related to the divalent state, the electrode potential for the reduction of Fm^{2+} to Fm^0 has been measured by Samhoun and David (18). Over a period of years, they developed and refined a radiopolarographic technique for determining half-wave potentials at a dropping-Hg cathode. In addition to Fm, they have measured either the III → 0 or II → 0 potential for all transplutonium actinides except No and Lr (18,24). The polarograph for Fm is shown in Figure 1. The electrochemical reaction taking place at a reversible electrode can be deduced from the slope of the polarographic wave. Specifically, the number of electrons exchanged at the electrode, based on the Nernst equation, is obtained from this slope. From their analysis of the polarograms, there were three electrons involved in the electrochemical reduction of the trivalent ions of the elements Am through Es and only two electrons for the reduction of Fm. This implies that Fm^{3+} was first reduced to Fm^{2+} before being further reduced to the metal. The III → II reduction step is not detected by this radiopolarographic technique because both the III and II ions are in the solution phase; whereas, the measured parameter is the distribution of the tracer between the aqueous and Hg phase.

The half-wave potentials measured by this method include the amalgamation potential of the metal-mercury reaction. The potential for the overall process for Fm, i.e.

$$Fm^{2+} + 2e^- \; = \; Fm(Hg),$$

was found to be -1.474 V with reference to the standard hydrogen electrode. The amalgamation potential was estimated to be 0.90 V by using the metal radii as a correlating parameter and interpolating within a series of divalent elements with known amalgamation potentials (19). This correlation is shown in Figure 2. The standard electrode potential is then given as -2.37 V for the $Fm^{2+} + 2e^- = Fm^0$ reaction. The authors' estimated 5 mV accuracy for the measured half-wave potential seems reasonable, but there is a much larger uncertainty in the estimated amalgamation potential. Because the amalgamation potential represents a large correction in obtaining the standard potential, caution should be exercised in combining this standard potential with other data to calculate additional thermodynamic properties.

Mendelevium

The isotope ^{256}Md is nearly always employed for chemical studies of this element. Besides having a convenient half life

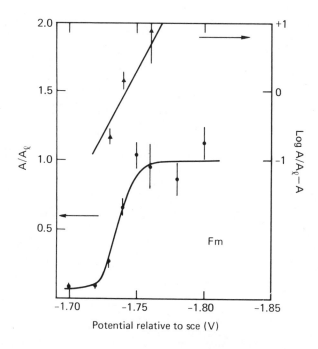

Figure 1.   Distribution of fermium as a function of applied voltage between mercury in a dropping mercury cathode and 0.1M tetramethyl ammonium perchlorate at pH = 2.4. The slope of the logarithmically transformed line indicates the number of electrons exchanged in the electrolysis reaction (24).

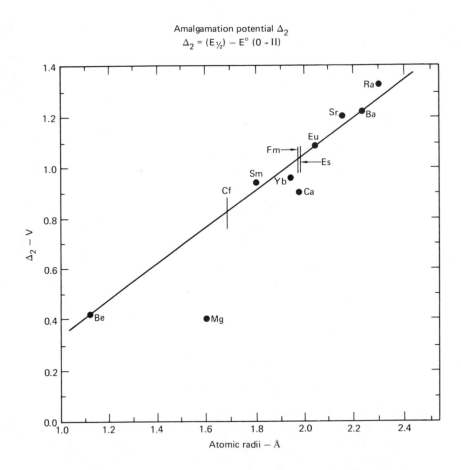

Amalgamation potential Δ_2

$$\Delta_2 = (E_{1/2}) - E^\circ (0 \to II)$$

Figure 2. Amalgamation potentials, Δ_2, derived from experimental data are plotted as a function of the atomic (metallic) radii. The amalgamation potential for fermium is obtained by using an estimated radius (19).

of 77 min, this nuclide can be made with millibarn cross sections
by a number of nuclear reactions between light and heavy ions
with actinide target nuclei. We have found that the bombardment
of fractions of a microgram of ^{254}Es with intense alpha-particle
beams will produce ~10^6 atoms of ^{256}Md in one to two hours of
irradiation time. The ^{256}Md is most easily detected through spontaneous fission arising from the ingrowth of its electron-capture
daughter ^{256}Fm. A difficulty with using spontaneous-fission
counting to determine the Md content of samples is that the
growth and decay of fission radioactivity in each sample must be
followed with time in order to resolve the amounts of Md and Fm
initially present. However, alpha-particles of a distinctive
energy coming from a 10% alpha-decay branch can also be used to
identify ^{256}Md in a mixture of actinide tracers.

Mendelevium metal was found to be more volatile than other
actinide metals as described in the section on fermium (7).
There are no experimental verifications of the electronic structure of Md, but this has been calculated by several methods to be
$5f^{13}7s^2$ in which the ground state level is $^2F_{7/2}$ (5).

The separation of Md from the other actinides can be accomplished either by reduction of Md^{3+} to the divalent state (20)
or by chromatographic separations with Md remaining in the tripositive state. Historically, Md^{3+} has been separated in columns
of cation-exchange resin by elution with α-hydroxyisobutyric acid
solutions (9). This method is still widely used even though
extraction chromatography requires less effort and attention to
technique. Horwitz and coworkers (10) developed a highly-effici-
ent and rapid separation of Md^{3+} by employing HNO_3 elutions of
columns of silica powder saturated with an organic extractant,
bis(2-ethylhexyl)phosphoric acid. The separation of Md from Es
and Fm could be completed in under 20 minutes and had the advantage of providing final solutions of Md free of complexing agents
that might be an interference in subsequent experiments.

When the divalent state of Md was first discovered, extraction chromatography was used to prove that the behavior of Md^{2+}
was dissimilar to that of Es^{3+} and Fm^{3+} (20). The extractant,
bis(2-ethylhexyl)phosphoric acid (HDEHP), has a much lower affinity for divalent ions than it does for the tri- and tetravalent
ones. Thus, the extraction of Md^{2+} is much poorer than the
extraction of the neighboring tripositive actinides as indicated
by the results shown in Table 2. This became the basis for a
separation method in which tracer Md in 0.1 M HCl is reduced by
fresh Jones' Reductor in the upper half of an extraction column
containing HDEHP absorbed on a fluorocarbon powder in the lower
half. Mendelevium, in the dipositive state, is rapidly eluted
with 0.1 M HCl whereas the other actinides are retained by the
extractant. The separation is quickly performed, but the Md contains small amounts of Zn^{2+} from the Jones' Reductor and also
Eu^{2+}, which was added prior to the elution to prevent reoxidation
of Md^{2+} by the extractant.

Table 2. Comparison of the extraction behavior of tracer einsteinium, fermium, and mendelevium after treatment with various reducing agents. The column-elution method of extraction chromatography was used with the extractant HDEHP adsorbed on a column bed of a fluoroplastic powder (20)

CONDITIONS FOR REDUCTION	STANDARD POTENTIAL OF REDUCING AGENT (volts)	% NON-EXTRACTED BY HDEHP COLUMN	
		Md	Es–Fm
Zn(Hg) AMALGAM, 80° ~20 min, 0.1 $\underline{M}$ HCl; Zn(Hg) AMALGAM IN UPPER HALF OF EXTRAC- TION COLUMN	+0.763	77	<0.10
0.01 $\underline{M}$ Eu^{2+} , 0.1 $\underline{M}$ HCl, ~2-3 min, 80°; Zn(Hg) AMALGAM IN UPPER HALF OF EXTRACTION COLUMN	+0.43	75	<0.10
0.6 $\underline{M}$ Cr^{2+} , 0.1 $\underline{M}$ HCl, ~2 min, 25°C; EXTRAC- TION COLUMN PRE- WASHED WITH 0.6 M Cr^{2+} IN 0.1 M HCl	+0.41	99	0.56

The solution chemistry of the trivalent oxidation state has not been investigated beyond its behavior in the separation procedures described above. All observations indicate that Md^{3+} is a "normal" actinide with an ionic radius slightly less than that of Fm. As might be expected, attempts to oxidize Md^{3+} with sodium bismuthate failed to show any evidence for Md^{4+} (20).

The divalent oxidation state was the first found for any member of the actinide series (20,21) and, therefore, stirred a strong theoretical and experimental effort to establish the reasons for the unexpected stability of this state in Md, and subsequently, in the adjacent actinides. We shall summarize the interpretations for divalency in the heaviest actinides in a later section of this review, but in this section, only the known properties of Md^{2+} will be presented.

In the earliest experiments with Md^{2+}, rough measurements were made of the reduction potential for the half-reaction

$$Md^{3+} + e^{-} = Md^{2+}.$$

The first measurement gave a reduction potential of −0.2 V with respect to the standard hydrogen electrode (20). This value was obtained from determining the equilibrium concentration of each metal ion in the reaction

$$V^{2+} + Md^{3+} \rightleftharpoons V^{3+} + Md^{2+}$$

and then calculating the equilibrium constant. After entering the equilibrium constant into the Nernst equation, it was found that V^{3+} was a better reducing agent than Md^{2+} by about 0.07 V. In other experiments, Mály observed the complete reduction of Md^{3+} with V^{2+} but the reduction was incomplete when Ti^{3+} was used (21). From these observations, he concluded the standard reduction potential of Md^{3+} was close to −0.1 volt. The standard potentials obtained by both groups are in reasonable agreement and, most importantly, they conclusively show that the stability of Md^{2+} is greater than any lanthanide(II) ion. This finding was surprising since divalency in the lanthanides is mainly associated with the special stability given by the half-filled and fully-filled f-electron shell. Divalent Md ions are at least one electron short of the stable $5f^{14}$ configuration.

Additional experiments which may not be clearly relevant to the divalent oxidation state, include the reduction of Md^{3+} to Md(Hg) by sodium amalgams and by electrolysis (22). Both the extraction experiments with Na amalgams and the electrolysis at a Hg cathode indicated a large enrichment of Md in the Hg phase relative to that of Np, Pu, Am, Cm, and Cf. The percentages of Es and Fm in the sodium amalgam were not greatly different from the percentage of Md. But a clear enrichment of Md was obtained in the electrolysis experiments as shown in Figure 3. The initial rate of amalgamation is much larger for Md than for Es and Fm.

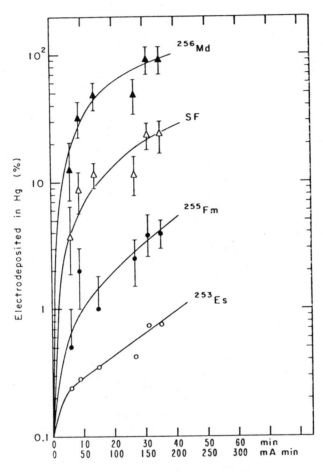

Journal of Inorganic and Nuclear Chemistry

Figure 3.   Percentage of actinide tracers electrodeposited in mercury as a function of time and passed charge. Current density used in this experiment was 5 mA/cm² (22).

Recently, new electrochemical experiments were carried out with Md in which controlled-potential electrolysis was used to study the reduction of Md^{3+} to the metallic state in a Hg amalgam (23,24). Half-wave potentials were measured by radiocoulometry and radiopolarography in the presence of noncomplexing and weak and strong complexing agents. The radiopolarogram obtained for Md in a noncomplexing media is presented in Figure 4. The half-wave potential for Fm was remeasured at the same time as that of Md because of its presence as a decay product of ^{256}Md. The results showed that the reduction potential of Md is about 10 mV more negative than Fm and that no significant difference is observed upon changing the medium from ClO_4^- to Cl^-. In citrate solutions, a shift of 90 mV was obtained for Md which is about the same shift seen with Fm and Ba ions in a citrate medium. The slope of the logarithmically transformed wave was 30 mV for Md and Fm and, for the reasons noted in the section on Fm, this slope corresponds to a two-electron exchange at the electrode. These results demonstrate that the electrochemical behavior of Md is very similar to that of Fm and can be summarized in the equation

$$Md^{2+} + 2e^- \;=\; Md(Hg); \qquad\qquad E^\circ = -1.50 \text{ V.}$$

If a 0.90 V amalgamation potential is assumed, then a standard reduction potential of -2.40 V is obtained.

In addition to the di- and trivalent ions of Md, a stable monovalent ion was reported by Mikheev et al. in 1972 (25). This oxidation state was indicated in the cocrystallization of Md with CsCl and RbCl after the coreduction of Md^{3+} and Sm^{3+} with Mg in an ethanol-7 M HCl solution. Mendelevium was also found enriched in Rb_2PtCl_6 precipitates, a specific carrier for the larger ions of the alkali metals. These results were explained by a stabilization of the monovalent ion due to completing the f shell which would give the $5f^{14}$ electronic configuration.

These experiments were recently repeated and a series of new ones were performed in which attempts were made to prepare Md^+ by reduction with $SmCl_2$ in an ethanolic or fused KCl medium (17). After the reductions, the coprecipitation behavior of Md was compared with the behavior of tracer amounts of Es, Fm, Eu, Sr, Y, and Cs. A large number of experiments showed that Md consistently followed the behavior of Fm^{2+}, Eu^{2+}, and Sr^{2+} rather than the behavior of Cs^+. The most telling experiment was the precipitation of Rb_2PtCl_6 after reduction of Md^{3+} with Sm^{2+}. The distribution of the tracer elements between the precipitate and an ~85% ethanol solution is given in the form of a ratio in Table 3. These results clearly demonstrate that Md did not coprecipitate with Rb_2PtCl_6, whereas virtually all of the Cs did so. The overall conclusion of this work was that Md cannot be reduced to a monovalent ion with Sm^{2+}, and therefore, the earlier claim for Md^+ was unsubstantiated.

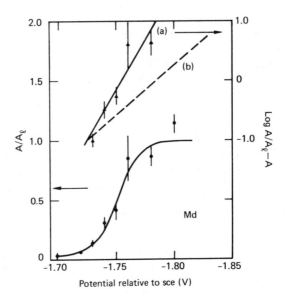

Figure 4.   Distribution of mendelevium as a function of applied voltage between mercury in a dropping mercury cathode and 0.1M tetramethyl ammonium perchlorate at pH = 2.4. The slope of the logarithmically transformed line indicates the number of electrons exchanged in the electrolysis reaction. The slope of line a is 30 mV and b is 60 mV, which corresponds to a one-electron reduction (24).

Table 3. Distribution of tracer elements after reduction with Sm^{2+} and coprecipitation with Rb_2PtCl_2 in ~85% ethanol. (Ref. 17).

		Distribution ratio for				
Fm	Md	Eu	Sr	Y	Es	Cs
0.004	0.005	0.006	0.012	0.017	0.033	110

This same conclusion was reached also by Samhoun et al. (23) and David and coworkers (24) on the basis of their electrochemical investigations of Md, which we described earlier. If the potential for the reaction $Md^+ + e^- \rightarrow Md$ was more positive than -1.5 V, it would have been observed in the electrochemical reductions. Furthermore, the logarithmic slope of the Md reduction waves could not be fitted to a slope of 60 mV expected for a one-electron change. And lastly, the shifts in potential caused by complexing Md with either citrate or chloride ions were consistent with it being a divalent ion and not with it being either a cesium-like or silver-like ion.

The attempts to produce a monovalent state have the positive effect of setting limits on its stability. From the limits obtained, we can then make an estimate of the stability of the $5f^{14}$ configuration relative to the $5f^{13}7s$. Presumably, the $f^{13}s$ configuration lies lower in energy than the f^{14} because there is no obvious stabilization of a monovalent state due to a possible closing of the 5f shell. The divalent ion is at least 1.3 V more stable than the monovalent.

Nobelium

The principle isotope of nobelium produced for investigations of its chemical properties is 3.1-min ^{255}No. In the earliest studies (26), this nuclide was synthesized by irradiating ^{244}Pu with 97-MeV ^{16}O ions, but larger yields were later obtained in bombardments of ^{249}Cf targets with 73-MeV ^{12}C ions (27). From the latter nuclear reaction, about 1200 atoms of ^{255}No were collected every ten minutes. Of these 1200 atoms, only 3 to 20% were detected after the chemical experiments because of losses by radioactive decay, losses in the experiments, and a 30% geometry for counting alpha particles emitted in the decay of this isotope. To obtain results that were statistically significant, the experiments were repeated until the required accuracy was attained.

Future work with No may require techniques or procedures of greater complexity than the one-step chemical methods used in past studies. The author believes that ^{259}No, because of its 1-h half life, would permit these more extensive investigations of No chemistry. Approximately 700 atoms can be made in a two-hour irradiation of ^{248}Cm with 96-MeV ^{18}O ions. In combination with the long half life, this number of atoms is sufficient to permit a broader range of experiments to be performed.

A central feature in the chemistry of No is the dominance of the divalent oxidation state (26). In this respect, No is unique within the lanthanide and actinide series, since none of the other twenty-seven members possess a highly-stable divalent ion. The electronic configuration of the neutral atom obtained from relativistic Hartree-Fock calculations is $5\underline{f}^{14}7\underline{s}^2$ (5). Clearly, the special stability of No^{2+} must arise from the difficulty in ionizing an $\underline{f}$ valence electron from the completed 5f shell. Thus, pairing of the last electron, to close the shell, results in the $\underline{f}$ electron levels taking a rather abrupt drop in energy below the Fermi surface.

The separation of No from other actinide elements is based entirely on the dissimilar behavior of No^{2+} in comparison with the tripositive actinide ions. Without the addition of strong oxidants, No will be present as No^{2+} in acidic solutions and will have the general chemical properties of Group IIA elements in the Periodic Table. We have found that the extraction chromato-graphic method described in the section on Md provides an effec-tive separation from all other actinides and lanthanides. In contrast to Md, reducing agents are unnecessary in separating No by this extraction chemistry.

The solution chemistry of No was explored shortly after the discovery of divalent Md (26). Subsequent studies include an estimation of the III → II reduction potential (28), aqueous complexing with carboxylate ions (29), and a determination of No^{2+} extraction and ion-exchange behavior in comparison with the alkaline earths (27). The first studies (26) indicated that the normal state of No in aqueous solution was that of a divalent ion. Nobelium was coprecipitated with BaSO$_4$ but not with LaF$_3$. After oxidation with Ce^{4+}, a large fraction of the No coprecipitated with LaF$_3$. This behavior is consistent with a change in oxida-tion state from (II) to (III). An elution position of No rela-tive to tracer quantities of Es, Y, Sr, Ba, and Ra (Figure 5) showed that No did not elute before Es as would be expected of a tripositive actinide ion.

The standard reduction potential of the No^{3+}/No^{2+} couple in aqueous solution was estimated by Silva and coworkers (28) from the extractibility of No after treatment with a variety of oxid-ants. The distinction between No^{2+} and No^{3+} was made on the basis of the affinity of the extractant, bis(2-ethylhexyl) phosphoric acid, for highly-charged cations. In 0.1 $\underline{M}$ acid, mono- and dipositive ions are poorly extracted, whereas the tri- and tetrapositive ions are strontly absorbed in the extractant. In comparison with the behavior of the tracer ions of Ra, Tl, Ce, Cm, and Cf it was shown that No was not fully extracted until

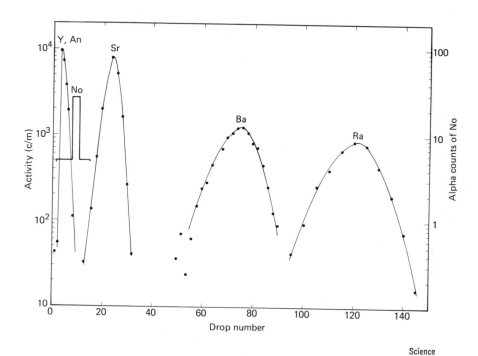

Figure 5.   Elution of nobelium from a heated (80°C) Dowex 50 × 12 cation-exchange column with 1.9M ammonium α-hydroxyisobutyrate (pH 4.8) (26)

H_5IO_6 (standard potential = 1.6 V) was used as an oxidant. Chromate and $HBrO_3$ partially oxidized No^{2+} to No^{3+}. From these observations, a potential of 1.4 to 1.5 V was estimated for the couple.

The extraction behavior of No^{2+} in a tri-n-octylamine–HCl system was compared with that of divalent Hg, Cd, Cu, Co, and Ba (27). This experiment provided a test of the chloride complex strength of No^{2+} because the amine anion-exchanger will only extract anionic species. It was found that Ba^{2+} and No^{2+} were not extractable over a range of 0.2 to 10 M HCl, while the other divalent ions of Hg, Cd, Cu, and Co were strongly extracted. This implies noncomplexing of No in the chloride medium which is a characteristic of the alkaline earths.

The elution position of No^{2+} from a cation-exchange resin with 4 M HCl eluant was compared with the elution positions of Be^{2+}, Mg^{2+}, Ca^{2+}, Sr^{2+}, Ba^{2+}, and Ra^{2+} (27). The No^{2+} ions eluted at exactly the Ca^{2+} position. Similar column elution experiments using bis(2-ethylhexyl)phosphoric acid adsorbed on an inert support material and 0.025 M HCl as the eluting acid, showed No^{2+} eluting between Ca^{2+} and Sr^{2+}. These elution curves are illustrated in Figure 6. With the same extractant, the distribution coefficients for No^{2+} were measured as a function of hydrogen-ion concentration. From the mass-action expression for the ion exchange, a slope of +2 was obtained from the line describing the log of the distribution coefficients vs. pH. The extraction of No^{2+} is second power with respect to the H^+ concentration, thus indicating a charge state of two for No because of the cation-exchange mechanism for extraction in this system.

Silva and coworkers (27) noted that other investigators had shown a linear correlation between the log of the distribution coefficients of the alkaline earths and their ionic radii. This appeared to be the case wherever a pure cation-exchange mechanism governed the distributions between phases, and hence, was applicable to distributions obtained with either cation-exchange resins or extractants. In Figure 7, log D is plotted as a function of ionic radius for the extraction of various dipositive cations into 0.1 M HDEHP from aqueous solutions. The measured distribution coefficient for No^{2+}, when placed on the correlation line, gave an ionic radius of 1.1 Å. If the distribution coefficients from their ion-exchange elutions were used, the ionic radius of No^{2+} would be the same as that of Ca^{2+} (1.0 Å), since both ions have the same elution position. An ionic radius of 1.1 Å was also obtained by applying Pauling's correction to the radius of the outermost, $6p_{3/2}$ shell, which was calculated from a relativistic radial wave function (Hartree–Fock–Slater). The calculated ionic radius is in agreement with the radii derived from their solvent-extraction and ion-exchange results. A radius of 1.1 Å for No^{2+} can be compared with 1.03 Å found for Yb^{2+} (30), the lanthanide homolog of No. Insertion of the No^{2+} ionic radius into an empirical form of the Born equation gave a single-ion heat of hydration of -355 kcal(g-atom)$^{-1}$.

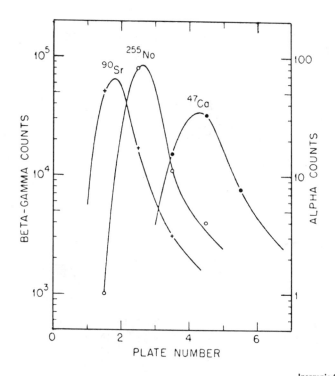

Inorganic Chemistry

Figure 6.   Elution of No²⁺, Ca²⁺, and Sr²⁺ with 0.025M HCl from a column of HDEHP on an inert support (27)

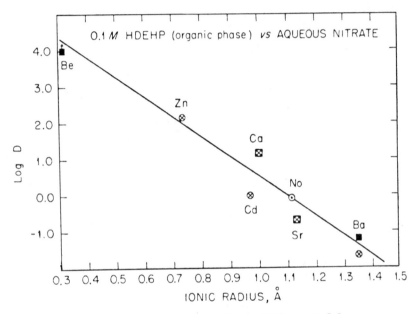

Inorganic Chemistry

Figure 7. *Log* D *vs. ionic radius for typical divalent cations in the HDEHP–aqueous nitrate system* (27)

The ability of No^{2+} to form complexes with citrate, oxyalate, and acetate ions in an aqueous solution of 0.5 M NH_4NO_3 was investigated by McDowell and coworkers (29). The complex strengths of Ca^{2+} and Sr^{2+} with these carboxylate ions were measured under the same conditions for comparison with the No results. The formation constants they obtained are given in Table 4 and indicate for each anion, the complexing tendency of No^{2+} is between that of Ca^{2+} and Sr^{2+} with Sr^{2+} being slightly more favored.

Table 4.   Complex formation constants for No^{2+}, Ca^{2+}, and Sr^{2+} from distribution data. (Reprinted with the permission of O. L. Keller, Jr. and Pergamon Press (Ref. 29)).

| System | | Formation Constants | |
Cation	Ligand	β_1	σ^a
No^{2+}	Cit	151.9	18.5
	Ox	48	5.6
	Ac	− 5	5
Ca^{2+}	Cit	333	11.2
	Ox	88.9	2.1
	Ac	5.5	0.7
Sr^{2+}	Cit	96.7	1.7
	Ox	25.3	0.5
	Ac	0.58	0.12

[a] Standard deviation of fitting of β_1.

The standard potential for the reduction of No^{2+} to $No(Hg)$ was measured by a modified radiopolarographic technique (31). Usually, the half-wave potential is determined by measuring the distribution of an element between the mercury and aqueous phases as a function of applied voltage. The half-life of ^{255}No is too short to allow time for the recovery of No from the Hg phase for assay, therefore Meyer et al. measured the depletion of No in the aqueous phase as a function of a controlled potential. They assumed that equilibrium was reached in 3 min of electrolysis and that the electrode reaction was reversible. A sharp drop in No concentration in the aqueous phase occurred between −1.8 and −1.9 V vs. the saturated calomel electrode or −1.6 V vs. the standard hydrogen electrode. Thus, their best estimates are summarized in the following equation.

$$No^{2+} + 2e^- = No(Hg); \qquad E^0 = -1.6 \pm 0.1 \text{ V}$$

If this potential is reduced by about the 1 V estimated for the amalgamation potential, then a value of about −2.6 V would be given for the II → 0 couple.

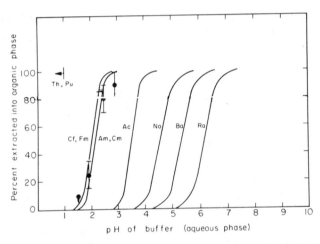

Inorganic and Nuclear Chemistry Letters

Figure 8. Percent extracted into the organic phase as a function of the pH of the aqueous phase: a summary of earlier data by the same authors (1).

Lawrencium

Element 103, lawrencium, is the last member of the actinide series and its chemical nature should be similar to its counterpart in the lanthanide series, Lu. However, confirming experimental information is nearly nonexistent because of the 35-s half life of ^{256}Lr and the great difficulty in producing a useful quantity for experiments. The bombardment of ^{249}Cf with ^{11}B ions is probably the most favorable nuclear reaction for producing ^{256}Lr. Even so, only about ten atoms have been made in each short irradiation and of these, only one or two were detected after completion of the chemical tests (1).

Lawrencium is expected to have a $5f^{14}6d7s^2$ electronic configuration (32) although Brewer computed a $5f^{14}7s^27p$ configuration (5). The energy difference between the two possible ground states is no more than a few thousand wave numbers, which is slightly greater than the errors in the extrapolations. There appears to be no method for resolving this question by direct experiment.

The ionization of Lr would be expected to stop with the f^{14} core intact because of the enhanced binding energy of possible valence electrons in the filled f shell. The stable valence state of Lr would then be the (III) state. Experiments to confirm this oxidation state of Lr were undertaken by Silva and coworkers (1). They compared the extraction behavior of Lr with several tri- and tetravalent actinides and with Ba^{2+}, Ra^{2+}, and No^{2+}. A chelating extractant, thenoyltrifluoroacetone dissolved in methyl isobutyl ketone, was employed to extract the tracer ions from aqueous solutions that had been buffered with acetate anions. Their results, shown in Figure 8, very clearly demonstrate that Lr is extracted within the same pH range as the trivalent actinides, and therefore, proves that Lr is trivalent.

Further studies of Lr have not been attempted.

Acknowledgment

Work performed under the auspices of the U. S. Department of Energy by the Lawrence Livermore Laboratory under contract number W-7405-ENG-48.

LITERATURE CITED

1. R. Silva, T. Sikkeland, M. Nurmia, and A. Ghiorso, Inorg. Nucl. Chem. Lett. 6, 733 (1970).

2. A. Ghiorso, B. G. Harvey, G. R. Choppin, S. G. Thompson, and G. T. Seaborg, Phys. Rev. 98, 1518 (1955).

3. R. J. Borg and E. K. Hulet, Lawrence Livermore Laboratory, private communication (1979).

4. L. S. Goodman, H. Diamond, H. E. Stanton, and M. S. Fred, Phys. Rev. A 4, 473 (1971).

5. L. Brewer, J. Opt. Soc. Amer. 61, 1101 (1971).

6. F. T. Porter and M. S. Freedman, Phys. Rev. Lett. 27, 293 (1971).

7. I. Zvara, V. Z. Belov, V. P. Domanov, B. L. Zhookov, T. Ryotts, Z. Chubener, M. R. Shalaevski, Joint Institute for Nuclear Research, Dubna, USSR, Preprint P6-10334 (in Russian) 1976.

8. E. P. Horwitz and C.A.A. Bloomquist, J. Inorg. Nucl. Chem. 35, 271 (1973).

9. G. R. Choppin, B. G. Harvey, and S. G. Thompson, J. Inorg. Nucl. Chem. 2, 66 (1955).

10. E. P. Horwitz and C.A.A. Bloomquist, Inorg. Nucl. Chem. Lett. 5, 753 (1969).

11. S. Hubert, M. Hussonnois, L. Brillard, G. Gody, R. Guillaumont, J. Inorg. Nucl. Chem. 36, 2361 (1974).

12. S. Hubert, M. Hussonnois, L. Brillard, R. Guillaumont, in Transplutonium Elements, W. Müller and R. Lindner, Eds., (North-Holland, Amsterdam, 1976) p. 109.

13. M. Hussonnois, S. Hubert, L. Aubin, R. Guillaumont, G. Boussiéres, Radchem. Radioanal. Lett. 10, 231 (1972).

14. N. B. Mikheev, V. I. Spitsyn, A. N. Kamenskaya, B. A. Gvozdev, V. A. Druin, I. A. Rumer, R. A. Dyachkova, N. A. Rozenkevitch, L. N. Auerman, Inorg. Nucl. Chem. Lett. 8, 929 (1972).

15. N. A. Rozenkevich, N. B. Mikheev, I. A. Rumer, L. N. Auerman, B. A. Gvozdev, A. N. Kamenskaya, Radiokhimiya 17, 441 (1975).

16. N. B. Mikheev, V. I. Spitsyn, A. N. Kamenskaya, N. A. Konovalova, I. A. Rumer, L. N. Auerman, A. M. Podorozhnyi, Inorg. Nucl. Chem. Lett. 13, 651 (1977).

17. E. K. Hulet, R. W. Lougheed, P. A. Baisden, J. H. Landrum, J. F. Wild, and R.F.D. Lundqvist, J. Inorg. Nucl. Chem. (in press, 1979).

18. K. Samhoun and F. David, in Transplutonium Elements,
    W. Müller and R. Lindner, Eds., (North-Holland, Amsterdam,
    1976) p. 297.

19. L. J. Nugent, J. Inorg. Nucl. Chem. 37, 1967 (1975).

20. E. K. Hulet, R. W. Lougheed, J. D. Brady, R. E. Stone, and
    M. S. Coops, Science 158, 486 (1967).

21. J. Mály and B. B. Cunningham, J. Inorg. Nucl. Chem. Lett. 3,
    445 (1967).

22. J. Mály, J. Inorg. Nucl. Chem. 31, 741 (1969).

23. K. Samhoun, F. David, R. L. Hahn, G. D. O'Kelley, J. R.
    Tarrant, and D. E. Hobart, J. Inorg. Nucl. Chem. (in press,
    1979).

24. F. David, K. Samhoun, E. K. Hulet, P. A. Baisden, R. Dougan,
    J. H. Landrum, R. W. Lougheed, J. F. Wild, and G. D.
    O'Kelley, to be published.

25. N. B. Mikheev, V. I. Spitsyn, A. N. Kamenskaya, I. A. Rumer,
    B. A. Gvozdev, N. A. Rosenkevich, and L. N. Auerman, Dokl.
    Akad. Nauk SSSR 208, 1146 (1973).

26. J. Mály, T. Sikkeland, R. Silva, and A. Ghiorso, Science 160,
    1114 (1968).

27. R. J. Silva, W. J. McDowell, O. L. Keller, Jr., and J. R.
    Tarrant, Inorg. Chem. 13, 2233 (1974).

28. R. J. Silva, T. Sikkeland, M. Nurmia, A. Ghiorso, and E. K.
    Hulet, J. Inorg. Nucl. Chem. 31, 3405 (1969).

29. W. J. McDowell, O. L. Keller, Jr., P. E. Dittner, J. R.
    Tarrant, and G. N. Case, J. Inorg. Nucl. Chem. 38, 1207
    (1976).

30. H. P. Beck and H. Bärnighausen, Z. Anorg. Allg. Chem. 386,
    221 (1971).

31. R. E. Meyer, W. J. McDowell, P. F. Dittner, R. J. Silva,
    J. R. Tarrant, J. Inorg. Nucl. Chem. 38, 1171 (1976).

32. L. J. Nugent, K. L. Vander Sluis, B. Fricke, J. B. Mann,
    Phys. Rev. A 9, 6, 2270 (1974).

RECEIVED March 4, 1980.

ELECTRONIC STRUCTURE AND SPECTROSCOPY

Hypersensitive Transitions in f-Electron Systems

B. R. JUDD

Physics Department, The Johns Hopkins University, Baltimore, MD 21218

The sensitivity of the optical absorption spectra of rare-earth ions to their environment was discussed as long ago as 1930 by Selwood (1). It was not until the spectral lines could be interpreted as transitions between well-defined states of the rare-earth ions that the key feature of the phenomenon could be established: the intensities of a few lines are exceptionally sensitive, and for these so-called hypersensitive lines the selection rules on J, the total angular momentum of the f electrons of the rare-earth ion, are the same as those for quadrupole radiation. That is, the change in J can be at most 2. The effect was first noticed when the absorption spectra of aqueous solutions of rare-earth nitrates and chlorides, as found by Hoogschagen (2), were compared. For example, the only transition of Er^{3+} exhibiting an intensity difference is $^4I_{15/2} \rightarrow \,^2H_{11/2}$. Again, the largest intensity change for solutions of Nd^{3+} occurs for $^4I_{9/2} \rightarrow \,^4G_{5/2}$ (3). A couple of years later, several more transitions for rare-earth ions were noted (4).

Hypersensitivity is not confined to rare-earth ions. Pappalardo, Carnall, and Fields (5) have identified two hypersensitive transitions in AmI_3 that correspond to a change of two units in J. It is evident that the relevant features of the 4f shell carry over to the 5f shell, and so it is convenient to consider both the rare-earth and the actinide ions at the same time.

1. Ions at Sites of Non-Vanishing Electric Field

The natural reaction to the experimental results is to conclude that the hypersensitive transitions are electric quadrupole. Such transitions involve no change in parity and are thus allowed within configurations of the type f^N. The sensitivity to environment could conceivably depend on the variability (due to covalency) of

0–8412–0568–X/80/47–131–267$05.00/0

the average value of r^2 for an f electron, which plays a key role in the quadrupolar intensities. Unfortunately, an actual calculation shows that this mechanism falls short of experiment by roughly five orders of magnitude [4].

It is now well established that almost all transitions within the f shell are electric dipole in nature. The breakdown of the Laporte parity rule is brought about by non-centro-symmetric terms in the crystal-field Hamiltonian V, which have the effect of mixing d and g states into the f shell. Transitions which are nominally f to f take place because of the permitted transitions $f \to d$ and $f \to g$. An early attempt [6] to explain the hypersensitivity used the fact that for rare-earth or actinide ions at site symmetries of the types

$$C_s, C_1, C_2, C_{2v}, C_3, C_{3v}, C_4, C_{4v}, C_6, C_{6v}$$

there are terms in V that correspond to the spherical harmonics Y_{1m}. We can now use Y_{1m} to connect an initial state of f^N to a virtual state of $f^{N-1}d$, $f^{N+1}d^{-1}$, or $f^{N-1}g$; and the electric-dipole operator, itself a harmonic of rank 1, completes the linkage to the final state of f^N. Since the crystal splittings of the terms of the excited configurations are small compared to the energies of excitation, we can contract the two harmonics Y_1, with the result that we get an operator whose transformation properties (when a scalar is subtracted out) are identical to those of Y_2. Such an operator would yield selection rules on J identical to those for quadrupole radiation. Any structural change that produces any of the site symmetries listed above would provide a source for an enhanced transitions probability.

Plausible though this mechanism is, it came under criticism [7] because, _inter alia_, it could not account for the intense hypersensitive transitions of the gaseous rare-earth trihalides [8]. However, there is recent evidence that the halides are not planar [9, 10], as had been previously supposed. If this is in fact the case, the importance of the mechanism based on Y_1 terms in V remains undecided.

It should be pointed out that the derivatives of the Y_1 terms do not all vanish at the origin. There appears to be an electric field acting on the nucleus of the rare-earth or actinide ion – an impossibility if the ion is in equilibrium. The resolution of this paradox lies in the admixtures of d and g states into the f shell. The redistribution of electronic charge of the rare-earth or actinide ion produces a second electric field at the nucleus that exactly cancels the first.

2. Inhomogeneous Dielectric and Dynamic Coupling

An early survey of the possible sources for the hypersensitivity concluded that the most likely candidate was a mechanism based on the inhomogeneities of the dielectric surrounding the rare-earth or actinide ion (4). It runs as follows. The radiation field induces sinusoidally fluctuating dipole moments in the ligands surrounding the ion. These induced dipoles necessarily radiate, and the emitted fields impinge on the rare-earth or actinide ion. Because of the proximity of source and receiver, the plane-wave condition no longer applies; the wave fronts are sufficiently distorted to produce substantial quadrupole components. It should be stressed, however, that the radiation field seen by the central ion is a superposition of the fields produced by all the ligands, and will, in general, be quite different from the quadrupole field produced by a single, distant, source. Although the quadrupole selection rules hold for J, we cannot make comparable statements for the corresponding magnetic quantum numbers M until an analysis of the crystal structure is carried out.

The original estimate of the importance of the above mechanism falls short of experiment by a factor of 30. An apparently more successful mechanism has been introduced by Mason, Peacock, and Stewart (11, 7). In their model, the f electrons of the rare-earth or actinide ion produce a field that polarizes the ligands. If the complex of central ion plus ligands does not possess a center of inversion, the complex exhibits an electric dipole moment that can interact directly with the electric vector of the radiation field. This mechanism has been referred to as dynamic coupling (7, 11). Although it seems to be different from the mechanism based on an inhomogeneous dielectric, the two are, in fact, identical (12). They are simply different verbalizations of the same mathematics. The good agreement that Mason, Peacock, and Stewart find with experiment is simply a reflection of better polarization data for the ligands and more accurate structural information, both of which substantially reduce the discrepant factor of 30 mentioned above.

However, neither the inhomogeneous dielectric mechanism nor its equivalent, the dynamic-coupling mechanism, makes allowance for the polarizability of the outer shells of the rare-earth or actinide ion. For an external quadrupole field to penetrate to the f electrons, we must include a screening factor $(1 - \sigma_2)$; the same factor must be introduced if we take the point of view of dynamic coupling and ask what reduction the quadrupole field of the f electrons experiences as it penetrates out to the ligands. For a free rare-earth ion, σ_2 can be as large as 0.7 (13, 14). When the screening factor is squared to obtain the reduction in intensity, it is found that the hypersensitive transition intensities are reduced by an order of magnitude.

Although this has the effect of putting in doubt all the good fits between theory and experiment found by Peacock and his collaborators, it removes the somewhat embarrassing result that hypersensitive lines in the rare-earth trichlorides, which the theory would otherwise predict to be intense, are experimentally unexceptional.

It could well turn out that for ions in contact with ligands, σ_2 should be substantially reduced. However that may be, it seems clear that we can no longer be certain that the inhomogeneous dielectric mechanism (or its equivalent, the dynamic-coupling mechanism) plays a dominant role.

3. Vibronic Contributions

A common feature of rare-earth and actinide spectra is the appearance of vibronic companions to the lines that represent purely electronic transitions. In solutions, vapors, and melts, these two types of lines may become blended, and separate intensity measurements become impossible. For ions at sites possessing a center of inversion, all the intensity resides in the vibronic lines. It is clear that any analysis of hypersensitivity should include the vibronic components.

Octahedral complexes have been studied by Faulkner and Richardson (15, 16). The vibronic analogs of the mechanisms of Secs. 2 and 3 both give significant contributions to the intensities. That is, an odd vibrational mode, when excited, leads to electric fields at the nucleus of the central ion that permits the mechanism of Sec. 2 to function: similarly, such a mode allows the six ligands to take up instantaneous positions in which the quadrupole radiation that they produce (as described in Sec. 3) can coherently combine rather than exactly cancel, as it would for a regular octahedron. These two mechanisms have been re-examined with the aid of more elaborate tensorial techniques (17), but the general conclusions of Faulkner and Richardson remain unaffected. Both mechanisms can be expected to play a role, their relative importance depending upon the specific system under investigation.

One feature of the analysis deserves mention. In the limit of large ion-to-ligand distances compared to ligand displacements from equilibrium and to f-electron radii, all the vibronic intensities should reside in just those transitions associated with hypersensitivity. A good example of this is provided by $Nd_2Mg_3(NO_3)_{12} \cdot 24H_2O$. The fact that twelve oxygen ions are packed around the neodymium ion makes the ion-to-ligand separation unusually large. A glance at the absorption spectrum (18) reveals that most sharp electronic lines do not appear to have noticeable vibronic components: the only case where well-defined vibronic lines appear is in group D_4 which corresponds to precisely the hypersensitive transition $^4I_{9/2} \to {}^4G_{5/2}$ at 5700Å.

4. Crystal Structures

Work is being carried out in collaboration with Dr. W.T. Carnall in an effort to connect hypersensitivity to the structures of the crystal lattices where it is exhibited. The mechanism based on an inhomogeneous dielectric leads to an expression for the intensities of hypersensitive lines that is proportional to (12)

$$Q_3 = \sum_{L,L'} a_L \, a_{L'} \, R_L^{-4} \, R_{L'}^{-4} \, P_3(\cos w_{LL'}),$$

where a_L and $a_{L'}$ are the polarizabilities of ligands L and L', situated at distances R_L and $R_{L'}$ from the nucleus of the rare-earth or actinide ion and subtending an angle $w_{LL'}$ to it. The function $P_3(z)$ is the Legendre polynomial $\frac{1}{2}(5z^3 - 3z)$. For identical ligands at equal distances from the central ion, it is convenient to define

$$Z_3 = a_L^{-2} \, R_L^8 \, Q_3 = \sum_{L,L'} P_3(\cos w_{LL'}).$$

Some correlation can be established between Z_3 and hypersensitivity. Thus, for a nearly planar molecule NdI_3, we find Z_3 is roughly 6, while for the crystal $NdCl_3$, it is only 0.9. Intense hypersensitive transitions occur in the molecule but not in the crystal. However, the pyramidal NdI_3 produces an electric field at the Nd^{3+} site, while the crystal does not; so the intensity correlation does not establish the predominance of the inhomogeneous-dielectric mechanism. The issue is further clouded by the vibronic contributions that undoubtedly occur in the molecule. Comparisons between different crystals are even more ambiguous. Thus Er^{3+} in YCl_3 and Am^{3+} in AmI_3 exhibit hypersensitivity, while $PuBr_3$ does not (19). The corresponding Z_3 values (for the relevant lattice structures YCl_3, BiI_3, and $TbCl_3$) are roughly 0.5, 0.0, and 0.4. All three systems permit electric fields at the rare-earth or actinide site.

It is important to recognize that it is experimentally much easier to compare intensities than make absolute measurements. It could easily happen that a spectrum dominated by a single hypersensitive line might have a total intensity less than a spectrum in which no hypersensitive lines stand out. The octahedral systems studied in acetonitrile solutions by Jørgensen and Ryan seem to fall into this category [20, 21]. The case of Er^{3+} in YCl_3 mentioned above may do so too.

5. Distinguishing Mechanisms

Setting aside vibronic transitions, we have the two competing mechanisms described in Secs. 2 and 3. Although they provide additive contributions to the intensities of transitions between level J and level J', interference effects take place if transitions between different M components are considered. In principle, it is possible to use this fact to separate the two mechanisms. A detailed analysis would take us too far afield: but a feeling for what happens can be gained by noting that the effective dipole operator for the electric field mechanism depends on the coupled tensor $(C_L^{(1)} U^{(2)})^{(1)}$, while that for the inhomogeneous dielectric mechanism depends on $(C_L^{(3)} U^{(2)})^{(1)}$ [12]. The tensor $U^{(2)}$ acts on the f electrons and provides the quadrupole character of the hypersensitive lines; the tensor $C_L^{(1)}$ and $C_L^{(3)}$ are proportional to spherical harmonics of ranks 1 and 3 in which the polar angles leading out from the nucleus of the rare-earth or actinide ion to ligand L appear.

As an example, consider the transitions $^7F_0 \rightarrow {}^5D_2$ for three ligands providing a site symmetry at a Eu^{3+} ion of the type C_{3v}. For various M components of the final state we obtain the following intensities:

$$M = 0: \quad ((9/5)^{\frac{1}{2}} C_0^{(3)} - p(2/5)^{\frac{1}{2}} C_0^{(1)})^2 \qquad (\pi \text{ polarization})$$

$$M = \pm 1: \quad ((3/5)^{\frac{1}{2}} C_0^{(3)} + p(3/10)^{\frac{1}{2}} C_0^{(1)})^2 \qquad (\sigma \text{ polarization})$$

$$M = \pm 2: \quad -C_3^{(3)} C_{-3}^{(3)} \qquad (\sigma \text{ polarization})$$

In these expressions, p is an arbitrary coefficient that measures the strength of the electric-field mechanism compared to one based on an inhomogeneous dielectric. The values of $C_q^{(k)}$ can be worked out once the polar angles of any one of the three ligands is specified. For example, $C_0^{(1)} = \cos \theta$. The components for which $M = \pm 1$ are mixed with those for which $M = \mp 2$ in C_{3v} symmetry, but if the relative mixtures are known (as they could be from Zeeman-effect data), a measurement of the three intensities would, in principle, determine p and also check the expressions above for consistency.

It is interesting to note that the sum of the five intensities (for $M = -2, -1, \ldots, 2$) above does not involve a term linear in p. For total intensities, the two mechanisms combine without interference.

6. Conclusion

After almost twenty years of work on the hypersensitive transitions, their origins are far from clear. In addition to the mechanisms discussed here, others have been introduced from time to time. The effect of charge transfer may not be negligible (22), though its importance has been discounted by Peacock (23). If one were to hazard a guess, it would be that different mechanisms play roles of varying importance from system to system; but until more experimental and theoretical work is carried out the question must remain open.

Acknowledgements

Dr. W.T. Carnall is thanked for many stimulating conversations. The work was supported in part by a grant from the National Science Foundation.

Literature Cited

1. Selwood, P.W. J. Amer. Chem. Soc., 1930, 52, 4308.
2. Hoogschagen, J. "De Absorptiespectra van de Zeldzame Aarden"; Noord-Hollandsche Uitgevers Maatschappij, Amsterdam, 1947.
3. Judd, B.R. Phys. Rev., 1962, 127, 750.
4. Jørgensen, Chr. K.; Judd, B.R. Mol. Phys., 1964, 8, 281.
5. Pappalardo, R.G.; Carnall, W.T.; Fields, P.R. J. Chem. Phys., 1969, 51, 1182.
6. Judd, B.R. J. Chem. Phys., 1966, 44, 839.

7. Mason, S.F.; Peacock, R.D.; Stewart, B. Mol. Phys., 1975, 30, 1829.

8. Gruen, D.M.; DeKock, C.W.; McBeth, R.L. Adv. Chem., 1967, 71, 102.

9. Giricheva, N.I.; Zasorin, E.Z.; Girichev, G.V.; Krasnov, K.S.; Spiridonov, V.P. Zh. Strukt. Khim., 1967, 17, 797.

10. Charkin, O.P.; Dyatkina, M.E. Zh. Strukt. Khim., 1964, 5, 921.

11. Mason, S.F.; Peacock, R.D.; Stewart, B. Chem. Phys. Lett., 1974, 29, 149.

12. Judd, B.R. J. Chem. Phys., 1979, 70, 4830.

13. Newman, D.J.; Price, D.C. J. Phys. C: Solid St. Phys., 1975, 8, 2985.

14. Ahmad, S.; Newman, D.J. J. Phys. C: Solid St. Phys., 1978, 11, L277.

15. Faulkner, T.R.; Richardson, F.S. Mol. Phys., 1978, 35, 1141.

16. Faulkner, T.R.; Richardson, F.S. Mol. Phys., 1978, 36, 193.

17. Judd, B.R. Physica Scripta, 1979, 20,

18. Dieke, G.H.; Eds. Crosswhite, H.M.; Crosswhite, H. "Spectra and Energy Levels of Rare Earth Ions in Crystals" Interscience (John Wiley), New York, 1968.

19. Carnall, W.T. private communication.

20. Ryan, J.L.; Jørgensen, C.K. J. Phys. Chem., 1966, 70, 2845.

21. Ryan, J.L. Inorg. Chem., 1969, 8, 2053.

22. Henrie, D.E.; Fellows, R.L.; Choppin, G.R. Coord. Chem. Rev., 1976, 18, 199.

23. Peacock, R.D. Mol. Phys., 1977, 33, 1239.

RECEIVED December 26, 1979.

Lanthanide and Actinide Lasers

MARVIN J. WEBER

Lawrence Livermore Laboratory, University of California, Livermore, CA 94550

Lanthanides and actinides were among the very first elements used to demonstrate laser action. Although the first laser used an iron group element, Cr^{3+} in Al_2O_3 (ruby), laser action from an actinide ion (U^{3+}) was also reported ([1]) in the same year, 1960. In the following year stimulated emission from both divalent (Sm^{2+}) ([2]) and trivalent (Nd^{3+}) ([3]) lanthanide ions was observed. The following two decades witnessed an astonishing proliferation of lasing ions and media. Elements of the lanthanide series contributed to this proliferation and in one case, that of solid-state lasers, dominate the field.

To date stimulated emission has been obtained from eleven trivalent and three divalent lanthanide ions; in hosts including crystalline and amorphous solids, metallo-organic and inorganic aprotic liquids, and neutral and ionized gases and molecular vapors; at wavelengths ranging from the infrared to the ultraviolet; from lasers operating pulsed and continuously; and from lasers ranging in size from thin films and small fibers for integrated optics applications to large disks for high-power Nd:glass lasers for inertial confinement fusion experiments.

The versatility and wide applicability of lanthanide ions for lasers arises from several desirable spectroscopic features. The electronic states of the ground $4f^n$ configurations provide complex and varied optical energy level structures, thus many different lasing schemes are possible. The large number of excited states suitable for optical pumping and the subsequent decay to metastable states having high quantum efficiencies and narrow f→f emission lines are favorable for achieving laser action. Because the locations of the energy levels do not change greatly with host, a given ion can be lased in many different hosts. The host can therefore be selected to optimize performance for a specific application.

The spectroscopic properties of the lanthanides and actinides as they relate to laser action are the principal

0–8412–0568–X/80/47–131–275$05.00/0

topic of this article. After a brief review of laser
fundamentals (4), the extent to which these elements have
been employed for lasers in various media are surveyed.
A comprehensive listing of all lanthanide laser ions and hosts
is beyond the scope of this paper, however references are
given to tabulations containing references to the original
work. The particular transitions used for lasing are shown
and discussed. These illustrate how and why laser action is
obtained and form the basis for considering possible
stimulated emission involving other ions and transitions.
Recent work, current activities, and future directions are
also noted. Because of space limitations, engineering details
and applications of lanthanide and actinide lasers are not
discussed, but are well covered in a book by Koechner (5).

Laser Fundamentals

To obtain stimulated emission between two energy levels,
a population inversion is necessary. This is usually achieved
by excitation into a third level or levels which rapidly and
efficiently transfer energy to a metastable upper laser level.
A generalized energy level scheme for laser action is shown in
Fig. 1. If the terminal laser level is the ground state and
the initial and final laser states have equal degeneracies,
then more than one-half of the ions must be excited to obtain
an inverted population and 3-level laser action. If, instead,
the terminal level 2 is above the ground state, then only an
excited-state population in level 3 sufficient to overcome
the Boltzmann population in level 2 is needed for population
inversion. This drastically reduces the pumping requirements.
Phonon-terminated or vibronic lasers are those in which level 2
is a vibrational-electronic state.
When it is difficult to excite ions into level 3 or the
level decays very rapidly, a population inversion and oscillation may be obtained using a cascade laser scheme involving
two consecutive lasing transitions. An example is shown at
the right of Fig. 1. To lase the 3→2 transition, ions are
first pumped into level 4 and then stimulated to emit to
level 3, thereby creating a population inversion with respect
to level 2. Here one relies on a stimulated rather than a
spontaneous emission rate for the 4→3 transition. The rate can
therefore be made very fast and controlled via the beam
intensity or the Q of the resonant optical cavity. Cascade
lasing schemes are also useful when the terminal level of a
lasing transition relaxes so slowly that oscillation self-
terminates because the population inversion and associated
gain decrease to a value insufficient to overcome the losses.
If the terminal state is stimulated to emit to a lower level,
oscillation can be maintained and the total energy stored in
the first upper laser level extracted

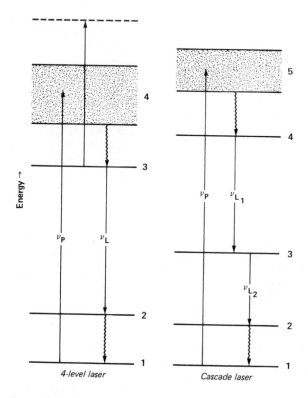

Figure 1.   Representative energy-level diagram and transitions for four-level and cascade lasing schemes: ν_P and ν_L are the pump and laser frequencies; wavy lines denote nonradiative transitions.

Cascade lasing requires that the host be transparent and there be no deleterious ground or excited-state absorptions at either laser wavelength. Transitions L_1 and L_2 may be associated with two different ions. In this case the energy in the terminal level 3 of the first lasing ions must be quickly and efficiently transferred to the upper laser level 3 of the second ion.

The threshold condition for laser oscillation is given by

$$R_1R_2\exp(2G\ell) = 1,\tag{1}$$

where R_1 and R_2 are the reflectivities of the mirrors in the optical resonator cavity, G is the gain per unit length, and ℓ is the length of the active lasing medium. The gain is determined by

$$G = \alpha - \alpha_s - \alpha_I,\tag{2}$$

where α is the gain coefficient of the lasing medium, and α_s and α_I are loss coefficients due to scattering and impurity absorption. If N_3 and N_2 are the populations in the upper and lower laser levels in Fig. 1, the net gain coefficient of the laser medium is

$$\alpha = N_3\sigma_{32} - N_2\sigma_{23} - N_3\sigma_{esa}.\tag{3}$$

In Eq. (3), σ_{32} and σ_{23} are the cross sections for stimulated emission and absorption. For narrow-line absorption and emission spectra, these two cross sections are equal. For broadband spectra with emission bandwidth greater than kT, the cross sections are connected by a generalized Einstein relation (6). The final term in Eq. (3) accounts for possible excited-state absorption from the upper laser level to higher excited-states indicated by the dashed level in Fig. 1. If $\sigma_{esa} > \sigma_{32}$, absorption from level 3 dominates stimulated emission and laser action is not possible.

The gain, from Eq. (3), is governed by a product of the stimulated emission cross section and the population inversion (N_3-N_2). The latter is dependent upon the absorption spectrum and its spectral match with the pump source, the lifetime of the metastable level 3 which determines the pumping rate required, and the quantum efficiency. The last quantity includes the fluorescence conversion efficiency (the number of ions excited to the fluorescing level per incident pump photon) and the quantum efficiency of the fluorescing state (the fractional number of photons emitted per excited ion in the upper laser level).

The peak cross section is determined by the oscillator strength and the linewidth of the transition. The linewidth may be (1) the natural or homogeneous width, governed by radiative and/or nonradiative transitions between Stark levels, or (2) inhomogeneously broadened in the case of disordered media such as glasses or mixed crystals. In the latter case the rate of energy extraction varies from ion to ion and spectral hole burning in the gain profile may occur.

Both f-f and f-d transitions have been used for lanthanide and actinide lasers. The spectroscopic properties of these transitions are compared in Table I. Since the d states have shorter lifetimes, faster pumping as well as higher energies are required for excitation. Possible pumping sources include ultrafast flashlamps, other lasers, electron beams, or synchrotron radiation. With one exception, all lanthanide and actinide lasers have been optically pumped.

Table I.   Comparison of spectroscopic properties of f-f and
f-d transitions of lanthanide ions in solids at 300 K.

	$4f \rightarrow 4f$	$5d \rightarrow 4f$
Oscillator strength	$\sim 10^{-6}$-10^{-8}	$\sim 10^{-1}$-10^{-2}
Ion-lattice coupling	weak	intermediate — strong
Fluorescence wavelength	~ 200-5000 nm	~ 150-1000 nm
Fluorescence linewidth	~ 10 cm^{-1}	$\gtrsim 1000$ cm^{-1}
Fluorescence lifetime	10^{-5}-10^{-2}s	10^{-8}-10^{-5} s
Excited-state absorption	$f \rightarrow f$ $f \rightarrow d$	$d \rightarrow d$ $d \rightarrow$ higher configurations

Table II lists all lanthanide and actinide laser ions and types of transitions.

Table II.   Electronic transitions and ions used for lanthanide
and actinide lasers.

Transition	Ions
$4f \rightarrow 4f$	Pr^{3+}, Nd^{3+}, Sm^{2+}, Eu^{3+}, Gd^{3+}, Tb^{3+}, Dy^{2+}, Dy^{3+}, Ho^{3+}, Er^{3+}, Tm^{2+}, Tm^{3+}, Yb^{3+}
$5d \rightarrow 4f$	Ce^{3+}, Sm^{2+}
$5f \rightarrow 5f$	U^{3+}

Lasing Media

Stimulated emission has been observed from lanthanide elements in gases, liquids, and solids. The lanthanides used and the number of ion-host combinations lased in each medium are given in Table III. Figure 2 shows the spectral ranges covered by lanthanide lasers in the different media. Gas lasers operating from the far infrared to the vacuum ultra- violet are known (7), thus the lanthanides cover only a modest range for this medium. Liquid laser action from lanthanide ions or organic dye molecules is limited to wavelengths between the infrared and ultraviolet transmission cut-offs, therefore the spectral coverages of both are comparable. In solids, lanthanides dominate both the number of lasers and

Table III. Lasing media and number of ion-host combinations used for lanthanide lasers.

	Lanthanides	Total
Gases		
Metal vapors	Sm, Eu(I,II), Tm, Yb(I,II)	6
Molecular vapors	Nd^{3+}	1
Liquids		
Chelates	Nd^{3+}, Eu^{3+}, Tb^{3+}	28
Aprotic solvents	Nd^{3+}	8
Solids		
Glasses	Nd^{3+}, Ho^{3+}, Er^{3+}, Tm^{3+}, Yb^{3+}	⩾100
Crystals	Divalent: Sm^{2+}, Dy^{2+}, Tm^{2+}	
	Trivalent: all except Pm^{3+}, Sm^{3+}	>200

the wavelength range covered. The only other ions used for solid-state lasers are a few iron group ions (Cr^{3+}, V^{2+}, Ni^{2+}, Co^{2+}); semiconductor and color-center lasers complete the category of solid-state lasers.

Gases. Gas lasers are attractive for high-power, high- efficiency systems and offer advantages of low materials cost, ability to flow the lasing medium to remove heat, and low susceptibility to damage and distortion due to high intensity optical fields. Two approaches to obtaining lanthanide laser action in a gaseous media are (1) lanthanide metal vapors excited in a gas discharge tube, and (2) lanthanide molecular vapors excited optically or with an electron beam.

Stimulated emission in the infrared wavelength region has been observed from neutral and/or singly-ionized atoms of four lanthanides: samarium, europium, thulium and ytterbium. A listing of rare-earth vapor lasers, wavelengths, and references is given in Ref. 7. Because the energy level structures for lanthanide vapors are complex and comprehensive spectroscopic data is not always available, identification of some of the laser transitions and details of the mechanisms for population inversion are uncertain. Experimentally, the lanthanide metal vapor together with a buffer gas is excited in a standard gas discharge tube equipped with windows and placed within an optical resonator cavity. Population inversion is obtained using current pulses up to several hundred amperes and durations of a few microseconds or longer. The quantum efficiency of transitions used for stimulated emission to not exceed 40 percent. The overall electrical efficiency of the laser is considerably less. Recently an average power of 2W was reported (8) for a He-Eu ion laser operating on the 1.0019-μm Eu II line at a pulse repetition frequency of 10 kHz.

Another approach to gas laser action is to use f-f transitions of optically-excited lanthanide molecular vapors. The spectroscopic properties of several rare-earth trihalide-aluminum chloride complexes and various rare-earth chelates has been studied (9) and optical gain observed for a Nd-Al-Cl vapor complex (10). Measurements of the fluorescence kinetics show evidence of strong excited-state excited-state quenching. This plus the low molecular densities achievable reduce the attractiveness of these systems for practical laser applications.

Liquids. Lanthanide laser action has been obtained for two groups of liquids: metallo-organic and inorganic aprotic liquids. The first group includes chelate lasers (Nd, Eu, Tb) which are reviewed by Lempicki and Samelson (11); research on aprotic materials and systems for high-power, pulsed liquid lasers are reviewed by Samelson and Kocher (12). Stimulated emission in both liquids occurs between 4f states of trivalent lanthanides. The tuning ranges of these lasers are small compared to that obtainable from organic dye lasers (13).

The spectroscopic properties of lanthanides in liquids are characterized by broad absorption and emission bands with line-widths that approach those in glasses. Lanthanide fluorescence in liquids is less prevalent than in solids because high frequency vibrations associated with the solvent cause non-radiative relaxation of excited electronic states. In chelates, the lanthanide ion is complexed to several organic groups or ligands. Chelates are soluble in many organic solvents. Lempicki (14) lists several ligands, cations, and solvents commonly used for rare-earth chelate lasers. As in glasses, the wavelengths of transitions exhibit small shifts with changing ligand or cation (15).

Stimulated emission is achieved by optical pumping with xenon-filled flashlamps in optical cavities and resonators similar to those used in solid-state lasers. The principal pumping for Eu and Tb chelate lasers is ascribed to absorption into the singlet state of the ligand followed by intersystem crossing to the triplet state and subsequent intermolecular transfer to an excited state of the lanthanide (11). Because the singlet absorption is very strong at the concentration necessary for lasing, only small volumes of active material can be pumped effectively. For the Nd chelate laser, many absorption bands of Nd^{3+} are below the ligand bands and are utilized for optical excitation.

The fluorescence quantum efficiency of excited lanthanides in most liquids is very low. To reduce fluorescence quenching due to interactions with high-frequency vibrations in liquids, solvent molecules should have no tightly bonded atoms of low atomic mass (16). Solvents containing hydrogen or other light atoms are therefore undesirable. Aprotic liquid laser materials consist of solutions of a rare-earth salt and an inorganic aprotic solvent.

The spectroscopic properties and chemistry of aprotic Nd^{3+} laser liquids plus references to earlier studies are discussed by Brecher and French (17). The oscillator strengths and fluorescence lifetimes are comparable to those in solids with quantum efficiencies near unity. Since fluorescence linewidths are smaller than in glasses, the stimulated emission cross sections are larger (18), although still less than in crystals. Aprotic liquid laser materials and references are listed in Ref. 19. Thus far only Nd^{3+} has been used as the laser ion although other lanthanide ions could also be used.

Solids. Solids are the most widely used host for lanthanide and actinide laser action. Hosts include over 200 different ion-crystal combinations and numerous glasses. The number of ion-crystal laser combinations for each ion is shown in Fig. 3. Lanthanide ions are generally introduced into solids as a substitutional impurity in concentrations of ≈1%. Oscillation has also been obtained with the lanthanide as a stoichiometric component of the host. Among the desired properties of a laser host, in addition to ability to incorporate the lanthanide ion with a homogeneous doping distribution, are high optical quality, transparency to the excitation and laser wavelengths, hardness sufficient for good optical finishing, resistance to damage by laser-induced electric breakdown, and, in the case of high repetition rate or continuous operation, good thermal conductivity and small stress-optic coefficients. Recent reviews of solid-state lasers are given in Refs. 20, 21, 22.

More than 140 different crystalline hosts have been used for lanthanide lasers. These include simple and mixed oxides and fluorides, and more complex compositions and structures (21).

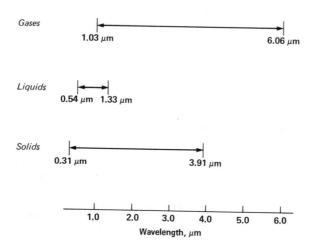

Figure 2. Spectral range of lanthanide lasers in various media

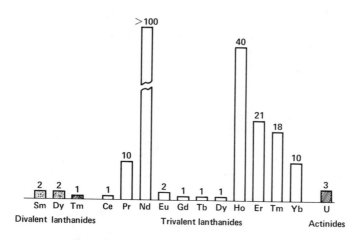

Figure 3. Number of different ion–crystal laser combinations grouped by lasing ion

Although a large number of crystalline lanthanide lasers have been explored, only a very few have achieved any practical acceptance, the prime example being the garnet $Y_3Al_5O_{12}$ (YAG). This material has a particularly favorable combination of being a very hard, optically isotropic crystal with sites suitable for trivalent lanthanide substitution without charge compensation. When the substitutional sites for trivalent lanthanide or actinide ions are divalent, such as in alkaline earth fluorides, excessive fluorine or other charge-compensating ions are added to maintain charge neutrality. A thorough discussion of the chemistry and growth of laser crystals is given by Nassau ([23]).

In glass, laser action has been observed only from trivalent lanthanides ([24], [25]). Hosts include oxide glasses (silicate, phosphate, borate, germanate, tellurite), one fluoride glass (beryllium), and mixed glasses such as borosilicate and fluorophosphate. For a given glass network former, changes in the number and type of network modifier ions affect the spectroscopic properties of the lanthanides. This feature is applied to tailor the glass compositions for specific laser applications. If each composition is defined as a new glass, then the number of ion-glass combinations lased becomes very large and the number in Table III is unknown and not very meaningful.

Glasses are inherently a disordered medium, therefore, the local environment at each lanthanide site is slightly different. This appears as spectral broadening and site-dependent transition probabilities. One manifestation of this inhomogeneity is spectral hole burning in the gain profile ([26], [27]). Because the stimulated emission cross sections are different at each site, the energy extracted from an inhomogeneous system is always less than that obtainable from a homogeneous system of the same average cross section ([28]).

The vibrational spectrum of the host is particularly important for determining the rate of nonradiative decay and fluorescence quantum efficiency of lanthanides ions. Studies show that in both crystals and glasses, the rate of multiphonon emission is determined principally by the size of the energy gap to the next lower level and the number of phonons required to conserve energy ([29]). Therefore hosts in which the maximum phonons energies are relatively small, e.g., $LaCl_3$, have more numerous and efficient fluorescing states. Unfortunately such materials frequently have poor physical properties for practical lasers. In glasses, the vibrational frequencies associated with the glass network former, e.g., the SiO_4 tetrahedra, are comparative large and the number of fluorescing states small. For this reason the number of lanthanide laser transitions in glasses is much less than in crystals.

The optical pumping efficiency and output power of many lasers is increased by codoping the medium with other ions which absorb pump radiation and effectively transfer the excitation to the upper laser level. This transfer may be either radiative or nonradiative. In general, sensitization schemes used for phosphors and other luminescence phenomena are also applicable to lasers (30). Requirements for the sensitizer ion include (a) no ground- or excited-state absorption at the laser wavelength, (b) absorption bands which complement rather than compete with absorption bands of the laser ion (because the fluorescence conversion efficiency usually is less for the former), (c) one or more metastable energy levels above the upper laser level, and (d) no other pairs of levels which can quench the activator fluorescence. In addition, for efficient transfer the concentration of sensitizer ions must be sufficiently high to provide significant transfer within the fluorescence lifetime of the activator.

Possible sensitizers for lanthanide and actinide ions include other lanthanide and actinide ions, other transition group ions, and molecular complexes. These may be present either as added impurities or as a component of the host. Of the many sensitization schemes reported, some offer only marginal improvement. The most efficient crystal laser is "alphabet" holmium: Ho^{3+} sensitized by Er^{3+}, Tm^{3+}, and Yb^{3+} (31). The absorption bands of these ions combine to form a quasi-continuous spectrum. Via a complex cascade, energy absorbed by the various ions is eventually transferred to the $^5I_7$ lasing level of Ho^{3+}.

The concept of upconversion (32) in which higher-lying states of an activator are excited by successive energy transfers from a less energetic sensitizer has also been applied to lanthanide lasers (33).

A list of sensitized lanthanide lasers is given in Table IV. The laser transitions are shown in the next section; for figures of the energy levels and transition of the sensitizer and activator ions and the original references see Refs. 21 and 34. Other sensitization schemes are known, but only those actually used for lasers are included. These have most commonly used f-f transitions of lanthanides. Possible d-d sensitization schemes have also been noted (35).

Survey of Lanthanide Ions

With the exception of promethium, stimulated emission has been reported for all of the lanthanides. The transitions used and the lasing characteristics of each are reviewed below. More detailed discussions of the spectroscopic features of the ion and the properties of the host that influence the potential for laser action are presented in a review article (19) and a book (36) devoted to rare-earth lasers.

Table IV. Ions used as sensitizers for optically-pumped lanthanide lasers.

Laser Ion	Laser transition	Sensitizer ion(s)
Nd^{3+}	$^4F_{3/2} \to {}^4I_{11/2}$	Ce^{3+}, Cr^{3+}, Mn^{2+}, UO^{2+}, $(VO_4)^{3-}$
Tb^{3+}	$^5D_4 \to {}^7F_5$	Gd^{3+}
Dy^{3+}	$^6H_{13/2} \to {}^6H_{15/2}$	Er^{3+}
Ho^{3+}	$^5I_7 \to {}^5I_8$	Er^{3+}, Tm^{3+}, Yb^{3+}, Cr^{3+}, Fe^{3+}, Ni^{2+}
	$^5S_2 \to {}^5I_8$	Yb^{3+} (*)
Er^{3+}	$^4I_{13/2} \to {}^4I_{15/2}$	Yb^{3+}
	$^4F_{9/2} \to {}^4I_{15/2}$	Yb^{3+} (*)
Tm^{3+}	$^3H_4 \to {}^3H_6$	Er^{3+}, Yb^{3+}, Cr^{3+}
	$^3F_4 \to {}^3H_5$	Cr^{3+}
Yb^{3+}	$^2F_{5/2} \to {}^2F_{7/2}$	Nd^{3+}, Cr^{3+}

*Multistep upconversion process

Energy level diagrams and lasing transitions for all trivalent lanthanide ions are shown in Fig. 4 and 5 (to simplify the diagrams, the extent of the crystalline Stark splitting, which varies with host, is not indicated). References to the original reports are given in Ref. 7 for gases, in Ref. 14 for liquids, and in Ref. 21 for solids.

Trivalent Ions. Energy levels associated with the $4f^n$ ground electronic configuration of the trivalent lanthanides are well understood for states up to $\approx$30,000-40,000 cm^{-1} both experimentally (37) and theoretically (38, 39). The ligand or crystal field of the host reduces the $(2J + 1)$- fold degeneracy of the free-ion states. Because the 4f electrons are shielded by the outer $5s^2$ and $5p^6$ electrons, the shift in the center of gravity of the free-ion energy levels and the extent of the crystalline Stark splitting are small, on the order of a few hundred cm^{-1}, and vary with the host. Levels of $4f^{n-1}$ 5d and other excited configurations are at higher energies and have been investigated for wide bandgap fluoride hosts (40, 41, 42). In many materials, however, the latter levels are near or above the fundamental absorption edge and therefore of limited usefulness for optical pumping or lasing.

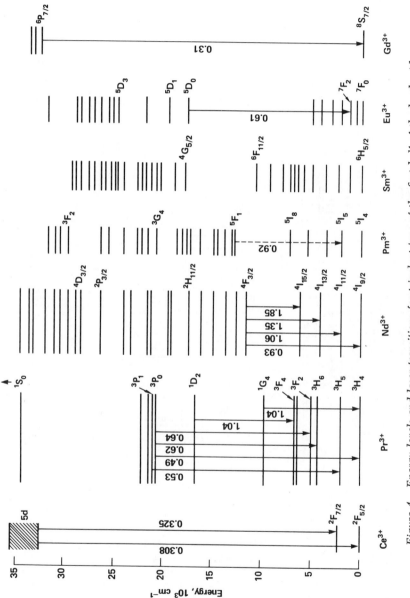

Figure 4.   Energy levels and laser transitions for trivalent ions of the first half of the lanthanide series. Approximate wavelengths of transitions are given in micrometers.

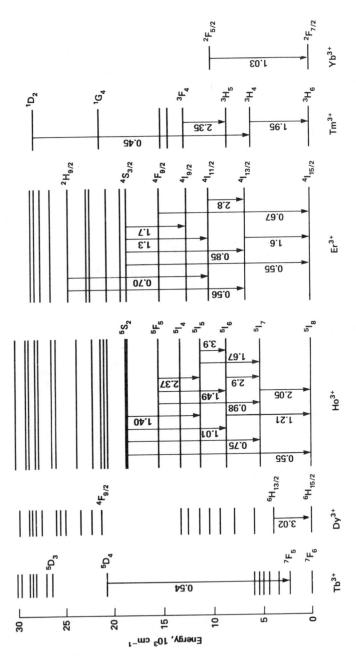

Figure 5.   Energy levels and laser transitions for trivalent ions of the second half of the lanthanide
series.   Approximate wavelengths of transitions are given in micrometers.

Cerium. Near-ultraviolet lasing from the lowest 5d band to states of the $4f^1$ ground configuration of Ce^{3+} in $LiYF_4$ was reported recently (43). Excitation was achieved by pumping into higher lying 5d bands at 248 or 193 nm using either a KrF or an ArF excimer laser, respectively. An efficient 4-level lasing scheme is formed by transitions terminating on the $^2F_{7/2}$ state. Oscillation also occurs to the $^2F_{5/2}$ ground state (44), the wavelength (308 nm) is the shortest of all lanthanide lasers. Because the Ce^{3+} fluorescence band is broad, the laser action is tunable. The tuning range achieved to date for both transitions in Ce:YLF is ≈ 500 cm^{-1} (44).

Although d→f lasing of Ce^{3+} and other lanthanides have been discussed for several years (45, 46), this was the first successful demonstration since the very early d→f lasing of Sm^{2+} in CaF_2. One difficulty in obtaining oscillation is possible excited-state absorption to higher lying states. This was shown to prevent oscillation of Ce^{3+}:YAG at room temperature (47). Since the nephelauxetic effect is different in oxide and fluoride hosts, the Ce^{3+} bands in $LiYF_4$ are shifted sufficiently to reduce excited-state absorption. Lasing from Ce^{3+} in other host crystals and glasses should be possible. If the d→f linewidth is predominantly homogeneous, hole burning and reduced energy extraction characteristic of glasses should not occur.

Praseodymium. The energy level scheme of Pr^{3+} includes several fluorescing and terminal states for 4-level operation. Absorption bands are few in number, hence thresholds are high for broadband optical pumps. Pulsed laser action has been observed from several excited states at 300 K and lower temperature (21, 34). Hosts include oxide and fluoride crystals. Lasing from the $^3P_0$ state should be possible from Pr^{3+} in glasses and liquids given adequate pumping.

The $^1S_0$ state located at $\approx 47,000$ cm^{-1} also exhibits fluorescence in wide bandgap hosts, such as fluoride crystals, and has been considered for laser action (48). Excitation into the 4f5d and higher lying bands rapidly decays to $^1S_0$. The $^1S_0 \rightarrow ^1G_4$ transition at ≈ 271 nm and the $^1S_0 \rightarrow ^3F_4$ transition at ≈ 250 nm are intense and could provide ultraviolet laser action if excited-state absorption is not dominant.

In $Y_3Al_5O_{12}$ the lowest 5d band is located below $^1S_0$ and fluoresces with high quantum efficiency at temperatures $\lesssim 300$ K (49). Several intense 4f5d→4f^2 transitions in the near-ultraviolet could provide tunable laser action barring strong excited-state absorption. Tunable laser action in the 215-260 nm range in $LiYF_4$ may also be possible (50).

Neodymium. As evident from Table III and Fig. 3, Nd^{3+} is the most extensively exploited laser ion and is the one trivalent lanthanide ion lased in all states of matter. The many optical absorption bands distributed through the visible and near-infrared spectral regions combined with rapid energy cascade to the metastable $^4F_{3/2}$ state provide good optical pumping efficiency for broadband sources. The $^4F_{3/2} \rightarrow ^4I_{11/2}$ transition is generally the most intense transition and forms a near-ideal 4-level lasing scheme at ambient temperatures. Pulsed laser action has been observed from $^4F_{3/2}$ to all ^4I_J states and cw laser action to the $^4I_{11/2}$ and $^4I_{13/2}$ states (22). Lasing involving 5d emission, which would be tunable in the region ≈172-195 nm in various fluoride hosts, has been mentioned (46) but not demonstrated.

Stimulated emission has been recorded for Nd^{3+} in more than 100 different crystals including doped single crystals, mixed crystals (solid solutions), and several crystals in which Nd is a stoichiometric component of the host (21). Because the spectroscopic properties are host dependent, the selection of materials provides variability with respect to cross sections and lifetimes as well as other physical properties.

The most widely used solid-state laser is Nd:YAG. The properties and operating characteristics of this laser have been thoroughly reviewed by Danielmeyer (51). All fluorescence transitions from $^4F_{3/2}$ to the ^4I_J states have lased in YAG. Cooling lowers the threshold for the $^4F_{3/2} \rightarrow ^4I_{9/2}$ transitions; the other transitions operate at ambient temperatures. Laser action was also observed from epitaxially-grown thin films (52) and single-crystal fibers (53) of Nd:YAG.

Other commercially available Nd-doped laser crystals include $YAlO_3$, $LiYF_4$, $La_2Be_2O_5$, and LaF_3.

Neodymium is also the most extensively developed glass laser (25). Systematic studies have shown that spectroscopic properties can be tailored, within limits, by selecting the glass network forming and network modifying ions (54). Many different oxide, fluoride, and mixed glass formers have been investigated (55). Thin film waveguides of Nd:silicate glass have been fabricated (56). At the other extreme, lasers for inertial confinement fusion experiments (57) employ long chains of Nd-doped glass disk amplifiers containing elliptical disks in sizes up to 50 x 600 nm.

Various broadband sources employed to optically pump Nd^{3+} include tungsten, mercury, xenon, and krypton lamps. The last source provides an especially good spectral match to the near-infrared absorption bands of Nd^{3+} in YAG. To reduce lattice heating resulting from the multiphonon emission decay to the $^4F_{3/2}$ state, semiconductor diodes and laser sources at 0.8-0.9 μm have pumped Nd lasers (58). Sun-pumped Nd and chromium-sensitized Nd lasers have been demonstrated and considered for space applications (59). Lasing of Nd^{3+} by electron beam excitation has also been reported (60).

Neodymium chelate laser action at 300 K was obtained, but to reduce nonradiative decay of $^4F_{3/2}$, a ligand containing

fluorine in place of hydrogen was used (61). Laser action has
been observed from several aprotic liquids consisting of a
solution of a Nd salt and an inorganic aprotic solvent. A list
of lasers and solvents is given in Ref. 19.

Gain was measured for the $^4F_{3/2} \to ^4I_{11/2}$ transition from one
molecular vapor, a $NdCl_3\text{-}AlCl_3$ complex (16). Intense excited
state-excited state quenching and low vapor pressures limit the
attractiveness of this lasing medium. The excited-state kinetics
for $Nd(thd)_3$ chelate vapors have also been investigated and the
prospects for laser action discussed (62).

Promethium. This ion has no stable isotopes. The isotope
Pm^{147} is a beta emitter (0.22 MeV) with a half-life of 2.6 years.
This radioactivity poses problems for the growth, fabrication,
operation, and lifetime of a solid-state laser. Stimulated
emission has not been reported for any host.

The energy level scheme of Pm^{3+} is very similar to that of
Nd^{3+} and hence is attractive for laser action. There are
numerous absorption bands for optical pumping and fluorescence
from $^5F_1$ state to levels of ^5I occurs at wavelengths ranging
from 0.81 to 1.72 μm. The large energy gap from $^5F_1$ to $^5I_8$
insures high quantum efficiency in most hosts. The most
promising transition for lasing is $^5F_1 \to ^5I_5$ which has a large
branching ratio and no competing excited-state absorption.
Krupke (63) calculated fluorescence intensities and the
radiative lifetime of $^5F_1$ for Pm:YAG using Judd-Ofelt intensity
parameters extrapolated from Nd^{3+}. The $^5F_1 \to ^5I_5$ transition at
0.92 μm had an oscillator strength within 70% of the value
for the $^4F_{3/2} \to ^4I_{11/2}$ transition of Nd:YAG. The $^5F_1 \to ^5I_6$ of the
transition and transitions from the thermally populated $^5F_2$
state are also intense and laser candidates.

The possibility of lasing Pm^{3+} in $LiYF_4$ has also been
considered (64). The greatest transition probability
corresponds to the π-polarized electric-dipole transition
between the $^5F_1$ Γ_1 level and the $^5I_5$ Γ_2 level at a wavelength
of 0.94 μm. Experiments suggest that the beta-ray activity
of Pm^{3+} may not have a severe effect on the optical properties
of $LiYF_4$ in the wavelength regions of interest (64).

Samarium. Stimulated emission has not been reported for
trivalent samarium in any medium. There are numerous absorption bands at wavelengths ≲500 nm and efficient fluorescence
occurs from the $^4G_{5/2}$ level in solids. The emission, however,
is divided among many terminal levels and several ion-ion
self-quenching transitions are possible. Fluorescence
transition having large branching ratios include $^4G_{5/2} \to ^6H_{7/2}$
and $^4G_{5/2} \to ^6H_{9/2}$ (65). Because of the high density of high-
lying states, the probability for excited-state absorption from
$^4G_{5/2}$ may, in many cases, be stronger than for stimulated
emission.

Europium. This is the most extensively studied chelate laser ion. Trivalent Eu has lased in 24 organic chelate solutions at temperatures ranging from -150 to 30°C. Some of the ligands, cations, and solvents used are given in Ref. 14. The principal optical pumping is ascribed to absorption into the singlet state of the ligand followed by intersystem crossing to the triplet state and subsequent intermolecular transfer to an excited state of Eu^{3+} (11).

In solids, only pulsed $^5D_0 \rightarrow ^7F_2$ laser action has been observed. Low temperatures were used to narrow the linewidths and reduce the population in the $^7F_2$ terminal state. Since there are no intense absorption bands in the visible, lasing thresholds were high. For efficient utilization of the higher-lying pumps bands, a rapid nonradiative cascade through the ^5D levels to $^5D_0$ is necessary to minimize fluorescence losses. Alternatively, lasing could be obtained from metastable $^5D_1$ and $^5D_2$ states to various levels of ^7F.

The absorption and emission cross sections of Eu^{3+} are relatively small. The $^5D_0 \rightarrow ^7F_2$ lasing transition is also a hypersensitive transition (66) and therefore very host dependent. No systematic effort appears to have been made to exploit this feature to improve lasing performance.

Because of the absence of $^5D_0$ quenching by simple ion pair interactions, high Eu concentrations and stoichiometric materials should be usable for laser action.

Gadolinium. Efficient fluorescence from the lowest excited state $^6P_{7/2}$ to the $^8S_{7/2}$ ground state occurs at ≈ 0.31 μm and forms a three-level lasing scheme. The high threshold characteristic of three-level operation and the requirements of good host transparency, low excited-state absorption, and an ultraviolet source (<0.3 μm) for optical pumping are all obstacles to obtaining stimulated emission. Only two cases of laser action have been reported; one in a crystal - $Y_3Al_5O_{12}$ (67) and one in a silicate glass (68). In both instances pumping was via a xenon flashlamp and the thresholds were very high.

Terbium. The $^5D_4 \rightarrow ^7F_5$ transition has the largest fluorescence branching from $^5D_4$ and forms a four-level laser scheme at ambient temperatures. However, stimulated emission has been observed in only one material, crystalline LiYF$_4$ (69). As in the case of Eu^{3+}, the principal absorption bands for optical pumping lie in the near-ultraviolet. If these are used to excite the $^5D_4$ level, the 6000 cm^{-1} $^5D_3 \rightarrow ^5D_4$ energy gap must be efficiently bridged. In LiYF$_4$ this was done by using a high Tb concentration ($\gtrsim 20\%$) so that there was rapid $^5D_3 \rightarrow ^5D_4$ decay by ion-ion interactions and energy-conserving $^7F_6 \rightarrow ^7F_{0,1}$ transitions. The 5d bands of Tb^{3+} are the lowest-lying of the trivalent lanthanides and, if too low, they can prevent lasing due to strong, $^5D_4$ 5d excited-state absorption. In LiYF$_4$, the nephelauxetic effect is small and the 5d bands are sufficiently high to avoid this difficulty; this is not the case in many other crystals, especially the oxides.

There is one report of optically-pumped Tb^{3+} chelate laser action at room temperature; the threshold was very high (70). THe excited-state kinetics of Tb^{3+} in vapor-phase terbium chelates (62) and terbium aluminum chloride complexes (71, 72) have been investigated but no laser action reported.

Dysprosium. Stimulated emission from Dy^{3+} in Er^{3+}-sensitized BaY$_2$F$_8$ at 3.02 μm is the longest wavelength non-cascade laser (73). Laser action was obtained at 77 K and involved a $^6H_{13/2} \rightarrow ^6H_{15/2}$ transition. Fluorescence also occurs from the $^4F_{9/2}$ level located at ≈21,000 cm^{-1} with intense emission to $^6H_{15/2}$ and $^6H_{13/2}$, but no laser action has been reported. The comments made earlier about the absence of stimulated emission from Sm^{3+} are again apropos. Several possible cascade laser schemes for Dy^{3+} have been described (74).

Holmium. In terms of the number of solid-state hosts, Ho^{3+} is the second most extensively exploited lanthanide laser ion; in terms of different transition lased, it is the most exploited. Stimulated emission is observed for 12 transitions with wavelengths ranging from 0.55 to 3.91 μm and in hosts including crystals, three stoichiometric materials (HoF$_3$-LiHoF$_4$, Ho$_3$Al$_5$O$_{12}$) (19, 21), thin films (52), and silicate glass (75). The most common laser transition, $^5I_7 \rightarrow ^5I_8$, has operated both pulsed and cw in crystals, however low temperatures are usually required. Phonon-terminated laser action has also been reported for Ho^{3+} in BaY$_2$F$_8$ (76).

Recently $^5S_2 \rightarrow ^5I_5 \rightarrow ^5I_6$ cascade laser action was observed for Ho^{3+} in LiYF$_4$ (77). The latter transition is at 3.914 μm and is the longest wavelength lanthanide laser reported to date. By using a 30-ns pump pulse and a high-Q cavity, $^5S_2 \rightarrow ^5I_5$ lasing was obtained within 0.5 μs. As a result, sufficient population buildup occurred in $^5I_5$ to achieve $^5I_5 \rightarrow ^5I_6$ oscillation before significant spontaneous decay from the $^5I_5$ state. Cascade lasing schemes of Ho^{3+} in Gd$_3$Ga$_5$O$_{12}$ crystals including $^5I_6 \rightarrow ^5I_8$ transitions were reported recently (78).

Erbium. The energy level diagram of Er^{3+} is similar to that of Ho^{3+} and stimulated emission involving a like number of transitions, wavelength range, and diversity of host materials is possible. The first cascade lasing scheme, $^4S_{3/2} \rightarrow ^4I_{13/2} \rightarrow ^4I_{15/2}$, was developed for Er^{3+} in a CaF$_2$-YF$_3$ crystal (79). Erbium was also part of a cascade laser scheme involving two different ions. In this scheme a $^4S_{3/2} \rightarrow ^4I_{13/2}$ lasing transition of Er^{3+} was followed by nonradiative Er^{3+}

$^4I_{13/2} \rightarrow {}^4I_{15/2}$: Tm^{3+} $^3H_6 \rightarrow {}^3H_4$ transfer and subsequent Tm^{3+} $^3H_4 \rightarrow {}^3H_6$ lasing (80).

Among particularly useful laser transitions of Er^{3+} are $^4S_{3/2} \rightarrow {}^4I_{13/2}$ at 0.85 μm and $^4I_{13} \rightarrow {}^4I_{15/2}$ at ≈1.6 μm. The latter wavelength is absorbed by the ocular media of the eye, thereby offering protection for the retina. Erbium-doped glass lasers were developed extensively for possible eye-safe applications (27). The spectroscopic properties and re-laxation of Er^{3+} in a $ErCl_3 \cdot (AlCl_3)_x$ vapor complex have been studied (81) with potential application for stimulated emission.

Possible 5d→4f lasing of Er^{3+} in $LiYF_4$ tunable from 165-172 nm has been mentioned (46).

Thulium. Stimulated emission has been obtained from three states of Tm^{3+}: $^3H_4$, $^3F_4$, and $^1D_2$. Other excited states having high quantum efficiency in most hosts include $^1G_4$ and $^1I_6$ (not shown in Fig. 5 but located at ≈34,000 cm^{-1}). With suitable pumping, oscillation should be readily obtainable from these states to various terminal states. The most intense transitions are $^1G_4 \rightarrow {}^3H_6$ and $^1I_6 \rightarrow {}^3H_4$. Emission from the lowest 5d band to 4f states in $LiYF_4$ has been suggested as a source of tunable coherent radiation in the range 165-172 nm (46).

The $^3H_4 \rightarrow {}^3H_6$ transition at wavelengths of 1.9 to 2.0 μm has been used for both pulsed and cw laser action in crystals (21). Pulsed $^3H_4 \rightarrow {}^3H_6$ lasing has also been observed in silicate glass (82). Tm^{3+} has only a few absorption bands in the visible and energy cascade is inefficient because of the large energy gaps between J states. The detrimental effects of these conditions on optical pumping efficient are ameliorated by co-doping the materials with fluorescence sensitizing ions (see Table IV).

Recently $^1D_2 \rightarrow {}^3H_4$ lasing of Tm^{3+} in $LiYF_4$ was obtained by direct excitation into the $^1D_2$ state using a XeF excimer laser (83). Because the $^3H_4$ state decays radiative, this lasing scheme results in minimal heating of the host by non-radiative transitions.

Ytterbium. There is only one absorption band, $^2F_{5/2}$, for optical-pumped $^2F_{5/2} \rightarrow {}^2F_{7/2}$ laser action (the 5d bands begin at energies ≳70,000 cm^{-1}). Therefore unless a narrowband resonant source such as a light-emitting semiconductor diode or fluorescence sensitization are used, the thresholds for oscillation are high. In addition, because the laser transition terminates on a Stark level of the ground-state manifold, low temperatures are required for low-threshold operation. In a silicate glass, lasing has been obtained at 1.015 μm at 77 K (84) and 1.06 μm at 300 K (85). Laser action should also be obtainable for Yb^{3+} in stoichiometric materials, because self-quenching is absent, and in aprotic solvents and chelates similar to those used for Nd^{3+} lasers.

Divalent Ions. The $4f^n$ and $4f^{n-1}5d$ energy levels of divalent lanthanides have been studied in alkaline-earth fluoride crystals (86, 87). The 5d levels occur at lower energies than for the isoelectronic trivalent state and in most cases extend into the visible. Because the spin-orbit parameters are smaller for the divalent ions, the separations of the J states of the $4f^n$ configuration are reduced.

Alkaline-earth fluorides have been the principal hosts for divalent lanthanide lasers. These are relatively soft, optically isotropic materials. Lanthanides enter the alkaline earth sites substitutionally without charge compensation. Because these sites have inversion symmetry, only magnetic-dipole or vibronic transitions are allowed between 4f states. These are weak and the resulting radiative lifetimes are long. In comparison, the radiative lifetimes of 5d→4f transitions, which are parity allowed, are short. The 4f→5d transitions are broad and thus provide good absorption bands for optical pumping.

Laser action has been reported for three divalent lanthanides (21, 34). Figure 6 summarizes the energy levels, transitions, and approximate wavelengths of these lasers. Only crystals have been used as hosts and reduced temperatures were used in all cases.

Of the lanthanides, Eu and Yb can be readily reduced to the divalent state and remain stable in many materials. This is true to a lesser degree for Sm and Tm. Special methods are usually required to reduce the remaining trivalent lanthanides to the divalent state (23). These include irradiation with x-rays, beta and gamma rays, metal diffusion, electrolysis, and photochemical reaction. Frequently, the resulting materials are not stable with respect to thermal and photochemical effects and the ions revert back to the trivalent state.

Samarium. Divalent Sm laser action has been demonstrated using both d→f and f→f transitions. The former was observed in CaF_2 (88, 89). At or below liquid nitrogen temperatures lasing occurs from 708 to 729 nm. For Sm^{2+} in SrF_2, the $^5D_0$ state is below the lowest 5d band and $^5D_0 \rightarrow {}^7F_1$ lasing has the lowest threshold at liquid helium temperatures (90). Samarium laser action was pulsed using xenon flashlamps or a ruby laser for excitation.

Europium. Broadband, Stokes-shifted 5d→4f emission is observed from Eu^{2+} in many hosts. The fluorescence occurs in the 400-500 nm region and has a lifetime of about 1-2 μs. Attempts to observe laser action from Eu^{2+} in a crystal

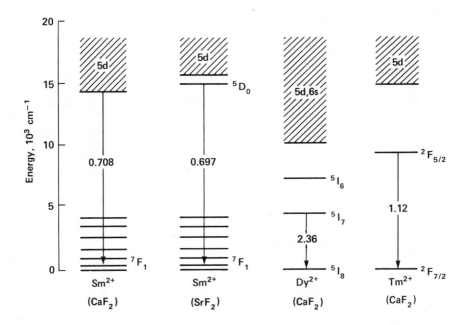

Figure 6. Energy levels and laser transitions for divalent lanthanide ions. Approximate wavelengths of transitions are given in micrometers.

(CaF$_2$) (91, 92) and a glass (93) have been unsuccessful. This
is due to losses by either excited-state absorption or color
centers because transient absorption has been observed (92, 94).
The excited-state absorption peak in CaF$_2$ is temperature
dependent.

Dysprosium. Both pulsed and cw infrared laser action are
reported at liquid nitrogen and helium temperatures (21, 34).
Pump sources include xenon, mercury, and tungsten lamps and
sunlight. Broad absorption bands throughout the visible and
near-infrared plus the long lifetime of the magnetic-dipole
$^5I_7 \rightarrow {}^5I_8$ transition in CaF$_2$ and SrF$_2$ ($\geq$10 ms) are favorable for
good energy storage.

Thulium. This ion has lased both pulsed (95) and cw in
CaF$_2$ (96), but the cw threshold are high even at 4 K. Excited-
state absorption from $^2F_{5/2}$ to 5d states, while energetically
comparable to the $^2F_{5/2} \rightarrow {}^2F_{7/2}$ laser transition, obviously is
not intense enough to prevent oscillation.

Survey of Actinide Ions

Qualitatively there are many similarities between the
energy levels and spectroscopic features of lanthanide and
actinide ions. Hence many of the earlier comments and discussions of lanthanide lasers are also apropos to possible
actinide lasers. As reviewed by Hessler and Carnall (97),
our knowledge and understanding of the energy levels and
spectral intensities of the 5f^n configurations has improved
significantly in recent years. Many of the interactions
governing the spectroscopic properties have been successfully
parameterized. Thus it is possible to make semi-quantitative
predictions about lasing prospects.

The positions of most of the lower J states of the ground
5f^n configurations of the trivalent actinides are known and
are given for LaCl$_3$ in Ref. 97. Figure 7 shows simplified
energy level diagrams for the trivalent actinides. The density
of states in the visible is very high. Because of the greater
degree of intermediate coupling, the J states for the actinide
ion order differently than for the corresponding 4f^n lanthanide
ion. Levels are frequently labeled by only the J quantum
number because the eigenstates have such mixed S,L character
that these are no longer meaningful quantum numbers. (We
will, however, sometimes label states using Russell-Saunders
designations for purposes of comparison with lanthanide
transitions.)

We consider only f-f transitions for lasing. The approximate positions of the 6d and charge transfer bands
throughout the actinide series are known (98, 99). The 6d
bands are lower than are the corresponding 5d bands of the

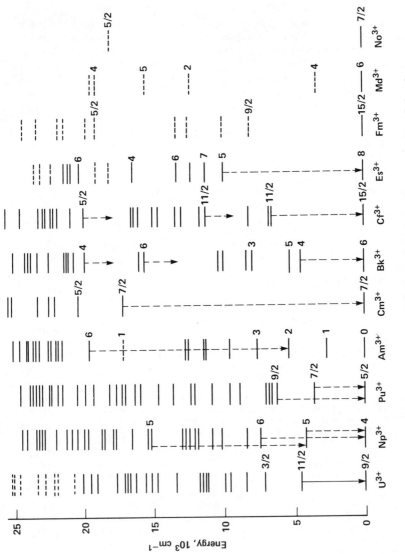

Figure 7. Energy levels of trivalent actinide ions: (– – –), predicted levels; (– →) fluorescence and possible laser transitions, discussed in the text.

lanthanides and they usually overlap the 5f levels. Hence excitation into the strong 6d bands would be followed by rapid nonradiative decay to lower 5f states as occurs for the analogous case of lanthanide 5d→4f relaxation (49). I am not aware of any reports of 6d→5f emission of actinide ions in solids. The presence of low-lying 6d bands should prevent lasing of many visible and shorter wavelength transitions because of intense excited-state absorption. Therefore most practical actinide laser action will be limited to the infrared spectral region.

The shielding of the 5f electrons of the actinides is less and therefore they are more sensitive to their environment than are the lanthanides. This has several consequences: (1) the electrostatic (Racah) parameters are smaller and the spin-orbit parameter $\zeta_{5f} \approx 2\zeta_{4f}$ (97), hence the energy gaps between J states are reduced. (2) The crystal-field parameters are approximately twice as large as for the lanthanides (97), therefore the Stark splitting is larger and the admixing of J states is greater (thereby reducing the effects of selection rules for transitions). (3) The admixing of opposite-parity states into the $5f^n$ configurations is larger and electric-dipole transitions are more probable. This is reflected in the actinide Judd-Ofelt intensity parameters which are larger because of this effect and the lower energies of the 6d bands. The probability of ion-ion energy transfer by electric dipole-dipole interactions (100) will also be greater. For self-quenching processes this is detrimental; for fluorescence sensitization it may be beneficial.

The dynamic crystal-field interactions and ion-phonon coupling are also expected to be stronger for the actinides. The natural linewidths of optical transitions are governed by one-and two-phonon transitions between Stark levels. Broader lines in crystals reduce the peak cross sections. In glasses, where inhomogeneous broadening predominates, an increased natural linewidth contributes to a more spectrally homogeneous transition. Only recently have studies of the homogeneous linewidths of actinides been made using fluorescence line narrowing techniques (101). Larger ion-phonon interaction strengths were observed. Increased ion-phonon coupling contributes, in addition, to increased probability for vibronic transitions.

Multiphonon processes which involve the ion-phonon coupling to higher order should also be more probable for actinide than for lanthanide ions. Systematic studies and quantitative data on the rates of these processes, such as exists for lanthanides (29), are still lacking for the actinides. Because of the larger probability for nonradiative decay, efficient laser action is further limited to transitions between the lower-lying J states which have large energy separations. These transitions are generally in the infrared.

The selection of host media having low vibrational frequencies is also more important for actinide fluorescence.

All of the actinides through Cf have isotopes with half-lifes of hundreds to thousands of years. The longest half-lifes of Es and Fm are measured in tens of days; Md and higher atomic number elements exist for only hours or less (102). Because of the radioactive decay, these latter elements are not considered for normal laser applications. Other concerns for practical lasers are (1) the presence of radioactive isotopes which may cause radiation damage and undesirable absorption bands in the host and (2) the cost and availability of adequate quantities of the required isotope.

An examination of the trivalent actinide energy level schemes reveals several possibilities for laser action. These are discussed in light of the general properties cited above. Only conventional broadband optical pump sources are considered. Obviously with selective laser excitation and cascade lasing schemes, stimulated emission from many more states should be possible, but these special situations are too numerous to be considered in detail here.

Actinide ions can be irradiated to achieve other valence states. In CaF_2 it was found the trivalent Am and Es could be reduced to the divalent state by gamma-ray irradiation; trivalent U, Np, Pu, and Cm, on the other hand, were converted to the tetravalent state (103). In the survey below, ions isoelectronic with the trivalent ion under consideration are included in parentheses; note, however, that depending upon the electrostatic and spin-orbit parameters, the ordering of the J states and possible lasing transitions may be different.

Uranium. (Np^{4+}). Uranium is the only actinide which has lased. The transition was $^4I_{11/2} \rightarrow {}^4I_{9/2}$ and oscillation occurs at 2.4-2.6 μm. Hosts included CaF_2, SrF_2 at temperatures ranging from 4-300 K. Both pulsed and cw oscillation were demonstrated (1, 104-109).

Trivalent uranium has many absorption bands in the visible and near-infrared suitable for xenon flashlamp pumping (see Fig. 7). The selection of host will govern how many excited-states fluoresce with high quantum efficiency. For example, fluorescence is observed from several excited states of U^{3+} in $LaCl_3$ (97). Therefore $^4F_{3/2} \rightarrow {}^4I_{9/2}$ oscillation may also be possible. Cascade $^4F_{3/2} \rightarrow {}^4I_{11/2} \rightarrow {}^4I_{9/2}$ lasing is another possibility, although f-f excited-state absorption may compete in the first step. In some hosts the 6d bands may occur at energies as low as 15,000 cm^{-1}, thus possible f→d excited-state absorption should limit oscillation to infrared wavelengths.

Neptunium. (Pu^{4+}). Emission from $^5I_6$ and $^5I_5$ to $^5I_4$ should occur with high efficiency in most crystalline hosts. The $^5I_6 \rightarrow {}^5I_4$ transition is particularly promising for lasing at

low temperatures sufficient to reduce the thermal population
in the terminal Stark levels of $^5I_4$. There are numerous
absorption bands throughout the visible and near-infrared
for optical pumping, but no bands for competing excited-
state absorption from $^5I_6$. $^5I_5 \rightarrow ^5I_4$ laser action is also
possible providing excitation into higher-lying pump bands
can rapidly bridge the $^5I_6 \rightarrow ^5I_5$ energy gap. Cascade $^5I_6 \rightarrow ^5I_5 \rightarrow ^5I_4$
lasing is another possibility.

Fluorescence is observed for several higher-lying excited
states of Np^{3+} in $LaBr_3$ (110). Intense emission at ≈505 nm
and 624 nm originates from the state at ≈19,800 cm^{-1} to $^5I_4$
and $^5I_5$ states, respectively. Intense fluorescence was also
observed from states at ≈17,300 cm^{-1} and 14,700 cm^{-1} to the
ground state. In all these cases, oscillation may be prevented by f→d absorption because the 6d bands begin at
≈25,000 cm^{-1}. A more promising possibility is the 4-level
lasing scheme from the J=5 ($^5F_5$) state at 15,000 cm^{-1} to $^5I_5$.
Concentration quenching of the $^5F_5$ fluorescence is probably
large.

When irradiated with ultraviolet light, a narrow emission
peak at 1.95 μm was observed from a radiation-damaged crystal
of ^{238}Pu-doped CaF_2. This feature was attributed to Pu^{4+}
(111).

Plutonium. (Am^{4+}). The energy level scheme and possible
lasing transitions for Pu^{3+} are very similar to those of Np^{3+}.
Prospective transitions include $^6H_{9/2} \rightarrow ^6H_{5/2}$, $^6H_{9/2} \rightarrow ^7H_{7/2}$,
and $^6H_{7/2} \rightarrow ^6H_{5/2}$. For efficient fluorescence and laser action
from either the $^6H_{9/2}$ or $^6H_{7/2}$ states, hosts should have low
phonon frequencies to reduce nonradiative decay by multiphonon
processes. Depending upon the host and the exact positions of
higher-lying states, excited-state absorption may reduce or
prevent net gain.

Emission was observed from Pu^{3+} in CaF_2 at 1.78 μm
corresponding to the $^6H_{9/2} \rightarrow ^6H_{5/2}$ transition (111). While
this transition satisfied the ΔJ = 2 rule for hypersensitive
transitions, the U(2) matrix element is not large (112) and
therefore should be less host dependent.

Americium. (Cm^{4+}). Attempts have been made to lase the
0.695 μm $^5L_6 \rightarrow ^7F_2$ transition of Am^{3+} in a $POCl_3$ liquid (113)
and in a $CaWO_4$ crystal (114), both at 300 K. No oscillation
was observed. The $^5L_6$ fluorescence lifetimes in $CaWO_4$ varied
from 50 to 80 μs; in the $POCl_3$ liquid the fluorescence was
weak and the decay time was very short which suggests the
presence of quenching. The 6d bands are located at energies
≈40,000 cm^{-1} and therefore f→d excited-state absorption should
not prevent oscillation, although f-f absorption could.

Although the 0.695 μm emission was attributed to the $^5L_6$
state and has been studied in $LaCl_3$ (115), nonradiative
decay to the J = 1 state at ≈17,000 cm^{-1} should occur in

hosts with high frequency vibrations. Transitions from
this level to levels of $^7F_1$ and $^7F_2$ are candiates for laser
action.

Transitions between the lower 7_F states are possible
infrared laser candidates in hosts with low vibrational
frequencies. The $^7F_1 \rightarrow ^7F_0$ transition is only magnetic-dipole
allowed and $^7F_2 \rightarrow ^7F_0$ should be a hypersensitive transition,
however selection rules are relaxed by J-state mixing.

Curium. (Am^{2+}, Bk^{4+}). Lasing of the $6(P,D)_{7/2} \rightarrow ^8S_{7/2}$
transition should be possible in a host where nonradiative
cascade to $^6P_{7/2}$ is efficient or, alternatively, from the
next higher excited state, J = 5/2. Pumping would be via
several visible and near-ultraviolet bands at $\approx 17,000$ cm^{-1}.
The bands around 25,000-27,000 cm^{-1} are particularly strong
because of the large matrix elements. In comparison to the
lanthanide analog Gd^{3+}, the oscillator strengths of Cm^{3+}
transitions are 10-100 times greater ([116]). The ground state
of Cm^{3+} is an S state; the splitting, while larger than for
Gd^{3+}, is still only a few tens of cm^{-1} ([117]) and low temperatures would be required for quasi-four-level lasing.

Radiation-reduced Am^{2+} in CaF_2 crystals shows absorption
bands beginning at $\gtrless 650$ nm in addition to residual Am^{3+}
lines ([118]). The crystal-field splitting of the ground
state is >5 cm^{-1}.

Berkelium. The two visible and one infrared emitting states
indicated in Fig. 7 are candidates for stimulated emission, but
in all cases the number of bands suitable for broadband pumping
sources is limited ([119]). Therefore lasing thresholds will be
high unless selective excitation is used. The lowest-lying
6d bands ($\approx 35,000$ cm^{-1}) and charge transfer states should not
cause excited-state absorption for most transitions ([99]).
Transitions from the J=4 level at $\approx 19,600$ cm^{-1} include 4-level
lasing schemes. Transitions from J = 6 level at 15,800 cm^{-1}
are also laser candidates, but pumping is again limited. THe
transition from the J=4 level at 4500 cm^{-1} to the $^7F_6$ ground
state should provide infrared lasing with no competing
excited-state absorption.

Californium. The transition from the J = 11/2 state at
≈ 6500 cm^{-1} to the $^6H_{15/2}$ ground state is a laser candidate.
There are several absorption bands in the visible and near-
infrared ([120]) which could be used for optical pumping in
hosts with strong ion-phonon coupling and efficient decay to
the metastable J=11/2 state. In some hosts, efficient
emission and possible laser transitions may also occur from
the higher J=11/2 state at $\approx 11,500$ cm^{-1} and the J=5/2 state
at $\approx 20,000$ cm^{-1}. Intense absorption bands begin at $\approx 33,000$
cm^{-1} and have been attributed to charge transfer states
rather than f-d transitions ([99]).

Einsteinium. The most promising lasing transition in most hosts is $^5I_5 \rightarrow {}^5I_8$ at $\approx$9800 cm^{-1}. Because the U(4) and U(6) matrix elements for this transition are not large, the radiative lifetime of $^5I_5$ should be relatively long. There are several pump bands in the near-infrared and visible (121). The J = 6 states at $\approx$13,150 cm^{-1} and 20,200 cm^{-1} have large U(2) matrix elements for absorption and are probably hypersensitive. The longest lived isotope ^{254}Es has a half-life of only 276 days.

Fluorescence Sensitization. The optical pumping efficiency and thereby the prospects for oscillation of several actinides may be improved by codoping with sensitizer ions using schemes similar to those employed for the lanthanides in Table IV. For example, the first excited J states of U^{3+}, Np^{3+}, and Pu^{3+} could be sensitized by energy transfer from lanthanide co-dopants such as Ho^{3+}, Er^{3+}, Tm^{3+}, and Yb^{3+} as in the "alphabet" Ho^{3+} scheme (31). Other schemes using actinides include Cf^{3+}-sensitized Bk^{3+} ($^7F_4 \rightarrow {}^7F_6$), U^{3+}-sensitized Np^{3+} ($^5I_5 \rightarrow {}^5I_4$) and U^{3+} plus Np^{3+}-sensitized Pu^{3+} ($^6H_{7/2} \rightarrow {}^6H_{5/2}$), where the potential lasing transition is in parentheses. Depending upon the host and the rates of nonradiative decay, energy transfer from U^{3+} could also lead to population of the Np^{3+} $^6H_{9/2}$ level. Other sensitization schemes involving lanthanide and other transition group ions are readily conceivable.

Conclusions

The energy levels and rates of radiative and nonradiative transitions important for achieving laser action are now well established for most lanthanide and actinide ions. With the experimental data base acquired over the past two decades, new lasing schemes can be predicted using a calculational approach. Krupke (74) has given examples of the use of the Judd-Ofelt theory and the phenomenological treatment of multiphonon relaxation to predict transitions for rare-earth quantum electronic devices in solids. Whereas our survey has been qualitative, as more data becomes available phenomenological parameters can be derived from which excitation and decay modes can be predicted and their rates estimated. This approach provides a valuable method for prescreening promising new ion-host combinations and lasing possibilities.

The field of lathanide lasers is mature but not exhausted. Additional laser schemes and materials will undoubtedly be exploited. There are 1639 free-ion energy levels associated with the 4f^n electronic configurations of the thirteen trivalent lanthanides. Yet, of the 192,177 possible transitions between pairs of levels, by mid-1979 only 41 had been used for lasers. It is certain that given suitable pump sources and materials, stimulated emission involving many

more transitions will be obtained. This is especially
true with the increasing availability of lasers at new wavelengths for pump sources and of tunable lasers for selective
excitation of levels.

The prospects for actinide lasers, based on available
spectroscopic data, is definitely more limited. Although
there are a few prospects for visible lasers, the presence
of low-lying 6d and electron transfer states can cause
intense excited-state absorption, thus limiting oscillation
principally to the infrared. Strong ion-host interactions
increase the probabilities for radiative and nonradiative
transitions and must be carefully considered with respect to
the overall operation and efficiency of any practical system.
In view of the ease and success of lasing lanthanide ions,
only some compelling reason such as the requirement of a
specific wavelength would warrant development of some of the
actinide lasing schemes discussed. Perhaps additional
spectroscopy will reveal advantages of using actinide ions
in other valence states and hosts for efficient laser action.

Achieving laser action is a result of a favorable
combination of many spectroscopic properties of an ion in a
given host. The ability to predict and demonstrate stimulated
emission is therefore a powerful confirmation of our understanding of the spectroscopy of lanthanide and actinide ions
and a motivation for further study of these ions.

Acknowledgments

I thank Drs. W. Carnall and J. Hessler for valuable
discussions and the opportunity to see their paper prior to
publication.

Work was performed under the auspices of the U.S.
Department of Energy and the Lawrence Livermore Laboratory
under contract number W-7405-Eng-48.

Abstract

Stimulated emission has now been observed from twelve
lanthanide ions and one actinide ion. Host media have included crystalline and amorphous solids, metallo-organic and
inorganic aprotic liquids, and metal and molecular vapors.
Laser action spans a spectral range from approximately 0.3 to
4.0 μm and involves electronic and phonon-assisted 4f-4f,
5d-4f, and 5f-5f transitions. The lanthanides have enjoyed
their greatest utilization in optically-pumped solid-state
lasers; sizes range from thin films and small fibers for
integrated optics applications to large rods and disks in
high-power glass lasers for fusion experiments. The spectroscopic properties which distinguish the operation of
lanthanide and actinide lasers in various hosts are reviewed.
Recent results and possible future directions to exploit the
unique characteristics of lanthanide and actinide elements for
lasers are also discussed.

Literature Cited

1. Sorokin, P. P.; Stevenson, M. J. Phys, Rev. Lett., 1960, 5, 557.

2. Sorokin, P. P.; Stevenson, M. J. IBM J. Rev. Dev., 1961, 5, 56.

3. Snitzer, E. Phys. Rev. Lett., 1961, 7, 444.

4. Yariv, A. "Introduction to Optical Electronics"; Holt, Rinehart, and Winston: New York, 1976.

5. Koechner,W. "Solid State Laser Engineering"; Springer-Verlag: Berlin, Heidelberg, New York, 1976.

6. McCumber, D. E. Phys. Rev., 1964 136, A954.

7. Beck, R.; English, W.; Gurs, K. "Table of Laser Lines in Gases and Vapors:, Second Edition; Springer-Verlag: Berlin, Heidelberg, New York, 1978.

8. Bokham, P. A.; Klimkin, V. M.; Prokopev, V. E.; Monastyrev, S. S. Sov. Tech. Phys. Lett., 1973, 3, 166.

9. Krupke, W. F.; Jacobs, R. R. in "The Rare Earths in Modern Science and Technology"; McCarthy, G. J.; Rhyne, J. J., Eds,; Plenum: New York, 1978; p. 519.

10. Jacobs, R. R.; Krupke, W. F. Appl. Phys. Lett., 1978, 32, 31.

11. Lempicki, A.; Samelson, H. in "Lasers", vol. 1; Levine, A. K., Ed.; Marcel Dekker, Inc: New York, 1966; p. 181.

12. Samelson, H.; Kocher, R. "High Energy Pulsed Liquid Laser - Final Technical Report, Contract N0014-68-C-0110", U.S. Office of Naval Research: Washington, D.C., 1974.

13. Peterson, O. G., in "Methods of Experimental Physics," vol. 15, part A; Tang, C. L., Ed. ; Academic Press: New York, 1979; p. 251.

14. Lempicki, A. in "Handbook of Lasers"; Pressley, R. J., Ed.; CRC Press: Cleveland, 1971; p. 355.

15. Schimitschek, E. J.; Nerich, R. B.; Trias, J. A. J. Chem. Phys., 1967, 64, 173.

16. Heller, A. Appl, Phys. Lett. 1966, 9, 106.

17. Brecher, C.; French, K. W. J. Phys. Chem., 1973, 77, 1370.

18. Samelson, H.; Heller, A.; Brecher, C. J. Opt. Soc. Am.,
     1968, 58, 1054.

19. Weber, M. J. in "Handbook on the Physics and Chemistry of
     Rare Earths", vol. 4; Gschneidner, K. A., Jr.; Eyring, L.,
     Eds.; North-Holland Publishing Co.: Amsterdam, 1979;
     p. 275.

20. Chesler, R. B.; Geusic, J. E. in "Laser Handbook", vol. 1;
     Arecchi, F. T., Schulz-Dubois, E. A., Eds; North-Holland
     Publishing Co.: Amsterdam, 1972; p. 325.

21. Kaminskii, A. A. "Crystal Lasers"; Springer-Verlag: Berlin,
     Heidelberg, New York, 1979.

22. Weber, M. J. in "Methods of Experimental Physics", vol. 15,
     part A; Tang, C., Ed.; Academic Press, Inc.: New York,
     1979; p. 167.

23. Nassau, K. in "Applied Solid State Science", vol. 2; Wolfe,
     R., Ed.; Academic Press: New York, 1971; p. 173.

24. Young, G. C. Proc. IEEE, 1969, 57, 1267.

25. Snitzer, E. Amer. Ceram. Soc. Bull., 1973, 52, 516.

26. Cabeza, A. Y.; Treat, R. P. J. Appl. Phys., 1966, 37, 3556.

27. Snitzer, E.; Young, G. C. in "Lasers", vol. 2; Levine, A. K.,
     Ed.; Marcel Dekker, Inc.: New York, 1968.

28. Brawer, S. A.; Weber, M. J. (to be published).

29. Riseberg, L. A.; Weber, M. J. in "Progress in Optics",
     vol. XIV; Wolf, E., Ed.; North-Holland Publishing Co.:
     Amsterdam, 1976; p. 91.

30. Van Uitert, L. G. in "Luminescence of Inorganic Solids",
     Goldberg, P., Ed.; Academic Press: New York, 1966; p. 465.

31. Johnson, L. F.; Geusic, J. E.; Van Uitert, L. G. Appl. Phys.
     Lett., 1966, 8, 200.

32. Auzel, F. Proc. IEEE 1973, 61, 758.

33. Johnson, L. F.; Guggenheim, H. J. Appl. Phys. Lett. 1971,
     19, 44.

34. Weber, J. J. in "Handbook of Lasers", Pressley, R. J., Ed.; The Chemical Rubber Co.: Cleveland, 1971; p. 371.

35. Yang, K. H.; DeLuca, J. A. Appl. Phys. Lett. 1979, 35, 301.

36. Reisfeld, R.; Jorgensen, C. K. "Lasers and Excited States of Rare Earths"; Springer-Verlag; Berlin, Heidelberg, New York, 1977.

37. Dieke, G. H. "Spectra and Energy Levels of Rare Earth Ions in Crystals"; John Wiley: New York, 1968.

38. Wybourne, B. G. "Spectroscopic Properties of Rare Earths"; Wiley Interscience: New York, 1965.

39. Hifner, S. "Optical Spectra of Transparent Earth Compounds"; Academic Press: New York, 1978.

40. Loh, E. Phys. Rev.,1966, 147, 332.

41. Heaps, W. S.; Elias, L. R.; Yen, W. M. Phys. Rev., 1976, B13, 94.

42. Yang, K. H.; DeLuca, J. A. Phys. Rev., 1978, B17, 4246.

43. Ehrlich, D. J.; Moulton, P. F.; Osgood, R. M., Jr. Optics Lett., 1979, 4, 184.

44. Ehrlich, D. J. (private communication).

45. Weber, M. J. "Proc. 11th Rare Earth Res. Conf.", 1974, 361, (available from National Technical Information Service, U.S. Dept. Commerce).

46. Yang, K. H.; DeLuca, J. A. Appl. Phys. Lett. 1977, 31, 594.

47. Jacobs, R. R.; Krupke, W. F.; Weber, M. J. Appl. Phys. Lett., 1978, 33, 410.

48. Elias, L. R.; Heaps, W. S.; Yen, W. M. Phys, Rev., 1973, B8, 4989.

49. Weber, M. J. Solid State Commun., 1973, 12, 741.

50. Yang, K. H.; DeLuca, J. A. Appl. Phys. Lett., 1976, 29, 499.

51. Danielmeyer, H. G.in "Lasers", vol. 4, Levine, A. K.; DeMaria, A. J., Eds.; Marcel Dekker, Inc.: New York, Basel, 1976; p. 1.

52. van der Ziel, J. P.; Bonner, W. A.; Kopf, L.; Singh, S.; Van Uitert, L. G. Appl. Phys. Lett., 1973, 22, 656.

53. Burrus, C. A.; Stone, J. Appl. Phys. Lett., 1975, 26, 318.

54. Jacobs, R. R.; Weber, M. J. IEEE J. Quantum Electron.,1976, QE-12, 102.

55. Brachkovskaya, N. B.; Grubin, A. A.; Lunter, S. G.; Przhevuskii, A. K.; Raaben, E. L.; Tolstoi, M. N. Sov. J. Quantum Electron., 1976, 6, 534.

56. Chen, Bor-Ue; Tang, C. L. Appl. Phys. Lett., 1976, 28, 435.

57. Glaze, J. A.; Simmons, W. W.; Hagen, W. F. Proc. Soc Photo-Opt. Instrum. Eng., 1976, 76, 7.

58. Saruwatari, M.; Kiumira, T.; Otsuka, K. Appl. Phys. Lett., 1976, 29, 291.

59. Falk, J.; Huff, L.; Taynai, J. D. IEEE J. Quantum Electron., 1975, QE-11, 14D.

60. Voron'ko, K. Yu.; Noelle, E. L.; Osiko, V. V.; Timoshechkin, M. I. JETP Letters, 1971, 13, 86.

61. Heller, A. J. Amer. Chem. Soc., 1967, 89, 167.

62. Jacobs, R. R.; Krupke, W. F. Appl. Phys. Lett., 1979, 34, 497.

63. Krupke, W. F. IEEE J. Quantum Electron., 1972, QE-8, 725.

64. Wortman, E. E.; Morrison, C. A. IEEE J. Quantum Electron., 1973, QE-9, 956.

65. Reisfield, R.; Bornstein, A.; Beohm, L. J. Solid State Chem., 1975, 14, 14.

66. Judd, B. R. (this volume and references cited therein).

67. Azamotov, Z. T.; Arsen'yev, P. A.; Chukichev, M. V. Opt. Spectrosc.,1970, 28, 289.

68. Gandy, H. W.; Ginther, R. J. Appl. Phys. Lett., 1962, 1, 25.

69. Jenssen, H. P.; Castleberry, D.; Gabbe, D.; Linz, A. IEEE J. Quantum Electron., 1973, QE-9, 665.

70. Bjorklund, S.; Kellermeyer, G.; Hunt, C. R.; Filipescu, N. Appl. Phys. Lett., 1967, 10, 160.

71. Hessler, J. P.; Wagner, F., Jr.; Williams, C. W.; Carnall, W. T. J. Appl. Phys., 1977, 48, 3260.

72. Jacobs, R. R.; Krupke, W. F. Appl. Phys. Lett., 1979, 35, 126.

73. Johnson, L. F.; Guggenheim, H. J. Appl. Phys. Lett., 1973, 23, 1973.

74. Krupke, W. F. in Proc. IEEE 1974 Region 6 Conf.; IEEE: New York, 1975; p. 17.

75. Gandy, H. W.; Ginther, R. J.; Weller, J. F. Appl. Phys. Lett., 1965, 6, 237.

76. Johnson, L. F.; Guggenheim, H. J. IEEE J. Quantum. Electron. 1974, QE-10, 441.

77. Esterowitz, L.; Eckardt, R. C.; Allen, R. E. Appl. Phys. Lett., 1979, 35, 236.

78. Kaminskii, A. A.; Fedorov, V.A.; Sarkisov, S. E.; Bohm, J.; Reiche, P.; Schultze, D. Phys, Stat. Sol. (a), 1979, 53, K219.

79. Kaminskii, A. A. Opt. Spectrosc., 1971, 31, 938.

80. Kaminskii, A. A. An SSN Izvest, Neorgan, mater., 1971, 7, 906.

81. Carnall, W. T.; Hessler, J. P.; Hoekstra, H. R.; Williams, C. W. J. Chem, Phys., 1978, 68, 4004.

82. Gandy, H. W.; Ginther, R. J.; Weller, J. F. J. Appl. Phys., 1966, 38, 3030.

83. Baer, J. W.; Knights, M. G.; Chicklis, E. P.; Jenssen, H. P. "Proc. Topical Meeting on Excimer Lasers"; IEEE-OSA: Sept. 1979.

84. Etzel, H. W.; Gandy, H. W.; Ginther, R. J.; Appl. Opt., 1962, 1, 535.

85. Snitzer, E. J. Quantum Electron., 1965, 2, 1675.

86. McClure, D. S.; Kiss, Z. J. Chem. Phys., 1963, 39, 3251.

87. Loh, E. Phys. Rev., 1968, 175, 533.

88. Sorokin, P. P.; Stevenson, M. J. IBM J. Res. Dev., 1961, 5, 56.

89. Kaiser, W.; Garrett, C. G. B.; Wood, D. L. Phys. Rev., 1961, 123, 766.

90. Sorokin, P. P.; Stevenson, J. J.; Lankard, J. R.; Pettit, G. D. Phys. Rev., 1962, 127, 503.

91. Jacobs, R. R.; Weber, M. J. Laser Program Annual Rpt.-1974, UCRL 50021-74; Lawrence Livermore Laboratory, 1975.

92. Owen, J. F.; Dorain, P. B.; Mroczkowski, S.; Chang, R. K. J. Opt. Soc. Am., 1979, 69, 1963.

93. Rapp, C. F.; Boling, N. L.; Carlen, C. M. J. Opt. Soc. Am., 1977, 67, 1425.

94. Boling, N. L. (private communication).

95. Kiss, Z. J.; Duncan, R. C. Proc. IRE, 1962, 50, 1532.

96. Duncan, R. C.; Kiss, Z. J. Appl. Phys. Lett., 1963, 3, 23.

97. Hessler, J. P.; Carnall, W. T. (this volume).

98. Nugent, L. J.; Baybarz, R. D.; Burnett, J. L.; Ryan, J. L. J. Phys. Chem., 1973, 77, 1528.

99. Nugent, L. J.; Vander Sluis, K. L. J. Chem. Phys., 1973, 59, 3440.

100. Kushida, J. Phys. Soc. Japan, 1973, 34, 1318, 1327, 1334.

101. Hessler, J. P.; Brundage, R. T.; Hegarty, J.; Yen, W. M. (to be published).

102. Weast, R. C., Ed. "Handbook of Chemistry and Physics"; CRC Press: Cleveland, 1978.

103. Stacy, J. J.; Edelstein, N.; McLaughlin, R. D. J. Chem. Phys., 1972, 57, 4980.

104. Porto, S. P. S.; Yariv, A. Proc. IRE, 1962, 50, 1542.

105. Boyd, G. D.; Collins, R. J.; Porto, S. P. S.; Yariv, A.; Hargreaves, W. A. Phys. Rev. Lett., 1962, 8, 269.

106. Porto, S. P. S.; Yariv, A. "Quantum Electronics: Proceedings of the Third International Congress", Grivet, P.; Bloembergen, N., Eds.; Columbia Univ. Press: New York, 1964; p. 717.

107. Wittke, J. P.; Kiss, Z. J.; Duncan, R. C.; McCormick, J. J. Proc. IEEE, 1963, 51, 56.

108. Porto, S. P. S.; Yariv, A. J. Appl. Phys., 1962, 33, 1620.

109. Porto, S. P. S.; Yariv, A. Proc. IRE, 1962, 50, 1543.

110. Krupke, W. F.; Gruber, J. B. J. Chem, Phys., 1967, 46, 542.

111. McLaughlin, R.; White, R.; Edelstein, N.; Conway, J. G. J. Chem. Phys., 1968, 48, 967.

112. Carnall, W. T.; Fields, P. R.; Pappalardo, R. G. J. Chem. Phys., 1970, 53, 2922.

113. Friedman, H. A.; Ball, J. T. Inorg. Nucl. Chem., 1972, 34, 3928.

114. Finch, C. B.; Clark, G. W. J. Phys. Chem. Solids, 1972, 33, 922.

115. Conway, J. G. J. Chem. Phys., 1956, 24, 1275.

116. Carnall, W. T.; Rajnak, K. J. Chem. Phys., 1975, 63, 3510.

117. Edelstein, N.; Easley, W. J. Chem, Phys., 1968, 48, 2110.

118. Edelstein, N.; Easley W.; McLaughlin, R. J. Chem, Phys., 1966, 44, 3130.

119. Carnall, W. T.; Fried, S.; Wagner, F., Jr. J. Chem. Phys., 1973, 58, 3614.

120. Carnall, W. T.; Fried, S.; Wagner, F., Jr. J. Chem. Phys., 1973, 58, 1938.

RECEIVED December 26, 1979.

Electronic Structure of Actinyl Ions

R. G. DENNING, J. O. W. NORRIS, I. G. SHORT, T. R. SNELLGROVE, and D. R. WOODWARK

Inorganic Chemistry Laboratory, South Parks Road, Oxford, OX1 3QR, United Kingdom

The covalent bond in actinide chemistry is seen in its simplest and most striking form in the actinyl ions, MO_2^{2+}. These ions, therefore, provide the most straightforward test of our understanding of the covalent bond in these elements. Although superficially similar to transition metal oxy-cations there are many striking differences. A useful example can be made of $MoO_2Cl_2(PPh_3O)_2$ and $UO_2Cl_2(PPh_3O)_2$ whose X-ray crystal structures have recently been reported ([1], [2]). The approximate geometries are shown in Figure 1. Apart from the larger radius of uranium, as observed in the metal-chlorine distances, the most striking point is the change from cis-dioxo geometry in the molybdenum compound to trans-dioxo geometry in the uranium compound. Actually these compounds are only prototypes of general stereochemical differences between dioxo compounds of the transition metals and of the actinides. From the examples in Table 1 it seems that the principal factor determining the geometry is the nature of the lowest energy metal valence shell and its occupancy. It is striking that the addition of 'd' electrons to the valence shell causes a change in geometry, whereas the addition of 'f' electrons causes no change in the actinyl ions.

Stereochemistry of dioxo compounds

The stereochemistry of the transition metal compounds can be rationalised in a simple way which is illustrated in Figure 2. If only the 'd' orbitals are considered to be important in the bond, the linear dioxo ions have metal orbitals of σ_g, π_g and δ_g symmetry while the oxygen bonding orbitals have $\sigma_g, \sigma_u, \pi_g$, and π_u symmetry. The argument may be illustrated by considering only the σ-orbitals. The upper part of Figure 2 shows the result of bending the MO_2 unit. In the d^0 ions the oxide orbitals are formally full and the metal orbitals vacant. The system is therefore stabilised on changing from the linear configuration, where there are two bonding and two non-bonding electrons to the bent geometry for which all four electrons are in bonding orbitals.

0–8412–0568–X/80/47–131–313$05.00/0

Table I. Geometry and Metal Oxygen Bond Lengths of Some Metal Dioxo Compounds.

d^0 cis		d^2 trans		$f^{0,n}$ trans	
$[VO_2(oxalate)_2]^-$	1.63 Å			$[UO_2Cl_4]^{2-}$	1.76 Å
$[MoO_2Cl_4]^{2-}$	1.71 Å	$[MoO_2(CN)_4]^{4-}$		$[NpO_2]^{2+}$	
$MoO_2Cl_2(PPh_3O)_2$	1.68 Å	$[RuO_2X_4]^{2-}$		$[PuO_2]^{2+}$	
$MoO_2(butanediolate)_2$	1.66 Å	$[OsO_2X_4]^{2-}$	1.73 Å	$[AmO_2]^{2+}$	
$[ReO_2F_4]^-$		$[ReO_2X_4]^{3-}$	1.87 Å		

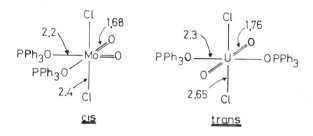

Figure 1. *The geometry of transition metal and actinide dioxo cations*

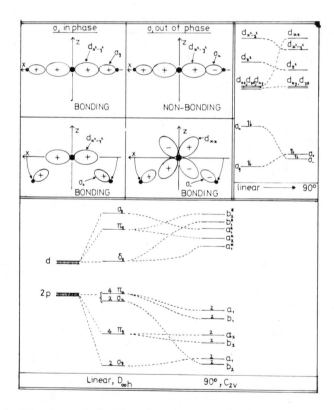

Figure 2. *The change in bonding characteristics and orbital energies in linear and bent transition metal dioxo species*

An identical argument applies to the π-orbitals and the lower part of Figure 2 shows a schematic energy level diagram (in the style of Walsh) for all the orbitals in the linear and bent configurations. When there are no metal electrons the bent configuration is favoured because no valence electrons are excluded, by virtue of their symmetry, from bonding. On the other hand the addition of two metal 'd' electrons occurs in non-bonding orbitals in the linear geometry but in anti-bonding orbitals in the bent geometry. In this way it is possible to understand the linearity of the d^2 ions. As it stands this argument obviously fails to predict the geometry of the actinyl ions and so it is necessary to know more about their electronic structure.

Energy levels in actinyl ions

The new feature in the actinyl ions is the probable importance of both 'f' and 'd' orbitals in the bonding. In the linear geometry the principal energy levels are those shown in the scheme in Figure 3. It is clear from many lines of evidence that the 'f' orbitals lie below the 'd' orbitals, and that the δ_g, δ_u and ϕ_u orbitals are excluded from bonding to oxygen by symmetry; but the ordering of the σ_u, π_u, σ_g and π_g bonding orbitals depends on the relative importance of σ and π-bonding on the one hand and 'd' and 'f' orbital bonding on the other. A great variety of ordering schemes have been suggested, both on experimental and theoretical grounds (3-8). Because of the lack of agreement on this important point and its relevance in characterising the bonding we have looked more closely at the experimental evidence and in particular at the electronic spectra.

Electronic spectra and structure

Figure 4 gives a survey of the polarised single crystal spectrum of $Cs_2UO_2Cl_4$ at 4.2K. Much of the detail in this spectrum has had to be omitted in the presentation but it should already be clear why the analysis of the spectrum has proved difficult. The techniques by which a spectrum like this can be analysed have been presented elsewhere (9-10) and are primarily of spectroscopic interest, but it is valuable to outline here the type of information that is obtainable. Two general observations can immediately be made from the data in Figure 4. First, the low intensity of the absorption suggests that the transitions are forbidden. Because the spin-orbit coupling is large in uranium it is not likely that a spin selection rule applies but rather a spatial selection rule must be operative. Second, the spectrum is clearly composed of progressions in the UO_2^{2+} symmetric stretching frequency ($\sim720cm^{-1}$). It is possible to disentangle the progressions based on different electronic origins by oxygen-18 substitution which markedly lowers the progression frequency. Vibrational features associated with pure electronic states are

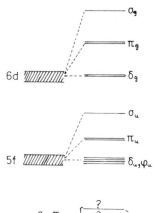

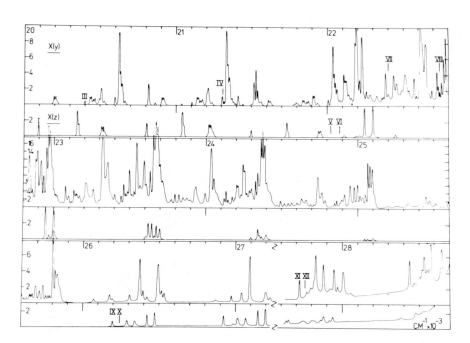

Figure 3. *Schematic orbital energies in actinyl ions*

Molecular Physics

Figure 4. *A single-crystal polarized absorption spectrum of $Cs_2UO_2Cl_4$ at 4.2K*
(10)

virtually unaffected by the substitution while those associated
with the progressions are strongly shifted (10).

Fortunately the crystal structure of $Cs_2UO_2Cl_4$ is particularly simple, there being only one molecule per unit cell, the
uranium atom lying at a C_{2h} site with inversion symmetry (11). In
the monoclinic system it proves possible to propagate the light in
three orthogonal directions X, Y and Z with respect to the molecular axis system (Figure 5) and to choose the electric vector of
the radiation (x), (y) and (z) in such a way as to define six
different experiments. The outcome is shown in Figure 6. By
comparing the X(y) and Z(y) spectra and the Z(x) and Y(x) spectra
the bands labelled I and II in the figure are seen to be magnetic-
dipole allowed, while a careful study of band III (9) shows it to
be electric-quadrupole allowed. Similar evidence shows that all
twelve electronic excited states observed in this spectrum (10)
are parity forbidden. Since the lowest energy empty orbitals are
ungerade 'f' orbitals it follows that the excitation must come
from either σ_u or π_u filled orbitals.

More evidence about the nature of the excited states comes
from Zeeman effect measurements. In the C_{2h} site in $Cs_2UO_2Cl_4$
there is no degeneracy possible so that all Zeeman effects are
second order, nevertheless the symmetry is sufficiently close to
D_{4h} that the second order effects are easily measured. Figure 7
shows some examples. The most important observation is that the
first excited state has a magnetic moment of 0.16 Bohr Magnetons.
Apparently the magnetic moment of the hole in the oxygen orbitals
almost cancels that of the 'f' electron. Two states, with the
Π_g ($D_{\infty h}$) symmetry implied by the magnetic dipole intensity, seem
possible, with the wavefunctions $|\bar{\sigma}_u\bar{\delta}_u{}^{2+}>$ and $|\bar{\pi}_u{}^{+1}\pi_u{}^{-1}\bar{\pi}_u{}^{-1}\phi_u{}^{3+}>$.
There is no simple choice at this point between these possibilit-
ies. Nevertheless the observed symmetries of the remaining
excited states are better described in terms of the former configuration. Figures 8 and 9 show the energies of the various
excited states arising from the $\sigma_u\delta_u$ and $\pi_u{}^3\phi_u$ configurations
using realistic spin-orbit coupling parameters and varying the
inter-electron repulsion parameters. Figure 8 predicts that the
second excited state will be of $\Delta_g(D_{\infty h})$ symmetry while Figure 9
predicts $\Gamma_g(D_{\infty h})$ symmetry. The electric quadrupole intensity of
band III in Figure 6 is only consistent with $B_{2g}(D_{4h})$ and $\Delta_g(D_{\infty h})$
symmetry suggesting that Figure 8 and the $\sigma_u\delta_u$ configuration give
the best description.

There are many additional pieces of evidence to support this
assertion, the most powerful of which is a theoretical argument
first advanced by Görller-Walrand and Vanquickenborne (12) and
slightly recast by us (13) which shows that in a strong axial
field it is not possible to observe first-order equatorial field
splittings in a two-open-shell system unless the configuration
is of the type $\sigma\gamma$, where γ is a general $D_{\infty h}$ representation. Since
there is ample evidence of first order equatorial field splittings
the excitation of a σ_u rather than a π_u electron is strongly

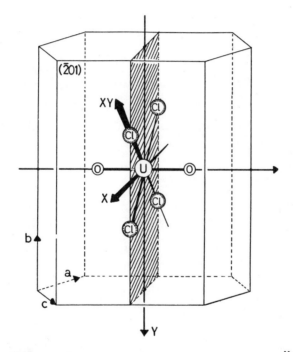

Molecular Physics

Figure 5. *Crystallographic axes, crystal habit, and molecular axes of* $Cs_2UO_2Cl_4$
(9)

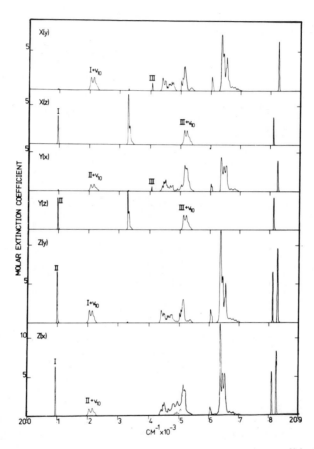

Figure 6.   *Absorption spectrum of single crystals of* $Cs_2UO_2Cl_4$ *at 4.2K in six different polarizations. Notation is explained in the text (9).*

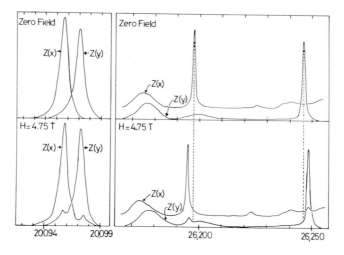

Molecular Physics

Figure 7. *The effect of a 4.75T magnetic field on orthogonally polarized single-crystal absorption spectra of $Cs_2UO_2Cl_4$ at 4.2K (10)*

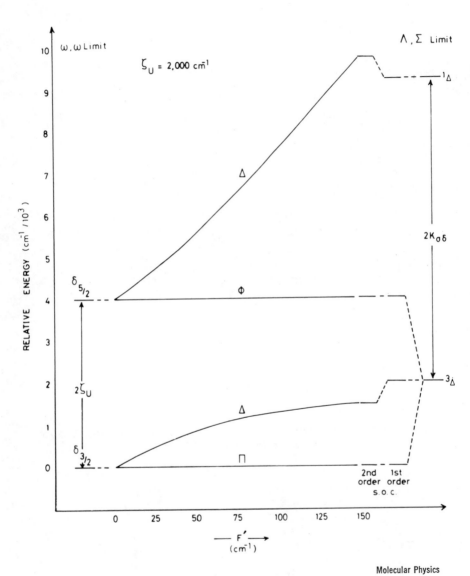

Figure 8.   Correlation diagram for the states arising from the $\sigma\delta$ configuration (13)

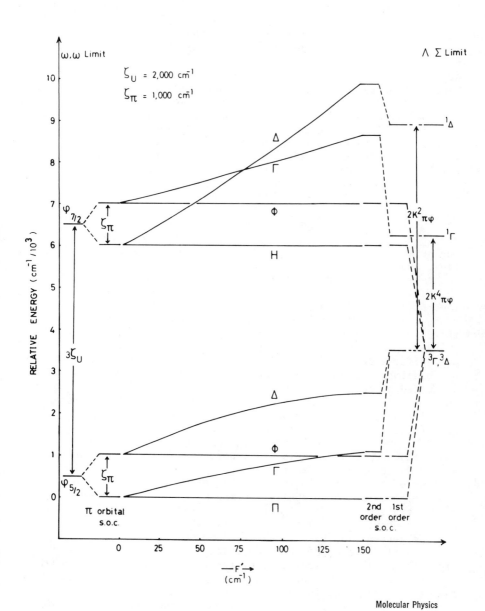

Figure 9.   *Correlation digram for the states arising from the $\pi^3\phi$ configuration* (13)

supported. For example, in the approximately D_{3h} site symmetry found in $CsUO_2(NO_3)_3$, axial field states of Φ_g symmetry (arising from a $\sigma_u\phi_u$ configuration) should be split by the equatorial field into A_1'' and A_2'' components (14). The absence of first order Zeeman effects makes these states difficult to identify. Nevertheless we have found that the nitrate internal modes couple appreciably to the electronic transitions in this compound (15). Figure 10 shows the nitrogen-15 isotope shift of one such feature. The magnitude of the shift identifies the mode, whose frequency is known from the pure vibrational spectrum, and the symmetry of the representations which it spans in D_{3h}. Taken with the polarisation data the symmetry of the electronic excited state to which this mode couples can then be constrained to either A_1'' or $E''(D_{3h})$. The absence of a magnetic moment narrows the choice to A_1''. The A_2'' component of the Φ_g ($D_{\infty h}$) state can also be identified via the similar isotope characterisation of a second nitrate internal mode.

Using a variety of experimental techniques of this kind we have been able to fix the energies and, with a few exceptions, the symmetries of twelve electronic excited states in $Cs_2UO_2Cl_4$ (10), and seven excited states in both $CsUO_2(NO_3)_3$ and $NaUO_2$ (acetate)$_3$ (14). Superficially the states appear to arise from the excitation of a σ_u electron and so we have tested a simple theoretical model based on the $\sigma_u\delta_u$ and $\sigma_u\phi_u$ excited configurations. To be convinced we should expect that similar values of the spin-orbit and inter-electron repulsion parameters should prevail in each compound but that the equatorial field parameters may differ widely. Figure 11 and 12 show the results of these calculations. Their significance and limitations have been discussed elsewhere (13) but it is easy to see that the agreement between the observed and calculated energies and magnetic moments is quite satisfactory for both $Cs_2UO_2Cl_4$ and $CsUO_2(NO_3)_3$, so that as a first approximation the model seems attractive.

The observation that the $\sigma_u\phi_u$ configuration lies 2900cm^{-1} above the $\sigma_u\delta_u$ configuration is important. This is not the same as the difference between the ϕ_u and δ_u virtual orbitals on account of the attraction between the electron in these orbitals and the hole in the σ_u shell. Making a reasonable estimate of this attraction sets the ϕ_u virtual orbital between 1500cm^{-1} and 2700cm^{-1} above the δ_u virtual orbital (13).

Jørgensen (16) takes the view, opposed to ours, that the first excited states of the uranyl ion stem from the $\pi_u^3\phi_u$ configuration. The implications for the relative ϕ_u and δ_u virtual orbital energies have not been investigated but it seems unlikely that this assignment is consistent with a ϕ_u orbital 2000cm^{-1} above the δ_u orbital. The simplest way to independently investigate the energies of these two orbitals is through the properties of the single 'f' electron in the neptunyl ion. To this end we have confirmed, by Zeeman effect measurements, the peculiar ESR results, due to Leung and Wong (17), that in $Cs_2U(Np)O_2Cl_4$ the

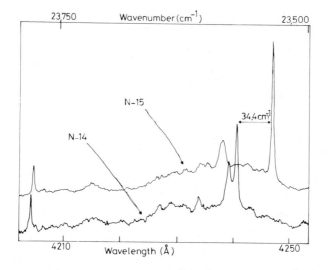

Figure 10.   Nitrogen-15 isotopic shift in the π-polarized, single-crystal absorption spectrum of $CsUO_2(NO_3)_3$ at 4.2K

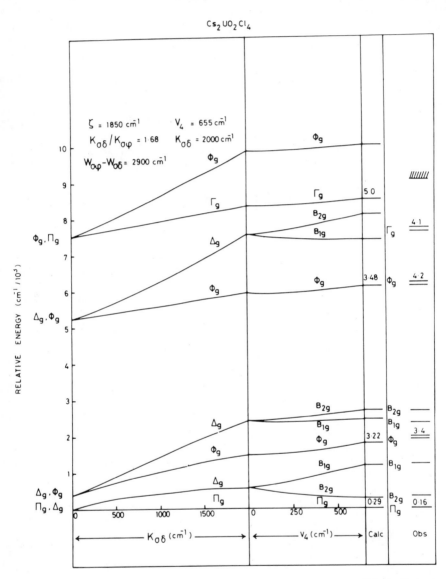

Figure 11. Calculated and observed energy levels for $Cs_2UO_2Cl_4$. Numbers on the diagram indicate magnetic moments (13).

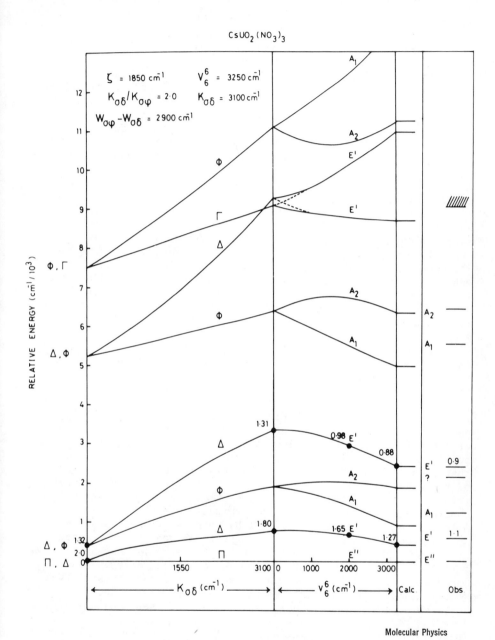

Figure 12. *Calculated and observed energy levels for CsUO₂(NO₃)₃. Numbers on the diagram indicate magnetic moments (13).*

ground state 'g' values are $g_\| \simeq g_\perp \simeq 1.32$. The apparent isotropy of
the 'g' value seems to contradict the extreme anisotropy of the
ligand field. The reason can be uncovered by a calculation of the
'g' values as a function of the energy difference of the ϕ_u and δ_u
orbitals (18). Figure 13 shows that when this difference is large
in either sense $g_\perp$ tends to zero and $g_\|$ to the appropriate value
for the ground state. Intermediate values can be seen to arise
because of the mutual interaction of the $\delta_{3/2}$ and $\phi_{5/2}$ states, as
a consequence of both the tetragonal field and the second-order
spin orbit coupling. The excellent agreement between the theoreti-
cal prediction that $g_\| \simeq g_\perp = 1.4$ and the experimental values sets
tight limits on the orbital energy difference at $2100 cm^{-1}$. This
is excellent support for the parameter choice used in our model of
the uranyl excited states.

   All in all our work implies that the highest filled orbitals
are of σ_u symmetry. To anyone reflecting on the electronic
structure of carbon dioxide it is extraordinary to find the σ_u
orbital above the π_u orbital, implying that the latter forms the
stronger bond. Nevertheless this state of affairs was anticipated
many years ago in the overlap calculations of Belford and Belford
(4). They pointed out that the angular nodal properties of the f_σ
and f_π orbitals are such that at short distances the f_σ-p_σ overlap
may actually be less than the f_π-p_π overlap; a result confirmed in
a calculation by Newman (5).

   The situation is, however, more complicated than this argument
implies because the $f(\pi_u)$ antibonding orbital energy is observed,
in the spectra of the neptunyl ion, to be about $15,000 cm^{-1}$ above
the ϕ_u and δ_u orbitals (18,19), while the $f(\sigma_u)$ orbital, presumably at much higher energy is not observed. It seems likely, from
recent comprehensive calculations (8), that the relatively high
energy of the filled (and empty) σ_u orbitals arises from the role
of the filled $6p(\sigma_u)$ orbital of the closed shell within the valence
shell; its interaction with oxygen orbitals being greater than
that of $6p(\pi_u)$. Whatever the explanation it is clear from the drop
of the uranyl symmetric stretching frequency in the excited states
(from $835 cm^{-1}$ to $710 cm^{-1}$) that the σ_u electron is quite strongly
bonding. Since the π_u, σ_g and π_g orbitals must all be placed
below the σ_u orbital they too must be seen as strongly bonding.

   The best evidence therefore suggests an energy level scheme
of the type shown in Figure 14. The implication is that all
twelve valence electrons are in bonding orbitals, offering an
explanation for the extraordinary stability and shortness of the
actinyl bond. Formally each metal oxygen bond is a triple bond.
Moreover because the σ_u and π_u orbitals are already bonding in the
linear geometry, by virtue of their interaction with 'f' orbitals,
there is no tendency for the linear dioxo unit to bend as is the
case in the transition metal oxy cations. Addition of further 'f'
electrons leads to the filling of orbitals which are non-bonding
towards oxygen so that the remaining actinyl ions are also linear.

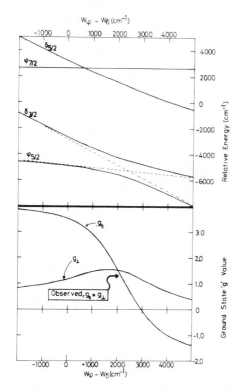

Figure 13. Ground-state "g" values and relative energies of the φ and δ states in $Cs_2U(Np)O_2Cl_4$ as a function of orbital energy difference

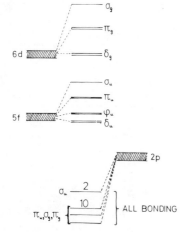

Figure 14. A possible energy-level scheme for actinyl ions

Summarising, there is clear evidence that both 'f' and 'd' orbitals participate in the actinyl bond and it is this joint participation which is responsible for both the stability and the linearity of the dioxo ions.

Literature Cited

1. Butcher, R. J.; Penfold, B. R.; Sinn, E. J. Chem. Soc.; Dalton Trans., 1979, 668.
2. Bombieri, G.; Forsellini, E.;Day, J. P.; Azzeez, W. I. J. Chem. Soc., Dalton Trans., 1978, 677.
3. McGlynn, S. P.; Smith, J. K. J. Mol. Spec., 1961, 6, 164.
4. Belford, R. L.; Belford, G. J. Chem. Phys., 1961, 34, 1330.
5. Newman, J. B. J. Chem. Phys., 1965, 43, 1691.
6. Boring, M.; Wood, J. H.; Moscowitz, J. W. J. Chem. Phys., 1975, 63, 638.
7. Yang, C. Y.; Johnson, K. H.; Horsley, J. A. J. Chem. Phys., 1978, 68, 1000.
8. Walch, P. F.; Ellis, D. E. J. Chem. Phys., 1976, 65, 2387.
9. Denning, R. G.; Snellgrove, T. R.; Woodwark, D. R. Mol. Phys., 1975, 30, 1819.
10. Denning, R. G.; Snellgrove, T. R.; Woodwark, D. R. Mol. Phys., 1976, 32, 419.
11. Hall, D.; Rae, A. D.; Water, T. N. Acta Cryst., 1966, 20, 160.
12. Görller - Walrand, C.; Vanquickenborne, L. G. J. Chem. Phys., 1972, 57, 1436.
13. Denning, R. G.; Snellgrove, T. R.; Woodwark, D. R. Mol. Phys., 1979, 37, 1109.
14. Denning, R. G.; Foster, D. N. P.; Snellgrove, T. R.; Woodwark, D. R. Mol. Phys., 1979, 37, 1089.
15. Denning, R. G.; Short, I. G.; Woodwark, D. R. (in preparation)
16. Jørgensen, C. K. J. Luminescence, 1979, 18, 63.
17. Leung, A. F.; Wong, E. Y. Phys. Rev., 1969, 180, 380.
18. Brown, D.; Denning, R. G.; Norris, J. O. W. (in preparation)
19. Stafsudd, O. M.; Leung, A. F.; Wong, E. Y. Phys. Rev., 1969, 180, 339.

RECEIVED December 26, 1979.

Synthesis and Characterization of Protactinium(IV), Neptunium(IV), and Plutonium (IV) Borohydrides

RODNEY H. BANKS and NORMAN M. EDELSTEIN

Materials and Molecular Research Division, Lawrence Berkeley Laboratory, Berkeley, CA 94720 and Department of Chemistry, University of California, Berkeley, CA 94720

Abstract

The actinide borohydrides of Pa, Np, and Pu have been prepared and some of their physical and optical properties measured.
X-ray powder diffraction photographs of $Pa(BH_4)_4$ have shown that
it is isostructural to $Th(BH_4)_4$ and $U(BH_4)_4$. $Np(BH_4)_4$ and
$Pu(BH_4)_4$ are much more volatile than the borohydrides of Th, Pa,
and U and are liquids at room temperature. Results from low-
temperature single-crystal x-ray diffraction investigation of
$Np(BH_4)_4$ show that its structure is very similar to $Zr(BH_4)_4$.
With the data from low-temperature infrared and Raman spectra, a
normal coordinate analysis on $Np(BH_4)_4$ and $Np(BD_4)_4$ has been
completed. EPR experiments on $Np(BH_4)_4/Zr(BH_4)_4$ and $Np(BD_4)_4/$
$Zr(BD_4)_4$ have characterized the ground electronic state.

Four of the seven known metal tetrakis-borohydrides--Zr, Hf,
Th, and U borohydrides (1,2)--were first synthesized about 30
years ago during the Manhattan project. They were found to be very
volatile and reactive compounds. In recent years, much structural,
spectroscopic, and chemical studies were done on these molecules.
New tetrakis-borohydrides of the actinides Pa, Np, and Pu have recently been prepared by analogous reactions used in the syntheses
of U and Th borohydrides (3). The Pa compound, $Pa(BH_4)_4$, is isomorphous to and behaves like $U(BH_4)_4$ and $Th(BH_4)_4$ while x-ray
studies on $Np(BH_4)_4$ and the isostructural $Pu(BH_4)_4$ have shown that
they resémble the highly volatile Zr and Hf compounds both in
structure and properties. All seven compounds contain triple
hydrogen bridge bonds connecting the boron atom to the metal.
The 14 coordinate Th, Pa, and U borohydrides (4), in addition,
have double-bridged borohydride groups that are involved in link-
ing metal atoms together in a low symmetry, polymeric structure.
The structures of the other four borohydride molecules are mono-

0–8412–0568–X/80/47–131–331$05.00/0

meric and much more symmetrical; the 12 coordinate metal is surrounded by a tetrahedral array of BH_4^- groups (5,6,7).

In an effort to understand the energy level structures of
actinide 4+ ions in borohydride environments, optical and magnetic
measurements have been initiated. Spectra of pure $Np(BH_4)_4$ and
$Np(BD_4)_4$, and these compounds diluted in single-crystal host
matrices of $Zr(BH_4)_4$ and $Zr(BD_4)_4$, respectively, have been
obtained in the region 2500–300 nm at 2K. The covalent actinide
borohydrides display rich vibronic spectra (8) and assignment of
the observed bands depends on a knowledge of the vibrational
energy levels of $M(BH_4)_4$ molecules. A normal coordinate analysis
derived from low-temperature infrared and Raman spectra of $Np(BH_4)_4$
and $Np(BD_4)_4$ was undertaken to elucidate the nature of their
fundamental vibrations and overtones. Electron paramagnetic
resonance (epr) spectra of $Np(BH_4)_4$ and $Np(BD_4)_4$ that characterize
the ground electronic state have been obtained in a number of host
materials. Optical spectra of $Pa(BH_4)_4$ and $Pa(BD_4)_4$ isolated in
an organic glass were obtained in the near infrared and visible
regions at 2K. This paper will summarize our progress to date
on these studies.

Experimental

Preparation of Borohydrides. Metal borohydrides are very
chemically reactive and most of them are pyrophoric in air. The
syntheses of the compounds and all manipulations with Al, Zr, Hf,
Np, and Pu borohydrides must therefore be performed in a grease-
free high-vacuum line. Work involving the less volatile Th, Pa,
and U borohydrides can also be done in argon-filled dryboxes.

All actinide borohydrides are made by reacting the anhydrous
actinide tetrafluoride with liquid $Al(BH_4)_3$ in the absence of a
solvent in a sealed glass reaction tube. The basic reaction
equation is:

$$AnF_4 + 2Al(BH_4)_3 \rightarrow An(BH_4)_4 + 2AlF_2BH_4.$$

Purification of the desired product is accomplished by sublimation
where only the unreacted $Al(BH_4)_3$ and $An(BH_4)_4$ are volatile. The
large difference in volatilities of these compounds permit easy
separation. $Th(BH_4)_4$ and $Pa(BH_4)_4$ are obtained on a $0°$ cold
finger by heating the solid reaction mixture to $120°$ and $55°$,
respectively. Uranium, neptunium, and plutonium borohydrides
sublime at room temperature and are collected in a dry ice trap
through which the $Al(BH_4)_3$ passes into a liquid nitrogen trap.
The stabilities of the actinide borohydrides dictate the type of
reaction conditions needed for successful preparation. The
polymeric compounds are stable at room temperature and their
syntheses are carried out at $25°$ for about five days. $Np(BH_4)_4$
and $Pu(BH_4)_4$ are unstable at room temperature and require that
the tetrafluorides react at $0°$ for only a few hours. These two

borohydrides must be stored at dry-ice or liquid-nitrogen temperature in a greaseless storage tube. $Zr(BH_4)_4$ used in experiments described here was prepared similarly to $U(BH_4)_4$ by reacting Na_2ZrF_6 with $Al(BH_4)_3$.

Preparation of Borodeuterides. All glassware which contacts the borodeuterides had been previously passivated with B_2D_6 or treated with D_2O and then baked out thoroughly under vacuum. The borodeuterides of Th, Pa, and U are prepared as described above using $Al(BD_4)_3$ as the source of BD_4^-. The high volatilities of the covalent borohydrides allow their deuterated analogs to be prepared by a more satisfactory method that utilizes the H ↔ D exchange property of these molecules with deuterium (9). If the D_2 gas is maintained in large excess, the extent of equilibrium will give the fully deuterated product in high yield. In a passivated glass bulb, a mixture of the borohydride vapor and 1 atm of D_2 gas was allowed to stand for a few days at room temperature. After freezing out the products at $-78°$ and evacuating, another volume of D_2 was added and the exchange reaction continued. Three cycles were sufficient to give the metal borodeuteride having an isotopic purity as high as that of the deuterium used (99.7%).

An attempt to prepare $Np(BT_4)_4$ using the above method resulted in the decomposition of the borohydride due to the extremely high radiation field of the T_2 gas (66 Ci) and no volatile Np compound was recovered.

The vapor pressure of $Np(BH_4)_4$ was determined as a function of temperature using a Bourdon gauge (5). The data for the liquid and solid shown in Figure 1 were used in calculating thermodynamic quantities of the actinide borohydrides given in Table 1. A single crystal x-ray study (5) was carried out for $Np(BH_4)_4$ at 130K. Its structure is shown in Figure 2.

Gas-phase infrared and low-temperature solid-state infrared and Raman spectra were obtained for $Np(BH_4)_4$ and $Np(BD_4)_4$ from 2.5 to 50μ. Assignments were made of the observed bands and the fundamental frequencies were fitted to calculated values in a normal coordinate analysis (10).

Electron paramagnetic resonance spectra were taken of $Np(BH_4)_4/Zr(BH_4)_4$ and $Np(BD_4)_4/Zr(BD_4)_4$ mixed crystals at X, K, and Q bands. Spin Hamiltonian parameters were found by a least-squares fit of the data.

Electronic spectra of $Pa(BH_4)_4$ and $Pa(BD_4)_4$ in an organic glass were obtained at 2K from 2200 nm - 300 nm.

Results and Discussion

The crystal structure of $U(BH_4)_4$ has been examined by single crystal x-ray (4b) and neutron diffraction techniques (4a). Much like the bonding in the well-known boron hydrides (11), this metal borohydride exhibits hydrogen bridge bonds that join the boron atom to the metal. In $U(BH_4)_4$, there are two tridentate and four

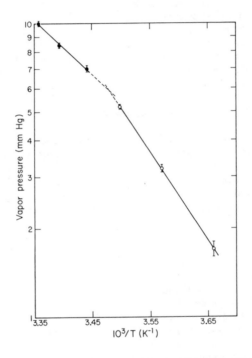

Figure 1. Vapor pressure vs. 1/T for Np(BH₄)₄: (○), data of the liquid; (●), data for the solid.

Table 1

Physical Properties of Metal Tetrakis–Borohydrides

Property	Th(BH$_4$)$_4$[a]	U(BH$_4$)$_4$[b]	Hf(BH$_4$)$_4$[a]	Zr(BH$_4$)$_4$[a]	Np(BH$_4$)$_4$[c]
Crystallographic space group	P4$_3$2$_1$2 (tetragonal)	P4$_3$2$_1$2[d] (tetragonal)	P$\bar{4}$3m[e] (cubic)	P$\bar{4}$3m[f] (cubic)	P4$_2$/nmc (tetragonal)
Solid-state structure	polymeric	polymeric	monomeric	monomeric	monomeric
Density in the solid state (gm/cc)	2.53	2.69	1.85	1.13	2.23
Melting point (°C)	203[g]	126[g]	29.0	28.7	14.2
Boiling point (°C) extrap.	-	-	118	123	153
Vapor pressure (mmHg/°C)	0.05/130	0.19/30	14.9/25	15.0/25	10.0/25
Liquid[h] A	-	-	2097	2039	1858
Liquid B	-	-	8.247	8.032	7.24
Solid[h] A	-	4264.6	2844	2983	3168
Solid B	-	13.354	10.719	10.919	11.80
Heat of sublimation (Kcal/mol)	21	19.5	13.0	13.6	14.5
Heat of vaporization (Kcal/mol)	-	-	9.6	9.3	8.5
Heat of fusion (Kcal/mol)	-	-	3.4	4.3	6.0
Entropy of sublimation (cal/mol°)	-	61.1	49.0	50.0	54.0
Entropy of vaporization (cal/mol°)	-	-	37.7	36.8	33.1
Entropy of fusion (cal/mol°)	-	-	11.3	13.2	20.9
Solubility in pentane	insol	slight	high	high	high

[a] Ref. 1
[b] Ref. 2
[c] Ref. 5
[d] Ref. 4
[e] Ref. 7
[f] Ref. 6
[g] With decomposition
[h] Log p(mmHg) = $-A/T + B$

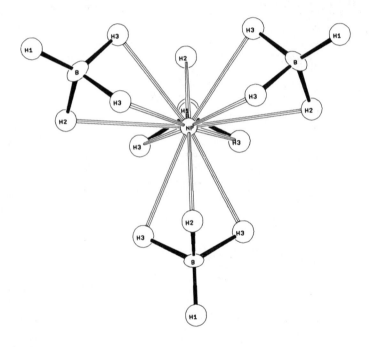

Figure 2.   ORTEP diagram of Np(BH₄)₄

bidentate BH_4^- groups. The tridentate bridge bond links the metal
atom to the boron atom through a triple-hydrogen-bridge bond
while the fourth hydrogen atom forms a terminal bond with the
boron atom. The bidentate bridge bond links one boron atom to two
metal atoms through two double-hydrogen-bridge bonds, resulting
in a helical polymeric structure.

Low-temperature x-ray powder diffraction photographs(3) of
$Np(BH_4)_4$ and $Pu(BH_4)_4$ revealed that they are isostructural and of
a unique structure type. The structure of $Np(BH_4)_4$ was determined
by a low-temperature, single-crystal x-ray study at 130K (5). The
borohydride molecules are monomeric and crystallize into the
tetragonal space group, $P4_2/nmc$, where a = 8.559(9) Å,
c = 6.017(9) Å, and Z = 2. The four terminal, triply-bridged
borohydride groups are bound to the Np atom with Np-B distances
of 2.46(3) Å. Although the hydrogen atoms were observed in the
Fourier maps and refined, values of the $Np-H_b$ bond lengths,
2.2(5) Å, had large standard deviations. No evidence was found
for symmetry lower than T_d for $Np(BH_4)_4$.

The molecular structure of $Np(BH_4)_4$ is illustrated in the
ORTEP diagram shown in Figure 2.

Structural studies on $Zr(BH_4)_4$ (6) and $Hf(BH_4)_4$ (7) have
shown that these molecules are monomeric and crystallize into a
cubic lattice with molecular structures very similar to those of
Np and Pu borohydrides.

Some of the physical properties of metal tetrakis-borohydrides, which are primarily determined by their solid-state
structure, are listed in Table 1. The polymeric Th, Pa, and U
borohydrides are of much lower volatility than the monomeric Zr,
Hf, Np, and Pu compounds. The intermolecular bonds connecting
molecules together decrease their volatility substantially since
these bonds break when the solid vaporizes (12). A plot of
log p(mmHg) vs 1/T yields the equation log p(mmHg) = -A/T + B,
where T is in K. Values of A and B allow the calculation of the
heats (ΔH) and entropies (ΔS) for phase-change processes as shown
in Table 1. The actinide ions in the polymeric compounds are 14
coordinate; however, in the gaseous state they are 12 coordinate
(12).

The free energy for the structure transformation at 290K
described by the equation

$U(BH_4)_4$ (solid, 14 coordinate, 4 double hydrogen bridge
      bonds, 2 triple-hydrogen-bridge bonds)

→ $U(BH)_4)_4$ (solid, 12 coordinate, 4 triple hydrogen
      bridge bonds)

can be estimated. ΔH and ΔS values for a 12 coordinate $U(BH_4)_4$
structure were obtained by an extrapolation of the measured
quantities for $Hf(BH_4)_4$ and $Np(BH_4)_4$ vs metal ionic radius.
Subtracting these derived $U(BH_4)_4$ values from the corresponding

measured ones gives the heat of transformation (4.5 Kcal/mol) and entropy of the transformation (6.5 cal/mol degree) of the 14 co-ordinate to the 12 coordinate structure for $U(BH_4)_4$. Using the equation $\Delta G = \Delta H - T\Delta S$, ΔG is found to be +2.6 Kcal/mol. This value can be compared to the free energy of an exchange process involving the bridge and terminal hydrogen atoms in solution for $(C_5H_5)_3UBH_4$ where $\Delta G^* \approx 5$ Kcal/mol at the coalescence temperature of $-140 \pm 20°C$ (13). The calculated value for the spontaneous transformation of the 14 coordinate structure to the 12 coordinate structure is ∿700K.

In addition to low vapor pressure, high melting points and low solubility in noncoordinating organic solvents are character-istic of the polymeric borohydrides. In contrast, Zr, Hf, and Np borohydrides melt around room temperature and are highly soluble in pentane.

Vibrational Spectroscopy. In spite of their complex molec-ular frameworks, the monomeric borohydrides display surprisingly simple vibrational spectra due to their high symmetry (T_d), which requires that many fundamental vibrations be degenerate. Normal coordinate analyses have been carried out for $Zr(BH_4)_4$ (14) and $Hf(BH_4)_4$ (15) and a similar study was completed for $Np(BH_4)_4$ in order to compare vibrational energy level structures and elucidate the nature of the fundamental vibrations of $Np(BH_4)_4$ (10). Table 2 lists the measured frequencies. The appendix compares our notation to that of earlier work (7b,15). Unless otherwise specified, the ^{11}B isotope is implied.

A tetrahedral $M(BH_4)_4$ molecule has 57 normal modes of vibration which are divided into five symmetry types, $4A_1 + A_2 + 5E + 5T_1 + 9T_2$. The nine T_2 modes are infrared active and the $4A_1$ (polarized), $5E$, and $9T_2$ modes are Raman active. Those vibrations of T_1 and A_2 symmetry are both infrared and Raman inactive.

νBH_t stretching motions transform as $A_1 + T_2$ and these are seen in the highest frequency regions 2600-2500 cm^{-1} (1950-1900 cm^{-1}) for the borohydride (borodeuteride). The 12 BH_b bonds (stretches) transform as $A_1 + E + T_1 + 2T_2$ and modes involving these coordinates occur at 2200-1900 cm^{-1} (1600-1500 cm^{-1}). The next region of fundamental activity, 1300-1100 cm^{-1} (1000-800 cm^{-1}), consists of normal modes composed of νMH_b stretches and δHBH bends, with each symmetry equivalent set classified as $A_1 + E + T_1 + 2T_2$. Modes consisting mainly of νMB stretches occur at lower energies, 600-450 cm^{-1} (500-400 cm^{-1}). The lowest frequency vibrations are observed below 200 cm^{-1} and involve δHMH and δBMB bends. In addition to the fundamental vibrations, over-tones and combination bands are also seen. Peaks due to ^{10}B and ^1H (in the deuteride spectra) are resolved in the solid-state, low-temperature spectra.

The gas-phase infrared spectra for $Np(BH_4)_4$ and $Np(BD_4)_4$ are shown in Figure 3. The frequencies and assignments for the

Table 2

Observed Bands in Gas-Phase IR Spectra of $Np(BH_4)_4$ and $Np(BD_4)_4$

Energy (cm^{-1})	Assignment	Internal coordinates	Comments
	$Np(BH_4)_4$		
2568	$\nu_1^{T_2}$	νBH_t	strong
2480	$2\nu_4^{T_2}, 2\nu_5^{T_2}$		weak, v.br.
2350	$\nu_{4,5}^{T_2} + \nu_6^{T_2}$		weak, br.
2155	$\nu_2^{T_2}$	νBH_b	strong
2130	$2\nu_3^{T_1}$		sh on $\nu_5^{T_2}$
2084	$\nu_3^{T_2}$	νBH_b	strong, sharp
1280			sh on $\nu_4^{T_2}, \nu_5^{T_2}$
1240	$\nu_4^{T_2}, \ \nu_5^{T_2}$	$\delta HBH, \nu MH_b$	strong, broad
1205			sh on $\nu_4^{T_2}, \nu_5^{T_2}$
1122	$\nu_6^{T_2}$	$\delta HBH, \nu MH_b$	medium, sl.br.
1080	$\nu_4^{T_2} - \nu_5^{E}$		sh on $\nu_6^{T_2}$
478	$\nu_8^{T_2}$	$\nu MB, \ \nu MH_b$	strong
	$Np(BD_4)_4$		
1930		$\nu^{10}BD_t$	sh on $\nu_1^{T_2}$
1922	$\nu_1^{T_2}$	νBD_t	strong
1605	$2\nu_3^{T_1}$ *		medium
1562		$\nu^{10}BD_b$	sh on $\nu_2^{T_2}$
1558	$\nu_2^{T_2}$ *	νBD_b	strong
1526	$\nu_3^{T_2}$	νBD_b	strong, sharp
1190		δHBD	v.weak, br.
928	$\nu_4^{T_2}, \nu_5^{T_2}$	$\delta DBD, \nu MD_b$	strong, sl.br
845	$\nu_6^{T_2}$	$\delta DBD, \nu MD_b$	weak, br.
437	$\nu_7^{T_2}$	$\nu MB, \ \nu MD_b$	strong

In the table: br = broad, sh = shoulder, sl = slightly, v = very, H_b = bridging hydrogen, H_t = terminal hydrogen. (See Appendix for description of notation)
*These two bands are apparently in Fermi resonance.

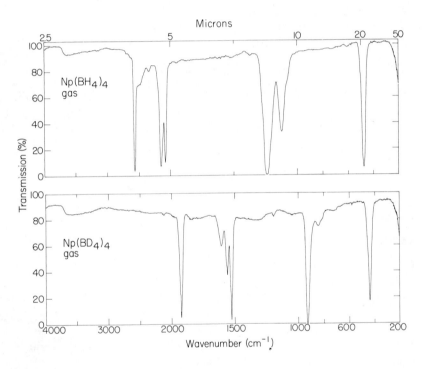

Figure 3.   Gas-phase IR spectra of $Np(BH_4)_4$ and $Np(BD_4)_4$

bands are given in Table 2. The solid-state spectra show many more bands and it is from these that a normal coordinate analysis was carried out.

A modified valence force field using the force constants and internal coordinates listed in Table 3 gave the calculated frequencies shown with the corresponding observed ones in Table 4. The force constants are very similar to those used in $Zr(BH_4)_4$ and $Hf(BH_4)_4$ even though the force fields are slightly different. The significant νMB force constant implies that there may be some direct metal-boron bonding in these borohydrides. The force field obtained for $Np(BH_4)_4$ is consistent for a molecule intermediate in covalency between that in diborane (16) and the alkali borohydrides (17).

Electron Paramagnetic Resonance. EPR spectroscopy involves transitions within the magnetic sublevels of the ground electronic state of a metal ion in the GHz energy region. The $^4I_{9/2}$ ground state of the Np^{4+} ion in a T_d crystal field of the form

$$V = B_4[C_0^{(4)} + (5/14)^{\frac{1}{2}}(C_4^{(4)} + C_{-4}^{(4)})] + B_6[C_0^{(6)} - (7/2)^{\frac{1}{2}}(C_4^{(6)} + C_{-4}^{(6)})]$$

splits into an isotropic Γ_6 doublet and two anisotropic Γ_8 quartets. Isotropic spectra for $Np(BH_4)_4$ and $Np(BD_4)_4$ establish the Γ_6 level as the ground state. The spin Hamiltonian describing the system is

$$\mathcal{H} = AI \cdot S' + g\beta H \cdot S' - g_I\beta H \cdot I ,$$

where A is the hyperfine coupling constant, $I = 5/2$ for ^{237}Np, and $S' = 1/2$. The calculations were carried out using $|F, m_F\rangle$ basis sets where $F = I + S'$. In zero magnetic field there are two states, $F = 2$ and $F = 3$, that are separated by $3A$. When the magnetic field is turned on, each of these two states splits into $(2F + 1) |m_F\rangle$ levels as shown in Figure 4 where A is assumed positive. The arrows in Figure 4 represent observed allowed transitions.

Results of least-squares calculations of the data to the spin-Hamiltonian above are shown in Table 5. The $Np(BH_4)_4/Zr(BH_4)_4$ spectra gave relatively broad resonances compared to the deuteride and a reliable g_I value could not be found. Inclusion of a non-zero g_I value in the calculations of the deuteride data improved the fit even though it was calculated to be very small. However, the significance of this improved fit must be tested further. Similar trials on the hydride data gave poorer fits.

The experimental g value is lower than calculated from LLW wavefunctions (17) (~ 2.7), which may indicate that covalency (19) or Jahn-Teller (20) effects may be important.

Electronic Spectra of Pa(BH_4)_4. Cary 17 spectra of $Pa(BH_4)_4$ and $Pa(BD_4)_4$ in an organic glass at 2K are shown in Figure 5.

Table 3

Best Fit Force Constants for Solid Neptunium Borohydride at 77K

Primary Force Constants		Interaction Force Constants	
Internal coordinate	Value	Internal coordinates	Value
νBH_t	3.51 md/Å	$\nu BH_b : \nu BH_b$ (intra)	.04 md/Å
νBH_b	2.36	$\nu MH_b : \nu MH_b$	.02
νMH_b	.37	$\nu MB : \delta H_b MH_b$	−.09 md/rad
νMB	1.28	$\nu BH_b : \delta H_b BH_b$ (intra)	.04
$\delta H_t BH_b$	.28 mdÅ/rad^2	$\nu MH_b : \delta H_b MH_b$	.04
$\delta H_b BH_b$	.36		
$\delta H_b MH_b (C_3)$	.26		
$\varepsilon H_b BMB$	.18*		

* This force constant was arbitrarily set at .18 since this depends almost solely on the A_2 torsion mode, which is not observed.

Table 4

Fundamental Vibrations (cm^{-1}) of $Np(BH_4)_4$ and $Np(BD_4)_4$

| Mode | $Np(BH_4)_4$ | | $Np(BD_4)_4$ | |
	Observed	Calculated	Observed	Calculated
$\nu_1{}^{T_2}$	2551	2557	1912	1911
$\nu_2{}^{T_2}$	2143	2144	1548	1603
$\nu_3{}^{T_2}$	2069	2078	1516	1485
$\nu_4{}^{T_2}$	1247	1266	926	897
$\nu_5{}^{T_2}$	1225	1223	917	895
$\nu_6{}^{T_2}$	1138	1104	860	824
$\nu_7{}^{T_2}$	--	575	437	447
$\nu_8{}^{T_2}$	475	488	--	415
$\nu_9{}^{T_2}$	130	156	112	139
$\nu_1{}^{A_1}$	2557	2554	1913	1905
$\nu_2{}^{A_1}$	2149	2147	1517	1523
$\nu_3{}^{A_1}$	1283	1284	955	953
$\nu_4{}^{A_1}$	517	517	475	466
$\nu_1{}^{E}$	2123	2117	1619	1589
$\nu_2{}^{E}$	1260	1270	905	899
$\nu_3{}^{E}$	1053	1089	795	807
$\nu_4{}^{E}$	--	571	--	413
$\nu_5{}^{E}$	168	142	154	125
$\nu_1{}^{T_1}$	--	2116	--	1587
$\nu_2{}^{T_1}$	--	1256	--	889
$\nu_3{}^{T_1}$	--	1084	--	810
$\nu_4{}^{T_1}$	--	565	--	405
$\nu_5{}^{T_1}$	--	405	--	288
ν^{A_2}	--	288	--	204

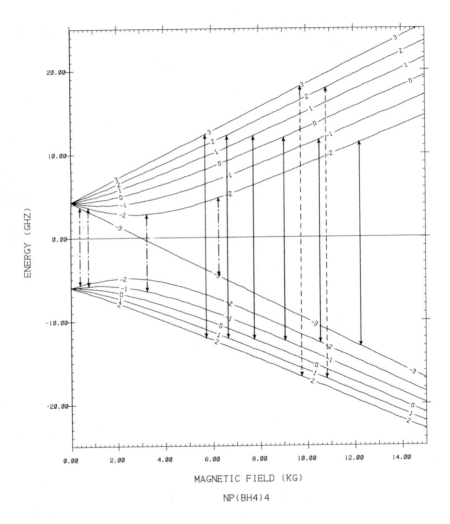

Figure 4.   *Observed allowed EPR transitions for* $Np(BH_4)_4$–$Zr(BH_4)_4$: (— · — ·),
X *band;* (——), **K** *band;* (– – –), **Q** *band.*

Table 5

Electron Paramagnetic Resonance of ^{237}Np(BH$_4$)$_4$ and ^{237}Np(BD$_4$)$_4$

$$\mathcal{H} = A\mathbf{I}\cdot\mathbf{S'} + g\beta\mathbf{H}\cdot\mathbf{S'} - g_I\beta\mathbf{H}\cdot\mathbf{I} \quad \begin{array}{l} I=5/2 \\ S'=1/2 \end{array}$$

Spin Hamiltonian Parameters

	$|A|$ (cm^{-1})	g	g_I
Np(BH$_4$)$_4$	.1140 ± .001	1.894 ± .002	∿0
Np(BD$_4$)$_4$	.1140 ± .001	1.892 ± .002	.0062 ± .002

Observed and Calculated Field Values (gauss) at K Band

Np(BH$_4$)$_4$ ν = 25.986GHz		Np(BD$_4$)$_4$ ν = 24.238GHz	
Observed	Calculated	Observed	Calculated
6355.6	6355.8	5683.1	5683.1
7295.0	7294.3	6596.0	6595.8
8406.0	8405.7	7695.0	7694.3
9700.0	9700.2	8991.0	8991.4
11177.0	11178.1	10487.6	10487.1
12829.6	12828.7	12167.9	12167.6

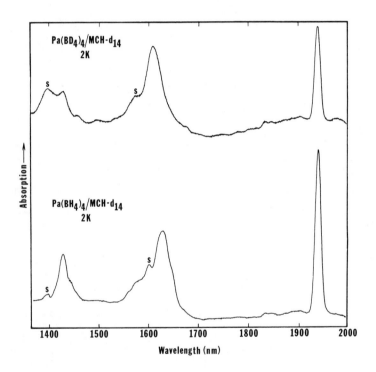

Figure 5.   Optical spectra of Pa(BH₄)₄ and Pa(BD₄)₄ in methylcyclohexane. **S** *above a peak represents a solvent band.*

In liquid solution at 25°C, the dissolved $Pa(BH_4)_4$ is monomeric and of T_d symmetry. Under these conditions there are five crystal field levels: Γ_7, Γ_8, and Γ_6, Γ_7', Γ_8' of the $^2F_{5/2}$ (ground) and $^2F_{7/2}$ levels. Point charge calculations (20) give the $\Gamma_8(^2F_{5/2})$ level as the ground state.

Keiderling (7b) has observed that when $U(BH_4)_4$ dissolved in an organic solvent is cooled to 2K, the monomeric T_d structure transforms back into the polymeric structure. Although it is tempting to assign the observed bands based upon the tetrahedral structure, definite conclusions must await comparison with pure $Pa(BH_4)_4$ spectra.

Near infrared and optical spectra have been obtained for $Np(BH_4)_4$ and $Np(BD_4)_4$ diluted in $Zr(BH_4)_4$, $Zr(BD_4)_4$ and methyl-cyclohexane at 2K. The spectra are dominated by vibronic transitions and the analysis of the data is now underway.

Summary

The actinide borohydrides $Pa(BH_4)_4$, $Np(BH_4)_4$, and $Pu(BH_4)_4$ have been synthesized. The structure of $Np(BH_4)_4$ has been studied by single-crystal x-ray diffraction and found to be similar in structure to $Hf(BH_4)_4$. A normal coordinate analysis on $Np(BH_4)_4$ was completed using IR and Raman spectra. The electronic ground state of $Np(BH_4)_4$ has been characterized by EPR spectroscopy. The electronic spectra of $Np(BH_4)_4$ and $Pa(BH_4)_4$ are under investigation.

Acknowledgment

This work was done with support from the Division of Nuclear Sciences, Office of Basic Energy Sciences, U.S. Department of Energy, under Contract No. W-7405-Eng-48.

Literature Cited

1. Hoekstra, H.R.; Katz, J.J. J. Am. Chem. Soc., 1949, 71, 2488.
2a. Schlesinger, H.I.,: Brown, H.C. J. Am. Chem. Soc., 1953, 75, 219.
2b. Katz, J.J.; Rabinowitch, E. "Chemistry of Uranium"; McGraw-Hill: New York NY, 1951.
3. Banks; R.H.; Edelstein, N.M.; Rietz, R.R.; Templeton, D.H.; Zalkin, A. J. Am. Chem. Soc., 1978, 100, 1975.
4a. Bernstein, E.R.; Hamilton, W.C.; Keiderling, T.A.; LaPlaca, S.J.; Lippard, S.J.; Mayerle, J.J. Inorg. Chem., 1972, 11, 3009.
4b. Bernstein, E.R.; Keiderling, T.A.; Lippard, S.J.; Mayerle, J.J. J. Am. Chem. Soc., 1972, 94, 2552.
5. Banks, R.H.; Edelstein, N.M.; Spencer, B.; Templeton, D.H.; Zalkin, A. J. Am. Chem. Soc., 1980, 102, 0000.
6. Bird, P.H.; Churchill, M.R. Chem. Comm., 1967, 403.

7a. Broach, R.S.; Chuang, I.S.; Williams, J.M.; Marks, T.J. Private communication, 1979.

7b. Keiderling, T.A. PhD thesis, Princeton University, 1974.

8. Bernstein, E.R.; Keiderling, T.A. J. Chem. Phys., 1973, 59, 2105.

9. Maybury, P.C.; Larrabee, J.C. Abstract of papers, 135th Meeting of the American Chemical Society, 1959. P-28M.

10. Banks, R.H.; Edelstein, N.M., to be published.

11. Muetterties, E.L. "Boron Hydride Chemistry"; Academic Press: New York NY, 1975.

12. James, B.D.; Smith, B.E.; Wallbridge, M.G.H. J. Mol. Struct. 1972, 14, 327.

13. Marks, T.J.; Kolb, J.R. J. Am. Chem. Soc., 1975, 97, 27.

14. Smith, B.E.; Shurvell, H.F.; James, B.D. JCS Dalton, 1978, 710.

15. Keiderling, T.A.; Wozniak, W.T.; Gay, R.S.; Jurkowitz, D.; Bernstein, E.R.; Lippard, S.J.; Spiro, T.G. Inorg. Chem., 1975, 14, 576.

16. Price, W.C. J. Chem. Phys., 1948, 16, 894.

17. Emery, A.R.; Taylor, R.C. J. Chem. Phys., 1958, 28, 1029.

18. Bleany, B. Proc. Roy. Soc. Lond., 1964, A277, 289.

19. Judd, B.R. 2nd Inter. Conf. on Electronic Structure of An, Warsaw, Poland, 1976.

20. Lea, K.R.; Leask, M.J.M.; Wolf, W.P. J. Phys. Chem. Solids, 1962, 23, 1381.

Appendix

In Tables 2 and 4, a fundamental or overtone is denoted by the symbol $n\nu_a^b$, where b is the Mulliken symbol for the irreducible representation of the mode and a is the number of the mode starting with 1 for the highest frequency, 2 for the second highest, etc. The n is omitted for fundamentals, equals 2 for first overtones, 3 for second overtones, etc.

The table given below relates our notation to that used in earlier work (7b,15).

This work	Literature
$\nu_1^{A_1} - \nu_4^{A_1}$	$\nu_1 - \nu_4$
ν^{A_2}	ν_5
$\nu_1^{E} - \nu_5^{E}$	$\nu_6 - \nu_{10}$
$\nu_1^{T_1} - \nu_5^{T_1}$	$\nu_{11} - \nu_{15}$
$\nu_1^{T_2} - \nu_9^{T_2}$	$\nu_{16} - \nu_{24}$

RECEIVED January 7, 1980.

Optical Properties of Actinide and Lanthanide Ions

JAN P. HESSLER and W. T. CARNALL

Chemistry Division, Argonne National Laboratory, 9700 South Cass Avenue, Argonne, IL 60439

The sharpness of many of the optical absorption and emission lines of the lanthanide ions in ionic crystals has intrigued scientists since 1908. We review some of the recent developments in this area of spectroscopy, emphasizing the optical properties of the tripositive lanthanide and actinide ions. In particular, we shall discuss the single ion properties of line position, intensity, width, and fluorescence lifetime. Such effects as the application of external electric and magnetic fields, hyperfine interactions, and cooperative effects such as long range ordering and energy transfer, although direct extensions of the above properties, must be excluded in such a short review.

The optical properties of the lanthanide and actinide ions are due to the unpaired electrons of the ion. The observed sharp transitions have been shown to be intraconfiguration transitions. The most widely studied systems have ground configurations (Xe, $4f^n$) and (Rn, $5f^n$) for the lanthanide and actinide ions respectively. The number of f-electrons, n, ranges from 1 to 13. The inert rare gas core allows us to discuss the systems in terms of the f-electrons only. Even such a conceptually simple system is complex enough to require a parameterization scheme. The physical significance of such a scheme and its role in developing an understanding of complex systems has been discussed by Newman (1). Our goal is not to uncritically accumulate parameters in some standard scheme which has limited utility, but instead to develop as comprehensive and universal a scheme as possible, one which can be applied to the energy level structure, radiative transition probabilities, temperature-dependent line widths, fluorescent lifetimes, electric and magnetic susceptibilities, hyperfine structure, and cooperative phenomena. In particular, the parameters we deduce should allow us to predict observables in an unmeasured region, be consistent with appropriate *ab initio* calculations, and be useful as input data into other parameterization schemes. An example of the last point is the analysis

0–8412–0568–X/80/47–131–349$05.00/0

of crystal-field parameters by an effective point charge or
other model.

We briefly summarize the parameterization schemes for f-
electron energy levels, intraconfiguration transition probabilities, and the electron-phonon interaction, and review the
current experimental situation for each area. We shall also
speculate on potentially fertile areas of future investigation.

I. Line Positions

A. Energy Level Parameterization Scheme. The parameterization scheme to derive the free-ion properties of the
f-electron system is based upon direct physical assumptions.
The approximate Hamiltonian describing N-electrons moving about
a nucleus of charge Ze, as discussed by Condon and Shortley (2),
can be written

$$\mathscr{H} = \sum_{i=1}^{N} \left\{ \frac{1}{2\mu} P_i^2 - \frac{Ze^2}{r_i} + \zeta(r_i)\vec{\ell}_i \cdot \vec{s}_i \right\} + \sum_{i>j=1}^{N} \frac{e^2}{r_{ij}} , \qquad (1)$$

where the symbols have their usual meaning. To treat a system
of several equivalent f-electrons, Racah developed the concepts
of tensor operators and the coefficients of fractional parentage.
These concepts have been reviewed by Judd (3).

The first approximation to paramterize equation (1) is to
assume that all electrons move in a central potential. If we
then limit the analysis to a single configuration, we need discuss only the Coulomb and spin-orbit interaction between the
equivalent f-electrons. With the aid of tensor operators the
Coulomb interaction can be expressed as

$$\mathscr{H}_{coul} = E^0 e_0 + E^1 e_1 + E^2 e_2 + E^3 e_3. \qquad (2)$$

The E^i's are parameters which may be expressed as a linear
combination of the Slater integrals, $F^{(k)}$: k = 0, 2, 4, and 6.
The e_i's are tensor operators. The spin-orbit interaction
within a single configuration may be parameterized by a single
spin-orbit radial integral, ζ_f, therefore

$$\mathscr{H}_{s.o.} = \zeta_f \sum_{i=1}^{n} (\vec{s}_i \cdot \vec{\ell}_i). \qquad (3)$$

The sum is over the n equivalent f-electrons.

Bethe (4) pointed out that when the free-ion is put into a
crystal the electric fields distort the isotropy of free space.
This causes a splitting in the free-ion energy levels, with any
residual degeneracy determined by the symmetry of the crystal.

In tensor operator notation the crystal-field interaction is written

$$\mathcal{H}_{C.F.} = \sum_{k,q} B_q^k \sum_{i=1}^{n} (C_q^{(k)})_i. \tag{4}$$

The B_q^k s are parameters and the $C_q^{(k)}$ s are tensor operators which are related to the spherical harmonics.

This basic parameterization scheme, used at the time of the last A.C.S. symposium on lanthanide and actinide chemistry (5), has been discussed in detail by Wybourne (6). In applying the scheme, the free-ion Hamiltonian was first diagonalized and then the crystal-field interaction was treated as a perturbation. This procedure yielded free-ion energy levels that frequently deviated by several hundred cm^{-1} from the observed energy levels.[a] In addition, the derived parameters such as the Slater radial integral, $F^{(2)}$, and the spin-orbit radial integral did not follow an expected systematic pattern across the lanthanide or actinide series (7).

These deviations are due to neglecting the Coulomb interaction between different configurations. To boldly proceed to enlarge the basis set of wave functions to include additional configurations would have resulted in an unmanageably large matrix. Instead, Rajnak and Wybourne (8) assumed that the Coulomb interaction between configurations was weak enough to be treated with perturbation techniques. They modified the Hamiltonian that operated within the ground configuration to include the greater part of the effects of all weakly perturbing configurations. This approach modifies the physical interpretation of the Slater radial integrals by introducing configuration interaction corrections and introduces additional parameters into the scheme. The additional parameters α, β, and γ are required to complete the description of two-body electrostatic configuration-interaction effects. The dominant contributions due to three-body interactions require an additional six parameters, T^k: $k = 2, 3, 4, 6, 7,$ and 8, defined by Judd (9).

Judd, Crosswhite, and Crosswhite (10) added relativistic effects to the scheme by considering the Breit operator and thereby produced effective spin-spin and spin-other-orbit interaction Hamiltonians. The reduced matrix elements may be expressed as a linear combination of the Marvin integrals, M^k: $k = 0, 2,$ and 4. They also considered the effect of additional configurations on the spin-orbit interaction to produce the electrostatically correlated spin-orbit interaction.

[a] Traditionally the spectroscopist has measured energy in cm^{-1}. In this system of units Plank's constant and the velocity of light are equal to 1. To convert to SI units, multiply values in cm^{-1} by $hc = 19.86484 \times 10^{-24}$ J-cm^{-1}.

Although this interaction has properties very similar to the spin-other-orbit interaction, it is distinct enough to require the additional paramters P^k: k = 2, 4, and 6.

This completes the current free-ion parameterization scheme. It involves twenty parameters which can be determined by comparison to experimental observations. The most important parameters are the four Slater radial integrals, the spin-orbit radial integral, and the three two-body configuration interaction parameters. With these eight parameters the free-ion levels can generally be fit to within a hundred cm^{-1}. The precise evaluation of the three-body configuration interaction parameters is critically dependent upon the observation of certain levels. Because it is often difficult to obtain a complete set of experimental levels, the three-body parameters are sometimes poorly defined. The Marvin integrals and the spin-other-orbit parameters produce changes in the free-ion levels which are on the order of the crystal-field splitting. Their evaluation, therefore, requires both extensive experimental data and an adequate model for the crystal-field interaction. Unfortunately, there has been no systematic evaluation of the effect of adding parameters to the scheme. More importantly, only the root-mean-squared deviation between observed and calculated energy levels has been used to test the quality of the theoretical predictions. No study of the correlation between fitted parameters has been undertaken. Such a study would be useful in establishing the importance of individual parameters and the overall adequacy of the scheme.

Ab initio calculations of the effective parameters are difficult because of the need to properly sum to infinite order the various configuration interaction contributions to the parameters. Morrison and Rajnak (11) used perturbation theory and graphical methods to correct Hartree-Fock theory and thereby calculated the parameters, α, β, γ, and corrections to the Slater radial integrals. Their work pointed out the need to properly include high angular momentum continuum states in any calculation of effective parameters. To include the continuum states, Morrison (12) used a perturbed-function approach to calculate the effect of core polarization on the two-body and Slater integrals. Newman and Taylor (13) modified the Hartree-Fock potential to change the form of the excited state spectrum and calculated Slater integrals and P^k parameters. Later, Balasubramanian, Islam, and Newman (14) introduced an infinitely deep potential well to calculate the three-particle correlation paramters, T^k. No systematic calculation has been published for either a finite number of parameters across an entire series or for all twenty parameters for a single ion.

With the significant improvements in high speed digital computers which have occurred within the last ten years, it is now possible to diagonalize a complete free-ion plus crystal-field Hamiltonian. This procedure reproduces the observed energy levels with a root-mean-squared deviation on the order of twenty-five cm^{-1}. Because the crystal field is not introduced as a perturbation, J-mixing of the wave functions is properly accounted for. This is especially important in studies of the actinides because J-mixing drastically alters the properties of the wave functions.

The virtues of the current scheme are relatively reliable predictions of the energy level positions, effective parameters that vary systematically across a series, and wave functions that may be utilized for additional calculations. The prediction of energy levels has aided the experimental study of new systems such as Gd^{3+} in CaF_2 (15). The systematic variation of parameters across a series has been used to estimate parameters for the initial analysis of an ion. The properly admixed wave functions will improve the transition probability analysis of the actinides.

B. <u>Current Status on Ln^{3+} and An^{3+} Ion Energy Levels.</u> The free-ion energy levels up to approximately 30000 cm^{-1} for all of the tripositive lanthanide ions in $LaCl_3$ single crystals are shown in Figure 1. Crosswhite (16) has recently tabulated and discussed the free-ion and crystal-field parameters needed to describe the lanthanide data. The tripositive actinide levels are shown in Figure 2. Table I summarizes the free-ion parameters for the actinides which have been studied in detail. The detailed analyses of the U^{3+} and Np^{3+} ions have recently been completed (17, 18). The analyses presented in Table I for Pu^{3+}, Am^{3+}, and Cm^{3+} are based on published spectra (19, 20, 21, 22) obtained by Conway and coworkers.

Crosswhite (23) has used the correlated multiconfiguration Hartree-Fock scheme of Froese-Fisher and Saxena (24) with the approximate relativistic corrections of Cowan and Griffin (25) to calculate the Slater, spin-orbit, and Marvin radial integrals for all of the actinide ions. A comparison of the calculated and effective parameters is shown in Table II. The relatively large differences between calculation and experiment are due to the fact that configuration interaction effects have not been properly included in the calculation. In spite of this fact, the differences vary smoothly and often monotonically across the series. Because the Marvin radial integral M^0 agrees with the experimental value, the calculated ratios $M^2(HRF)/M^0$ (HRF) = 0.56 and M^4 (HRF)$/M^0$(HRF) = 0.38 for all tripositive actinide ions, are used to fix M^2 and M^4 in the experimental scheme.

The analysis of crystal-field components has remained at the single-particle level introduced by Bethe (4). Crystal-field parameters for the actinide ions in lanthanum trichloride are shown in Table III. They are approximately twice as large as the values found for the lanthanides. Although the values

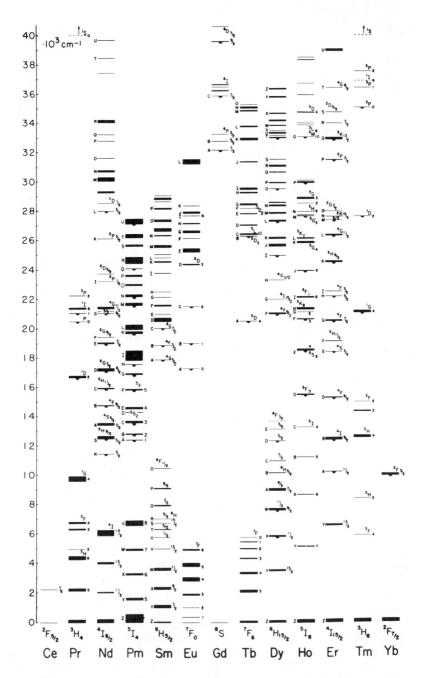

Figure 1.   Energy-level structure of the tripositive lanthanide ions in LaCl₃

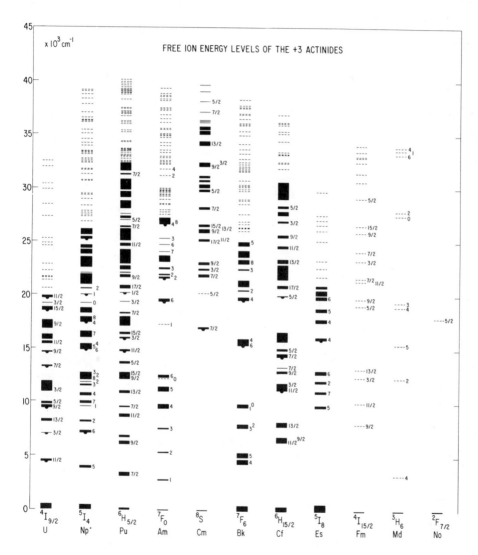

Figure 2.   *Energy-level structure of the tripositive actinide ions in LaCl₃*

Table I. Free-Ion Parameters for Trivalent Actinide Ions in Lanthanum Trichloride. Units are cm^{-1}.

Param.	U^{3+}	Np^{3+}	Pu^{3+}	Am^{3+}	Cm^{3+}
E_{AV}	19544	29999	39631	53700	64124
$F^{(2)}$	39715	44907	48670	[51800][a]	55109
$F^{(4)}$	33537	36918	39188	[41440]	43803
$F^{(6)}$	23670	25766	27493	[30050]	32610
ζ_f	1623	1938	2241	[2580]	2903
α	27.6	31.5	29.7	[29]	28.3
β	-772	-740	-671	[-660]	[-650]
γ	[1000]	899	1067	[1000]	825
T^2	217	278	186	[200]	[200]
T^3	63	44	48	[50]	[50]
T^4	255	64	38	[40]	[40]
T^6	-107	-361	-364	[-360]	[-360]
T^7	617	434	364	[390]	[390]
T^8	[350]	353	332	[340]	[340]
M^0 [b]	[0.67]	0.68	0.95	[0.99]	[1.09]
P^2 [c]	1276	894	822	[850]	912

[a] A [] indicates that the parameter was estimated and held constant for all fitting.

[b] For all cases: $M^2/M^0 = 0.56$ and $M^4/M^0 = 0.38$.

[c] For all cases: $P^4/P^2 = 0.75$ and $P^6/P^2 = 0.50$.

Table II.   Comparison of Relativistic Hartree-Fock Integrals for $(Rn, 5f^n)$ and Effective Parameters for Tripositive Actinide Ions in Lanthanum Trichloride.   Units are cm^{-1}.

Param.	U^{3+}	Np^{3+}	Pu^{3+}	Am^{3+}	Cm^{3+}
$\Delta F^{(2)a}$	31727	30037	29553	29546	29222
$\Delta F^{(4)}$	12833	11815	11754	11604	11246
$\Delta F^{(6)}$	10248	9918	9842	8855	7793
$\Delta \zeta$	275	244	238	212	216
$\dfrac{M^0 (exp)}{M^0 (HRF)}$	1.00	0.88	1.08	1.00	0.99

$^a \Delta F^{(k)} = F^{(k)} (HRF) - F^{(k)} (exp).$

Table III.   Crystal-Field Parameters for Trivalent Actinide Ions in Lanthanum Trichloride.   Units are cm^{-1}.

Param.	U^{3+}	Np^{3+}	Pu^{3+}	Am^{3+}	Cm^{3+}
B_0^2	260	163	226	$[230]^a$	246
B_0^4	-533	-632	-543	$[-610]$	-671
B_0^6	-1438	-1625	-1695	$[-1590]$	-1410
B_6^6	1025	1028	1000	$[980]$	921

^{a} A [] indicates that the parameter was estimated and held constant for all fitting.

are approximately constant across a series, there are significant variations that distort the monotonic behavior. This
may be indicative of an incomplete parameterization scheme for
the crystal-field interaction, just as the non-systematic
behavior of the Slater and spin-orbit integrals indicated the
need for the addition of the configuration interaction.

    C. Advances in Experimental Techniques. The ions U^{3+}
through Cm^{3+} have been studied by classical photographic
techniques, which may also be applied to the study of Bk^{3+} and
Cf^{3+}. The ion Es^{3+} is too radioactive to utilize these techniques. To overcome this problem and to extend the experimental
capabilities into the time domain, we have applied pulsed dye
laser technology. Selective excitation of a specific ion
within the background of daughter ions is used to discriminate
against the radioactive induced fluorescence. Time resolved
detection of fluorescence is used to identify groups of
fluorescing levels with a single upper level. By monitoring
a known fluorescence line as a function of the dye laser
wavelength, the equivalent of an absorption spectrum may be
obtained. With these techniques (26), both absorption and
fluorescence data may be obtained for Es^{3+}. The precision of
the data is comparable to that obtained with classical methods.
Similar techniques may also succeed in locating some levels in
Fm^{3+}.

II. Line Intensities

    A. Transition Probability Parameterization Scheme. As
early as 1937 Van Vleck (27) referred to the "puzzle of the
intensities of the absorption lines of the lanthanide ions".
Later Broer, Gorter, and Hoogschagen (28) showed that the
observed intensities were too large to be accounted for by
magnetic dipole or electric quadrupole radiation, but that
induced electric dipole transitions could account for the intensity. The central problem with electric dipole transitions
within a configuration is that they are LaPorte (or parity)
forbidden. To obtain non-vanishing matrix elements for the
electric dipole operator requires that opposite parity configurations be admixed into the states of the f^n configuration.

    Judd (29), in his classic paper of 1962, used the odd
parity terms of the ligand field to accomplish this admixture.
After applying second order perturbation theory and several
simplifying assumptions, he showed that the electric dipole
line strength between J-manifolds may be expressed as the sum
of three terms, each being the product of an intensity parameter
and a reduced matrix element of the tensor operator $U^{(\lambda)}$ of
rank λ. The electric dipole line strength, S_{ed}, can be written
in the form

$$S_{ed}(aJ,a'J') = e^2 \sum_{\lambda=2,4,6} \Omega_\lambda <f^n\gamma J||U^{(\lambda)}||f^n\gamma'J'>^2. \qquad (5)$$

The Ω_λ s are the intensity parameters. The line strength for the magnetic dipole transitions is given by

$$S_{md}(aJ,a'J') = \mu_B^2 <f^n\gamma J||\vec{L} + 2\vec{S}||f^n\gamma'J'>^2 \qquad (6)$$

where $\mu_B = eh/4\pi mc$.

The wave functions used in the expressions for the line strengths are precisely those deduced by an analysis of the free-ion energy level structure. Therefore, only three new parameters, the Ω_λ s, have been introduced to account for the line strengths. This scheme has been remarkably successful in modeling experimental observations in both crystal and solution environments. It also accommodates the existence of the "hyper-sensitive" transitions. Peacock (30) has recently reviewed the field with regard to lanthanide f-f transitions. The simplicity of this scheme has been utilized by Krupke (31) and Caird (32) to predict potential laser transitions in the lanthanides.

B. <u>Current Status of An^{3+} Ion Line Strengths</u>. As with the lanthanides, solution spectra were the first to be investigated in terms of the Judd parameterization scheme. The light actinides U^{3+}, Np^{3+}, and Pu^{3+} have a rather high density of states in the optical region, therefore the free-ion J-manifolds overlap and analysis is difficult. Am^{3+} is a special case. Only transitions between the ground J = 0 and even J-manifolds are allowed in the context of the free-ion approximation. For Cm^{3+} and the heavier actinides Bk^{3+}, Cf^{3+}, and Es^{3+} a number of the free-ion J-manifolds are well resolved as can be seen in the absorption spectra shown in Figure 3. The intensity parameters for these systems are given in Table IV. For Cm^{3+} and Cf^{3+} the parameterization scheme yields a good fit to the experimental observations. For the case of Bk^{3+} the large value of Ω_2 is not consistent with neighboring values of the series. The source of this discrepancy has not yet been identified. We note, if ligand-field interactions are not included in the determination of free-ion wave functions, then J-mixing between the manifolds will not occur. This J-mixing will be very important in the calculation of transition probabilities in the actinide systems.

C. <u>Line Intensities Between Individual Stark Components</u>. Simultaneously with Judd's work, Ofelt (33) independently

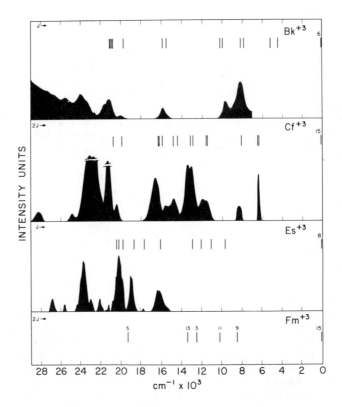

Figure 3.   Intraconfiguration absorption spectra of the heavier tripositive actinide ions in aqueous solution. The vertical lines are the calculated positions of the free-ion energy levels. A broad background absorption has been subtracted from the data. No measurements have been obtained for fermium.

Table IV.   Intensity Parameters for the Heavier Actinides in Dilute Acid Solution.   Units are pm^2, 1 $pm^2 = 10^{-20}$ cm^2.

Ion	Ω_2	Ω_4	Ω_6	No. Bands Fit
Cm^{3+}[a]	15.2	16.8	38.1	17
Bk^{3+}	100	40.4	12.4	6
Cf^{3+}	3.8	12.5	21.3	11
Es^{3+}[b]	Ω_2 <<	Ω_4 >	Ω_6	

[a] W. T. Carnall and K. Rajnak, *J. Chem. Phys.* 63, 3510–3514 (1975).

[b] W. T. Carnall, D. Cohen, P. R. Fields, R. K. Sjoblom, and R. F. Barnes, *J. Chem. Phys.* 59, 1785–1789 (1973).

analyzed the electric dipole coupling mechanism.   Ofelt actually parameterized the transition probability between the crystal-field split components of two J-manifolds.   In dealing with the individual Stark component transitions we can not sum over the odd crystal-field components to obtain the simple three param-eter scheme used for solutions.   Therefore, a rather large set of empirical odd crystal-field parameters, which are critically dependent on the point symmetry of the ion, remain to be deter-mined.   For C_{3h} symmetry there are six empirical parameters, but for D_{3h} there can be as few as three. Newman and Balasubramanian (34) have recently obtained the most general description of transition probabilities between crystal-field levels.   In general, the absorption and fluorescence spectra are used to deduce a set of empirical parameters such as for Eu^{3+} in euro-pium ethylsulfate (35) and in KY_3F_{10} (36).   Good, Jr. and his colleagues (37) have deduced the ratios of the odd crystal-field parameters for erbium ethylsulfate by the application of a transverse magnetic field.   Esterowitz *et al.* (38) have calculated the odd crystal-field parameters for Pr^{3+} in $LiYF_4$ using an effective point charge model and performing a lattice summation.   They then predict the relative magnitude of tran-sitions between two J-manifolds.   They studied both S_4 and D_{2d} point symmetries and obtained qualitative agreement.

III. Line Widths and Lifetimes

A. Electron-Phonon Interaction Parameterization Scheme. In observing the fluorescence decay rate from a given J-manifold, it is generally found that the decay rate is independent of both the crystal-field level used to excite the system and the level used to monitor the fluorescence decay. This observation indicates that the crystal-field levels within a manifold attain thermal equilibrium within a time short compared to the fluorescence decay time. To obtain this equilibrium, the electronic states must interact with the host lattice which induces transitions between the various crystal-field levels. The interaction responsible for such transitions is the electron-phonon interaction. This interaction produces phonon-induced electric-dipole transitions, phonon side-band structure, and temperature-dependent line widths and fluorescence decay rates. It is also responsible for non-resonant, or more specifically, phonon-assisted energy transfer between both similar and different ions. Studies of these and other dynamic processes have been the focus of most of the spectroscopic studies of the transition metal and lanthanide ions over the past decade. An introduction to the lanthanide work is given by Hüfner (39).

The simplest electron-phonon interaction is the direct one-phonon process that induces a transition between two electronic levels that are separated by less than the maximum phonon energy of the crystal. The first successful parameterization of the electron-phonon interaction was given by Orbach in his paper on the spin-lattice relaxation in rare-earth salts (40). McCumber and Sturge (41) extended these ideas to optical transitions in solids. The parameters describing single-phonon transitions between crystal-field levels are products of the phonon energy, the electron-phonon coupling constant, and a matrix element between the crystal-field levels. The details are given in a lecture by Orbach (42). Two-phonon processes are much more difficult to treat quantitatively. Both higher order perturbation theory and higher order terms in the electron-phonon interaction must be considered.

To directly measure the transition rate between various crystal-field components would be very difficult because of the relatively low fluorescence transition probabilities and the high phonon-induced transition rates. Instead, the temperature dependence of the homogeneous line width of optical transitions is measured and related to the transition rate by the Heisenberg uncertainty relation. By measuring the temperature dependence of the homogeneous line width for several transitions between two J-manifolds and modeling the results in terms of one- and two-phonon processes, the effective parameters involving the electron-phonon coupling may be deduced. Because these parameters depend on the details of the phonon density of states and the properties of the electronic states, detailed analysis of the parameters has not been undertaken.

B. Experimental Line Widths for Ln^{3+} and An^{3+} Transitions.
The first detailed quantitative study of the temperature dependence of the line width in a lanthanide system was done on Pr^{3+}
in LaF_3 by Yen, Scott, and Shawlow ([43]). Since that time line
widths as low as 15 kHz have been observed for the $^1D_2$ to $^3H_4$
transition of Pr^{3+} in LaF_3 at 2 K ([44]). To observe such narrow
homogeneous line widths within the strain broadened inhomogeneous line width found in solid systems, experimental techniques
involving tunable dye lasers are usually employed. These
techniques have been reviewed by Selzer ([45]) while specific
results for the lanthanide systems have been reviewed by Yen
([46]).

The first measurement of the temperature dependence of an
optical line width in an actinide system, Np^{3+} in $LaCl_3$, was
recently completed ([47]). The fluorescence transitions at 671.4
and 677.2 nm were studied from 10 to 200 K.   The
low temperature limit for the line width of the 677.2 nm transition is 16.5 GHz and is a measure of the width of the first
excited crystal-field level of the ground manifold. The 671.4
nm transition has a line width of 0.55 GHz at 10 K. Its temperature dependence is described in terms of an effective three-
level scheme for the excited manifold. The parameters are
comparable to those found for Pr^{3+} in LaF_3. Further comparison
depends upon the details of the phonon spectrum and the electronic states. At low temperatures, the residual width of the
671.4 nm transition was limited by the laser line width. This
is consistent with the very narrow line widths observed in Pr^{3+}.
Additional detailed studies of this type and proper contrast
and comparison between lanthanides and actinides may provide
the additional information needed to describe the electron-
phonon and electron-ligand interactions of the actinides.

C. Many-phonon Processes. The experimental observation
that only certain manifolds fluoresce and that the fluorescence
lifetime is temperature-dependent, indicates that there is nonradiative relaxation between manifolds. Because the energy
difference between manifolds greatly exceeds kT, the effective
energy range for one- and two-phonon processes, multiphonon
processes are required to account for the relaxation. From
measurements of the temperature dependence of excited state
lifetimes and quantum efficiencies, a theoretical model involving
multiphonon transition rates has been developed. The first
systematic study of the multiphonon orbit-lattice relaxation of
lanthanides in single crystals was given by Reisberg and Moos
([48]). The salient feature of the model and the experimental
results is that the spontaneous transition rate for multiphonon
excitation is independent of the particular lanthanide ion or
J-manifold of the ion and depends solely on the host crystal
and the energy gap between manifolds. This model, although
naive, works very well for weakly coupled systems. The results

for specific lanthanide ions in various hosts have been reviewed
by Riseberg and Weber (49).

The situation for actinide ions is ambiguous due to a lack
of experimental data. Because of the larger crystal-field
parameters of the actinides, one would anticipate that the ion-
lattice coupling is stronger. This assumption was not obviously
shown to be true in the recent line width measurements of
neptunium (47). Experimental measurements of the temperature
dependence of the fluorescence lifetimes and quantum efficiencies will provide a direct test for the multiphonon coupling
and the universality of the energy gap dependence of the multiphonon spontaneous transition rate.

IV. New Directions

At a symposium of this type it is appropriate to speculate
on where significant advances may be anticipated in the future.
The free-ion energy level structure of trivalent $4f^n$ and $5f^n$
configurations is fairly well understood. Some additional work
on the static crystal-field interaction is needed. Although
the original idea of Bethe (4) properly accounts for the number
and symmetry of the individual Stark levels, significant deviations remain between experimental and calculated energy levels.
One straightforward improvement is to introduce an additional
potential which represents the two-particle correlation induced
by the ligand fields (50). This unfortunately introduces as
many as 637 new parameters for very low point symmetry or 41
parameters for octahedral symmetry. The problem is then to
devise physical models to reduce the number of free parameters.
Examples of such reduction schemes have been discussed by Judd
(51, 52) and Newman (53). A systematic evaluation of such
schemes is needed along with experimental tests of the physical
mechanisms proposed to reduce the number of parameters.

It is well known that the first excited configuration of
the trivalent actinides, $(Rn, 5f^{n-1}, 6d)$, occurs at a much
lower energy than in a corresponding lanthanide. With new
laser techniques it is possible to investigate this configuration. Such studies will provide direct information on the
structure of the configuration and the more important information
on the interaction of the two configurations. The corresponding
two-photon studies can probe configurations of the same parity
as the ground configuration. This will provide a direct test
of the assumptions needed to formulate the effective Hamiltonian
scheme.

The very recent measurement of the electron-phonon interaction in actinide systems will be followed by additional
measurements along the lines developed for studies of the lanthanide and transition metal systems. Initial studies to
contrast the various sytems will be important in establishing
the relative magnitude of the electron-phonon coupling strength

in the actinides. If this coupling turns out to be intermediate
between the lanthanide and transition metal systems, as is
now assumed, this intermediate coupling may allow new studies
of the many phonon-induced reactions that govern the dynamical
properties of optically excited ions.

In an attempt to define a central issue for future investigation, we propose the following question, "to what extent
can the wave functions deduced in the effective Hamiltonian
approximation and constrained to produce only the energy level
structure of the ion, be used to predict and model other
physical properties of the ion?" In particular, why
is there such a large discrepancy between observed and calculated Zeeman splitting factors?   Hyperfine properties
may now be measured in metastable states
with the same precision as ground state measurements. Can
the same wave functions be used to parameterize the electric
and magnetic hyperfine interactions? The electron-phonon
interaction governs the dynamic processes. To what extent
can the static wave functions be used to discuss these dynamic
processes? Finally, the wave functions are single-ion wave
functions. What changes are required to discuss the ion-ion
interactions that lead to the energy transfer of an optically
excited ion? If a scheme can be obtained which is applicable
to all of the above properties, we will have achieved our
initial goal.

Acknowledgements

Professor B. R. Judd, Dr. K. Rajnak, and Dr. H. M.
Crosswhite are thanked for the many stimulating conversations
over the years. Dr. H. M. Crosswhite and Hannah Crosswhite are
thanked for computing the parameters of Tables I, II, and III.
This work was performed under the auspices of the Office
of Basic Energy Sciences, Division of Nuclear and Chemical
Sciences, U. S. Department of Energy.

Literature Cited

1. Newman, D. J., *Aust. J. Phys.* 1978, **31**, 489-513.
2. Condon, E. U. and Shortley, G. H., *The Theory of Atomic
    Spectra*, Cambridge University Press, New York, 1935.
3. Judd, B. R., *Operator Techniques in Atomic Spectroscopy*,
    McGraw-Hill Book Co., Inc., New York, 1963.
4. Bethe, H., *Ann. Physik* 1929, **3**, 133-208.
5. *Lanthanide/Actinide Chemistry*, Advances in Chemistry Series,
    Vol. 71, Fields, P. R. and Moeller, T., Symposium Charimen,
    American Chemical Society, Washington, D.C., 1967.
6. Wybourne, B. G., *Spectroscopic Properties of Rare Earths*,
    Interscience Publishers, New York, 1965.

7. Carnall, W. T. and Fields, P. R., *Lanthanide and Actinide Absorption Spectra in Solution*, Advances in Chemistry Series, Vol. 71, Fields, P. R. and Moeller, T., Symposium Chairmen, American Chemical Society, Washington, D.C., 1967.

8. Rajnak, K. and Wybourne, B. G., *Phys. Rev.* 1963, 132, 280-290.

9. Judd, B. R., *Phys. Rev.* 1966, 141, 4-14.

10. Judd, B. R.; Crosswhite, H. M. and Crosswhite, H., *Phys. Rev.* 1968, 169, 130-138.

11. Morrison, J. C. and Rajnak, K., *Phys. Rev. A* 1971, 4, 536-542.

12. Morrison, J. C., *Phys. Rev. A* 1972, 6, 643-650.

13. Newman, D. J. and Taylor, C. D., *J. Phys. B* 1972, 5, 2332-2338.

14. Balasubramanian, G.; Islam, M. M.; and Newman, D. J., *J. Phys. B* 1975, 8, 2601-2607.

15. Crosswhite, H. M.; Schwiesow, R. L.; and Carnall, W. T., *J. Chem. Phys.* 1969, 50, 5032-5033.

16. Crosswhite, H. M., *Systematic Atomic and Crystal-Field Parameters for Lanthanides in LaCl₃ and LaF₃*, Colloques Internationaux du C.N.R.S. -- Spectroscopie des Éléments de Transition et des Éléments Lourds dans les Solides, 28 Juin-3 Juillet 1976, Éditions du C.N.R.S., Paris, (1977) pp. 65-69.

17. Crosswhite, H. M.; Crosswhite, H.; Carnall, W. T.; and Paszek, A. P., to be published.

18. Carnall, W. T.; Crosswhite, H.; Crosswhite, H. M.; Hessler, J. P.; Edelstein, N.; Conway, J. G.; Shalimoff, G. V.; and Sarup, R., to be published.

19. Lämmermann, H. and Conway, J. G., *J. Chem. Phys.* 1963, 38, 259-269.

20. Conway, J. G. and Rajnak, K., *J. Chem. Phys.* 1966, 44, 348-354.

21. Conway, J. G., *J. Chem. Phys.* 1964, 40, 2504-2507.

22. Gruber, J. B.; Cochran, W. R.; Conway, J. G.; and Nicol, A. T., *J. Chem. Phys.* 1966, 45, 1423-1427.

23. Crosswhite, H. M., Argonne National Laboratory, personal communication (1979).

24. Froese-Fisher, C. and Saxena, K. M. S., *Phys. Rev. A* 1975 12, 2281-2287.

25. Cowan, R. D. and Griffin, D. C., *J. Opt. Soc. Am.* 1976, 66, 1010-1014.

26. Hessler, J. P.; Caird, J. A.; Carnall, W. T.; Crosswhite, H. M.; Sjoblom, R. K.; and Wagner, Jr., F., *Fluorescence and Excitation Spectra of Bk(3+), Cf(3+), and Es(3+) Ions in Single Crystals of LaCl₃*, The Rare Earths in Modern Science and Technology, McCarthy, G. J. and Rhyne, J. J., Eds., Plenum Publ. Co., New York, N.Y., (1978) pp. 507-512.

27. Van Vleck, J. H., *J. Phys. Chem.* 1937, 41, 67-80.

28. Broer, L. J. F.; Gorter, C. J. and Hoogschagen, J., *Physica* 1945, 11, 231-250.

29. Judd, B. R., *Phys. Rev.* 1962, 127, 750-761.
30. Peacock, R. D., *The Intensities of Lanthanide f-f Transitions*, Structure and Bonding, Vol. 22, Springer-Verlag, New York, N.Y. (1975) pp. 83-121.
31. Krupke, W. F., *IEEE Proc. Region 6 Conf.* IEEE, New York, N.Y. (1975) pp. 17-31.
32. Caird, J. A., *On the Evaluation of Rare Earth Laser Materials and the Matrix Elements of Orbital Tensor Operators*, Ph. D. thesis, University of Southern California, Los Angeles (1975), Available from Univeristy Microfilms, Ann Arbor, Michigan.
33. Ofelt, G. S., *J. Chem. Phys.* 1962, 37, 511-520.
34. Newman, D. J. and Balasubramanian, G., *J. Phys. C* 1975, 8, 37-44.
35. Axe, Jr., J. D., *J. Chem. Phys.* 1963, 39, 1154-11 .
36. Porcher, P. and Caro, P., *J. Chem. Phys.* 1978, 68, 4176-4182.
37. Syme, R. W. G.; Haas, W. J.; Spedding, F. H.; and Good, Jr., R. H., *Chem. Phys. Lett.* 1968, 2, 132-136.
38. Esterowitz, L.; Bartoli, F. J.; Allen, R. E.; Wortman, D. E.; Morrison, C. A.; and Leavitt, R. P., *Phys. Rev. B* 1979, 19, 6442-6455.
39. Hüfner, S., *Optical Spectra of Transparent Rare Earth Compounds*, Academic Press, New York (1978) pp. 115-135.
40. Orbach, R., *Proc. Roy. Soc. (London)* 1961, A264, 458-484.
41. McCumber, D. E. and Sturge, M. D., *J. Appl. Phys.* 1963, 34, 1682-1684.
42. Orbach, R., *Relaxation and Energy Transfer*, Optical Properties of Ions in Solids, DiBartolo, B., Ed., Plenum Press, New York (1975) pp. 355-399.
43. Yen, W. M.; Scott, W. C.; and Shawlow, A. L., *Phys. Rev.* 1964, 136, A271-A283.
44. Szabo, A.; DeVoe, R. G.; Rand, S. and Brewer, R. G., *Bull. Am. Phys. Soc., Series II* 1979, 24, 444.
45. Selzer, P. M., *General Techniques and Experimental Methods in Laser Spectroscopy of Solids*, Laser Spectroscopy of Ions and Molecules in Solids, Yen, W. M., Ed., Springer-Verlag, in press.
46. Yen, W. M., *Laser Spectroscopy of Ions in Crystals*, Laser Spectroscopy of Ions and Molecules in Solids, Yen, W. M., Ed, Springer-Verlag, in press.
47. Hessler, J. P.; Brundage, R. T.; Hegarty, J. and Yen, W. M., to be published.
48. Riseberg, L. A. and Moos, H. W., *Phys. Rev.* 1968, 174, 429-438.
49. Riseberg, L. A. and Weber, M. J., *Relaxation Phenomena in Rare-Earth Luminescence*, Progress in Optics, Vol. 14, Wolf, E., Ed., North-Holland Publ., Amsterdam, (1976) pp. 89-159.

50. Bishton, S. S. and Newman, D. J., *J. Phys. C* 1970, 3, 1753-1761.
51. Judd, B. R., *Phys. Rev. Lett.* 1977, 39, 242-244.
52. Judd, B. R., *J. Chem. Phys.* 1977, 66, 3163-3170.
53. Newman, D. J., *J. Phys. C* 1977, 10, 4753-4764.

RECEIVED December 26, 1979.

Photochemistry of Uranium Compounds

ROBERT T. PAINE and MARCIA S. KITE

Department of Chemistry, University of New Mexico, Albuquerque, NM 87131

The characterization and utilization of photochemical processes are rapidly developing into one of the major areas of
activity in modern inorganic and physical chemistry. In the past,
the photochemistry of classical metal coordination complexes has
received the greatest amount of attention, but recently the
photochemistry of organometallic compounds has attracted notice
(1,2,3). In particular, the photochemistry and photophysics of
uranyl compounds have been investigated for more than four
decades and a great deal has been learned about the primary
photoprocesses and the photo-induced reaction mechanisms displayed by these complexes (3,4). The popularity of uranyl compounds in photochemical studies is derived from their ready
availability and stability, their facile redox chemistry and
photosensitivity and their rich excited state chemistry. Since
current reviews of uranyl photochemistry are expected to appear
in the near future, vide infra, further discussion of this topic
here will be limited.
Instead, we wish to draw attention to the developing photochemistry of other classes of uranium compounds which, until
recently, have received relatively little notice. Historically,
much of the apparent lack of interest in the photochemistry of
uranium (nonuranyl) compounds has been a result of the diffi-
culties found in obtaining stable, well defined complexes. Few
nonuranyl uranium compounds known prior to 1970 are volatile or
stable in air and many compounds are not particularly soluble or
stable in common optically transparent solvents. The recent
surge in the development of new nonaqueous uranium coordination
chemistry and organouranium chemistry (5,6,7), the renewed
interest in photochemically driven isotope separation schemes
and the maturation of organotransition metal photochemistry (1,
2,8) have, however, provided stimuli to the initiation of a
broader spectrum of photochemical and photophysical studies of
uranium compounds. Here we will review the limited, unclassified
progress which has been made toward defining the scope of photochemistry involving uranium compounds. The subject is still in

0–8412–0568–X/80/47–131–369$05.00/0
© 1980 American Chemical Society

an early stage of development. At this point few quantitative photomechanistic studies have been reported and few systematic trends of synthetic utility have been derived. The progress which has been made, however, clearly indicates that further activity in the field will result in the description of useful, transferable photophysical data and a wealth of synthetic photochemical applications.

Uranium Hexafluoride

The photochemistry and spectroscopic properties of UF_6 have, of course, attracted much attention. The gas and liquid phase photochemistry were first explored in the presence of several fluorocarbon compounds during the Manhattan project (9). Mixtures were irradiated at 366 nm for 20 h and uranium products were produced with a reported quantum yield range of 2.4-3.4. Following this work, published activity on this molecule was dormant for some time until a study by Hartmanshenn and Barral appeared in a brief communication in 1971 (10). Uranium hexafluoride vapors and mixtures of UF_6 with H_2, CO, SO_2, O_2, and Xe were exposed to broadband UV radiation; β-UF_5 was formed as the photodecomposition product in each case. This work was followed by further studies of the photoreduction of gaseous UF_6 and gaseous mixtures of UF_6 and H_2 or CO (11). Together these studies led to the description of highly efficient photochemical syntheses of 5-10 g quantities of β-UF_5 in which H_2 or CO act as fluorine atom scavengers in the photoreduction reactions (12,13). The resulting UF_5 was reported to be quite soluble in several non-aqueous solvents including CH_3CN, DMF, and DMSO and the soluble UF_5 has been used as a starting material for the preparation of U(V) coordination compounds and organometallic compounds (14).

The early synthetic reports on the photoreduction of UF_6 led to a new flurry of gas and condensed phase spectroscopic and photophysical studies of UF_6 and its reduction products. McDowell and coworkers (15) studied the high resolution infrared spectrum of UF_6 at ambient and low temperatures. This work was followed by a series of vibrational and electronic spectroscopic studies of matrix isolated UF_6 (16,17,18,19,20). In the first experiments, UF_6 deposited in Ar or CO matrices was vibrationally characterized by infrared spectroscopy and then exposed to broadband UV radiation at 10°K. In argon, photoreduction proceeded rapidly; the 619 cm^{-1} UF_6 infrared peak decreased in intensity while two new peaks grew in at 584 cm^{-1} and 561 cm^{-1}. The new peaks were assigned to the expected UF_5 photolysis product and a tentative C_{4v} structure assignment was made. The wavelength dependence of the photoreduction was studied using a monochromatized UV source (1 kw Hg-Xe lamp, Schoeffel 6M-250 monochromator). The relative quantum efficiency of the UF_6 dissociation per unit absorbance of UF_6 was found to be relatively constant in the allowed B-X absorption band (250-300 nm) (17). Radiation in the forbidden A-$\tilde{X}$ band (340-410 nm) was found to be 10^{-4} as efficient in effecting conversion of UF_6 to UF_5 thereby explaining the low synthetic yields found in early photolysis studies (9). Photolysis of CO/UF_6 matrices (10°K) led to a very rapid production of UF_5 and the back reaction $UF_5 + F \rightarrow UF_6$ found in the argon matrix was retarded by CO scavenging of the fluorine atoms. Prolonged UV irradiation of these matrices led to the formation of a new, broad infrared band centered at 499 cm^{-1}. It was proposed that this band be assigned to matrix isolated UF_4 or polymerized UF_5. Jones (17) extended these matrix results by obtaining both high resolution infrared and Raman spectra. The improved spectroscopic data allowed Jones to firmly deduce the matrix isolated structure of UF_5 (C_{4v}) and determine the UF force constants.

More recent matrix studies (19,20) have attempted to extend the UF_6 photochemistry to include infrared stimulated photo-reduction. Catalano, et al. (19) reported no reaction between gaseous UF_6 and SiH_4 at 100°C; however, irradiation of a UF_6/SiH_4 matrix (12°K) with photons ($\sim$ 16 μm) produced by an incoherent broadband Nernst glower (10 μW cm^{-2}) or a tunable diode laser ($\sim$ 25 mW cm^{-2}) apparently induced a reduction reaction through excitation of the UF_6 ν_3 mode. Both UF_5 and UF_4 were identified in the matrix isolated products. Some doubt has been cast on the results of the broadband infrared photochemistry. Jones (20) has observed that UF_6/SiH_4 matrices are uneffected by infrared irradiation from a Nernst glower rigorously filtered of all UV components. These conflicting results require that additional work be completed before matrix photoreduction through infrared stimulation is verified.

The early broadband UV matrix photochemistry and the search for photochemical isotope separation schemes directed considerable renewed attention to the detailed investigations of gas phase UF_6 photochemistry. Farrar and Smith previously reviewed the progress, up to 1972, in photochemically induced uranium isotope separation (21). Proposed photochemical schemes were outlined in this report including two photon (infrared and UV) processes, but little specific photochemical or photophysical data were presented. In 1976 Eerkens published further spectroscopic data pertinent to laser-driven UF_6 isotope separation (22). The author also states that laser-driven photochemical uranium isotope separation using a UF_6 -HCl mixture was achieved in 1972-73 at AiResearch Manufacturing Company using a CO_2 laser line coincident with the $\nu_3 + \nu_4 + \nu_6$ combination band of UF_6. An enrichment factor of 1.1 was also claimed. To date a report verifying this work has not appeared in the unclassified literature.

Further literature descriptions of laser initiated photodissociation of UF_6 in the gas phase are few. Letokhov and Moore have outlined various general aspects and problems dealing with UF_6 laser photochemistry and isotope separation (23). In

1978, Wittig (24) described multiple photon photodissociation of
UF_6 initiated by the output from a CF_4 laser operating at 615
cm^{-1} (5-25 mJ fluence) and by the combined outputs from a CF_4
laser (615 cm^{-1}) and a CO_2 laser (1077 cm^{-1}, 0.7 J). The
experiments indicated that the one color CF_4 laser output at 5
mJ fluence was sufficient to cause photodissocation through
excitation of ν_3 alone. The addition of the second color (CO_2
laser) dramatically enhanced the dissociation rate. Based upon
these experimental results, Wittig concluded that the CF_4 laser
output leads to vibrationally excited UF_6 molecules having a
broadened and frequency shifted $\nu_2 + \nu_3$ combination band
($\sim$ 1157 cm^{-1}). It was proposed that the CO_2 laser effectively
drove the photodissociation of the excited UF_6 through irradiation of the broad combination band. Photodissociation was not
observed for UF_6 in its vibrational ground state using the one
color CO_2 excitation at 1077 cm^{-1}.
     At the same time Kaldor, et al. (25) reported the observa-
tion of infrared multiphoton photodissociation of UF_6 using a
one color CF_4 laser irradiation source (16 μm) coincident with
ν_3. A dissociation threshold and yield were estimated to be in
the range of those found for SF_6. The preliminary report was
recently followed by a more extensive paper from the Exxon group
(26) in which infrared multiphoton excitation and dissociation
in the ν_3 mode was further described. In addition, two color
excitation (16 μm) and dissociation (10.6 μm) experiments similar
to those described by Wittig were described.

Other Uranium Halides

     Little photochemical work has been accomplished for other
uranium halides. In 1954, Freed and Sancier (45) reported the
broadband UV photodissociation of UCl_4 dissolved in 10% n-pro-
panol and 1:1 propane/propene mixtures. The resulting purple
solution was assumed to contain UCl_3 based upon similar spectroscopic features in spectra of authentic samples of UCl_3 in the
same solvent. More recently, Donohue (46) reinvestigated the
photochemistry of UCl_4 in solution. Irradiation of alcoholic
solutions at 254 nm (low pressure Hg lamp) produced quantum
yields estimated to be in the range of 3 to 10%. Preliminary
studies using 308 nm (XeCl laser) excitation showed no change
in the vis-UV spectrum of UCl_4. Addition of 18-crown-6 poly-
ether to the alcoholic UCl_4 solutions followed by irradiation
at 254 nm resulted in an increased quantum yield and precipitation of a U(III) crown ether complex. The identities of the
U(IV) and U(III) species in this study are not yet clear, and
further work on this interesting system is needed.
     Moody (47) has recently reported thermal chemistry which is
related to Donohue's photochemical observations. It was found
that UCl_4 can be reduced to $UCl_3(THF)_x$ in THF solution and the
UCl_3 forms a THF insoluable UCl_3 crown ether complex. Attempts

to prepare a U(III) alkoxide have led to oxidized U(IV) products. In our own laboratory we have briefly investigated the photo-reduction of UCl_4 in THF solution. Under UV stimulation (medium pressure Hg lamp) small yields ($\sim$ 1%) of U(III) are identified by vis-UV spectrophotometry. We have also noted that addition of a crown ether enhances the U(III) production. Further work on this system will be forthcoming.

Condorelli, et al., (48) have studied the UV photochemistry of $[(C_2H_5)_4N]_2UCl_6$ in acetonitrile under a variety of conditions. Irradiation at 313, 333, 405, or 436 nm left the solutions uneffected; however, irradiation at 254 nm produced photochemical reactions. In the presence of air, the photoproduct was identi-fied as $UO_2Cl_4^{2-}$. In the absence of air more complex chemistry was found and the authors present a mechanism describing the overall chemistry.

Classical Coordination Complexes

Few photochemical investigations of classical nonuranyl complexes have been accomplished. Adams and Smith (49) reported photolyses of uranium(IV) citrate aqueous solutions. In the absence of oxygen, the complex was apparently stable toward photochemical oxidation; however, in the presence of oxygen, irradiation from a tungsten lamp source led to the formation of a uranyl citrate complex. The uranium(VI) citrate undergoes photoreduction to a U(IV) citrate species in the absence of air. The authors also claim that related chemistry is found for a uranium(IV) tartrate complex. Dainton and James (50) have reported photochemical electron transfer reactions involving U(III) and U(IV) in aqueous solutions. The results dealing with photosensitization reactions are interesting and they warrant further attention.

We have recently examined the photo-reactivity of $U[N(Si(CH_3)_3)_2]_3Cl$ in THF solution. Although the study is not complete at this date, the complex is reduced by broadband UV radiation from a medium pressure Hg lamp. The major product in each case appears to be $U[N(Si(CH_3)_3)_2]_3$ which has been synthe-sized by thermal techniques and subjected to extensive charac-terization by Andersen and coworkers (51). Detailed photochemical studies of this system are in progress.

Uranium Alkoxides

In 1976 Sostero, et al. (52) reported photochemical prepara-tions and reactions of several oxochlorouranium compounds. It was noted that photolysis of $(C_5H_5NH)_2UOCl_5$ in dry ethanol with a Hg UV source led to the formation of $U(OC_2H_5)_5$. The ethoxide was identified by its electronic spectral properties and a pro-posed mechanism for the photoreaction was presented. Quantum yield and product yield data were not given. Marks, et al. (53)

have recently reported a laser induced multiphoton gas phase
photodecomposition reaction of $U(OCH_3)_6$ using the output from a
CO_2 laser at the 00°1-10°0 transition which is nearly coincident
with an infrared absorption of the uranium methoxide at 931 cm^{-1}.
The authors report isotopic ^{235}U enrichment in the samples
treated under these conditions, but little specific information
is given regarding the observed photochemistry and photophysics.
Further studies of the photochemical reactions of uranium alko-
xides is certainly warranted based upon these two investigations.

Organometallic Compounds

Unlike organotransition metal chemistry, photochemical
techniques have not been widely applied to organoactinide
chemistry. Marks and coworkers (54) have observed that $Cp_3Th(i-C_3H_7)$ undergoes thermolysis at 170°C in toluene solution. The
products isolated from the reaction are C_3H_8 and $[Cp_2Th(C_5H_4)]_2$,
and no evidence is found for β-hydride elimination reaction
products. When the same compound is photolyzed at 5°C in a
benzene solution β-hydride elimination is observed. The products are propane, propene, and Cp_3Th. The mechanism of the
photochemical reaction is discussed in some detail by the
authors, and the system apparently represents the first example
of a photo-induced β-hydride elimination reaction which is
thermally blocked. This observation is in opposition to
observations in organotransition metal chemistry, where β-hydride
elimination is a thermally observed but photochemically hindered
process. A similar reaction occurs with the uranium analog (55).
Additional photochemical investigations of organouranium compounds are warranted. It may be hoped that whatever studies are
initiated will contain both synthetic and photophysical components.

Uranium Borohydride

Photochemical decomposition of $U(BH_4)_4$ was noted by investigators during the Manhattan project, but little detail is
available (19). In 1974 Engleman (56) noted that the UV flash
photolysis of $U(BH_4)_4$ produced a high density of emission lines
which could be assigned in part to UI, UII and perhaps BH. It
was then concluded in the preliminary study that $U(BH_4)_4$ could
serve as a convenient source of uranium atoms. Subsequently,
we have reported gas and solution phase photochemistry of $U(BH_4)_4$
and $U(BD_4)_4$ (57). Gaseous samples of both compounds were subjected to broadband UV photolysis for two hours using a 100 watt
Hanovia lamp. Typically 10-20% of the sample decomposed and
formation of $H_2(D_2)$ and $B_2H_6(B_2D_6)$ was noted along with a brown
solid. In dry THF or methylcyclohexane photodecomposition was
also realized and $U(BH_4)_3(U(BD_4)_3)$ was isolated and identified
as a product. The wavelength dependence of the UV promoted
photochemistry has not yet been determined.

Infrared induced photochemistry of $U(BD_4)_4$ was also explored by us using a CO_2 laser transition nearly coincident with the 924 cm^{-1} mode of the molecule (57). In a static system, the molecule was irradiated by one IR pulse (0.8 J/cm^2) and the gaseous products collected in a cold trap. After 25 pulse-collection sequences the vapors were found to contain B_2D_6 and D_2. An increase in the pulse repetition rate (0.5 Hz) without volatile product collection resulted in the appearance of an intense visible light emission after the second and succeeding pulses. The emission was analyzed and found to be a broad ($\sim$ 100 Å) band centered at 5914 Å. These results suggested that the first pulse leads to dissociation of the $U(BD_4)_4$ with the formation of $U(BD_4)_3$ which then interacts with the succeeding ir pulses. The second absorption leads to the observed emission. Attempts to measure the reaction threshold and quantum yields were unsuccessful. Further laser induced photochemistry on this molecule is warranted.

Uranyl Compounds

Although we have chosen to omit most of the large body of uranyl photochemistry from this short review, it is worth pointing out recent results from the Exxon group (58,59,60). The laser induced gas phase photodissociation of $[UO_2(hfacac)_2 \cdot THF]$ has been explored, and evidence for isotopic selectivity in both the uranium and oxygen isotopes was presented. These papers are particularly interesting, and they show an emerging trend toward detailed photochemical characterization of uranium complexes.

During the course of the collection of material for this review it was found that two other reviews are forthcoming which will summarize recent advances in the photochemistry of uranyl species. Interested investigators should look for papers by C. K. Jorgensen and R. Reisfeld in Structure and Bonding and H. Güsten in Gmelin.

Conclusions

Our examination of the photochemical literature of uranium clearly shows that extensive attention has been given to UF_6, while other compounds, until recently, have been almost ignored. The attention given to UF_6, of course, relates back to the great interest in achieving a low cost laser induced isotope separation process for uranium isotopes. The economics of isotope separation, which have been briefly discussed by Letokhov and Moore (61), have consequently dictated the direction of much of the applied photochemical research on uranium compounds. Nonetheless, from the existing spectroscopic and photochemical data outlined here it would be expected that coordination and organometallic complexes should display fundamentally interesting, facile photochemistry, and the few studies which have been completed indicate that this is true. It is apparent that this field needs additional fundamental photochemical attention. In particular, those interested in pursuing topics in uranium photochemistry should give attention to the following:

1. Optical spectroscopy: more detailed infrared, visible-UV and MCD studies, including assignments of observed transitions, are needed.
2. Synthesis: more wide spread application of photochemical techniques needs to be made in the synthesis and characterization of new compounds.
3. Systematics: attempts should be made to systematize photochemical reactions of uranium compounds.
4. Photophysics: the synthetic chemist should work in concert with photochemists in order that a greater characterization of photochemical reaction mechanisms occurs.

Investigators beginning new studies in this field should be rewarded with results of equal interest to data that have been obtained in organotransition metal photochemical research.

Literature Cited

1. Wrighton, M. S., Ed. "Inorganic and Organometallic Photochemistry," Adv. Chem. Ser., no. 168, American Chemical Society, Washington, DC, 1978.
2. Koerner von Gustorf, E. A.; Leenders, L. H. G.; Fischler, I.; Perutz, R. N. "Aspects of Organo-transition Metal Photochemistry and their Biological Implications," Adv. Inorg. Chem. Radiochem., 1976, 19, 65.
3. Balzani, V.; Carassiti, V. "Photochemistry of Coordination Compounds," Academic Press, NY, 1970.
4. Burrows, H. D.; Kemp, T. J. "The Photochemistry of the Uranyl Ion," Chem. Soc. Rev., 1974, 3, 139.
5. Tsutsui, M.; Ely, N.; Dubois, R. "σ-Bonded Organic Derivatives of f-elements," Acct. Chem. Res., 1976, 9, 217.
6. Marks, T. J. "Actinide Organometallic Chemistry," Acct. Chem. Res., 1976, 9, 223.
7. Marks, T. J. "Chemistry and Spectroscopy of f-element Organometallics. Part II: The Actinides," Prog. Inorg. Chem., 1977, 27, 223.
8. Wrighton, M. S. "The Photochemistry of Metal Carbonyls," Chem. Rev., 1974, 74, 401.
9. Urey, H. C. "Investigations of the Photochemical Method for Uranium Isotope Separation," Project SAM, Columbia University, July 1943 (A-750).
10. Hartmanshenn, O.; Barral, J. C. "Nouveaux Modes de Reduction de l'hexafluorure d'uranium," Compt. Rend. Ser. C, 1971, 267, 2139.

11. Asprey, L. B.; Paine, R. T. "One Electron Reduction Synthesis of Uranium Pentafluoride," Chem. Commun. 1973, 921.

12. Paine, R. T.; Asprey, L. B. "Metal Pentafluorides," Inorg. Syn., 1979, 19, 137.

13. Halstead, G. W.; Eller, P. G.; Asprey, L. B.; Salazar, K. V. "Convenient Multigram Synthesis of Uranium Pentafluoride and Uranium Pentaethoxide," Inorg. Chem., 1978, 17, 2967.

14. Halstead, G. W.; Eller, P. G.; Eastman, M. P. "Nonaqueous Chemistry of Uranium Pentafluoride," Inorg. Chem., 1979, 18, 2867.

15. McDowell, R. S.; Asprey, L. B.; Paine, R. T. "Vibrational Spectroscopy of Uranium Hexafluoride," J. Chem. Phys., 1974, 61, 3571.

16. Paine, R. T.; McDowell, R. S.; Asprey, L. B.; Jones, L. H. "Vibrational Spectroscopy of Matrix Isolated UF_6 and UF_5," J. Chem. Phys., 1976, 64, 3081.

17. Lewis, W. B.; Asprey, L. B.; Jones, L. H.; McDowell, R. S.; Rabideau, S. W.; Zeltmann, A. H.; Paine, R. T. "Electronic and Vibronic States of Uranium Hexafluoride in the Gas and in the Solid Phase at Very Low Temperature," J. Chem. Phys., 1976, 65, 2707.

18. Jones, L. H.; Ekberg, S. A. "Potential Constants and Structure of the UF_5 Monomer," J. Chem. Phys., 1977, 67, 2591.

19. Catalano, E.; Barletta, R. E.; Pearson, R. K. "Infrared Laser Single Photon Absorption Reaction Chemistry in the Solid State, I. The System SiH_4-UF_6," J. Chem. Phys., 1979, 70, 3291.

20. Jones, L. H.; Ekberg, S. A. "Photo-induced Reaction of UF_6 with SiH_4 in a Low Temperature SiH_4 Matrix," J. Chem. Phys. submitted for publication.

21. Farrar, R. L.; Smith, D. F. "Photochemical Isotope Separation as Applied to Uranium," USAEC K-L-3054 (Rev. 1), 1972.

22. Eerkens, J. W. "Reaction Chemistry of the UF_6 LISOSEP Process," Opt. Commun., 1976, 18, 32; Eerkens, J. W. "Spectral Considerations in the Laser Isotope Separation of Uranium Hexafluoride," App. Phys., 1976, 10, 15.

23. Letokov, V. S.; Moore, C. B. "Chemical and Biochemical Applications of Lasers," Vol. III, Moore, C. B. Ed.: Academic Press, NY, 1979.

24. Tiee, J. J.; Wittig, C. "The Photodissociation of UF_6 Using Infrared Lasers," Opt. Commun., 1978, 27, 377.

25. Rabinowitz, P.; Stein, A.; Kaldor, A. "Infrared Multiphoton Dissociation of UF_6," Opt. Commun., 1978, 27, 381.

26. Horsley, J. A.; Rabinowitz, P.; Stein, A.; Cox, D. M.; Brickman, R.; Kaldor, A. "Laser Chemistry Experiments with UF_6," J. Quantum Electron, submitted for publication.

27. Andreoni, A.; Bücher, H. "Fluorescence of Uranium Hexafluoride in the Gas Phase," Chem. Phys. Lett., 1976, 40, 237.

28. Benetti, P.; Cubeddu, R.; Sacchi, C. A.; Svelto, O.; Zaraga, F. "Fluorescence from Gaseous UF_6 Excited by a Near UV Dye Laser," Chem. Phys. Lett., 1976, 40, 240.

29. DeWitte, O.; Dumanchin, R.; Michen, M.; Chatelet, J. "Near UV Photophysics of Gaseous UF_6," Chem. Phys. Lett., 1977, 48, 505.

30. Andreoni, A.; Cubeddu, R.; de Silvestri, S.; Zaraga, F. "Temperature Dependence of the Fluorescence Lifetime of Gaseous UF_6 Excited at 374 nm," Chem. Phys. Lett., 1977, 48, 431.

31. Ambartzumian, R. V.; Zubarev, I. G.; Iongansen, A. A.; Kotov, A. V. "Investigation of the Kinetics of the Vibrational Excitation of the UF_6 Molecule by the IR-UV Resonance Method," Sov. J. Quantum Electron, 1978, 8, 910.

32. Grzybowski, J. M.; Andrews, L. "Ultraviolet Laser Induced Fluorescence of UF_6 Isolated in Argon Matrices," J. Chem. Phys., 1978, 68, 4540; Miller, J. C.; Allison, S. W.; Andrews, L. "Laser Spectroscopy of Matrix Isolated UF_6 at 12°K," J. Chem. Phys., 1979, 70, 3524.

33. Wampler, F. B.; Oldenborg, R. C.; Rice, W. W. "Laser Induced Time Resolved Emission of Electronically Excited UF_6," Chem. Phys. Lett., 1978, 54, 554.

34. Oldenborg, R. C.; Rice, W. W.; Wampler, F. B. "Laser Induced Fluorescence of Gaseous UF_6 in the A-X Band," J. Chem. Phys., 1978, 69, 2181.

35. Wampler, F. B.; Rice, W. W.; Oldenborg, R. C.; Akerman, M. A.; Magnuson, D. W.; Smith, D. F.; Werner, G. K. "Emission and Ionization Induced by Focused Kr-F Laser Irradiation of UF_6," Opt. Lett., 1979, 4, 143.

36. Rice, W. W.; Wampler, F. B.; Oldenberg, R. C.; Lewis, W. B.; Tiee, J. J.; Pack, R. T. "Spectra and Modeling of Laser Induced Emission for Multiple Photon (λ = 248.4 nm) Irradiation of UF_6," J. Chem. Phys., submitted for publication.

37. Lewis, W. B.; Wampler, F. B.; Huber, E. J.; Fitzgibbon, G. C. "Photolysis of Uranium Hexafluoride and Some Reaction Variables Affecting the Apparent Quantum Yield," J. Photochem., submitted for publication.

38. Kim, K. C.; Fleming, R.; Seitz, D.; Reisfeld, M., "Laser Flash Photolysis of UF_6: The 17 μm Infrared Spectrum of the Transient UF_5 Molecule," Chem. Phys. Lett., 1979, 62, 61.

39. Kim, K. C.; Fleming, R.; Seitz, D. "The Infrared Absorption Spectrum of the Isolated UF_5 Molecule in the 580-630 cm^{-1} Region," Chem. Phys. Lett., 1979, 63, 471.

40. Lucht, R. A.; Beardall, J. S.; Kennedy, R. C.; Sullivan, G. W.; Rink, J. P. "Multiple Photon Absorption of 16 μm Radiation in UF_6 at 300°K," Opt. Lett., 1979, 4, 216.

41. Wampler, F. B.; Oldenborg, R. C.; Rice, W. W. "Lifetimes of Electronically Excited UF_6 in the Presence of Inorganic Quenchers," J. Photochem., 1978, 9, 473, and references therein.

42. Kroger, P. M.; Riley, S. J.; Kwei, G. H. "Polyhalide Photofragment Spectra. II. Ultraviolet Photodissociation Dynamics of UF_6," J. Chem. Phys., 1978, 68, 4195.
43. Beauchamp, J. L. "Properties and Reactions of Uranium Hexafluoride by Ion Cyclotron Resonance Spectroscopy," J. Chem. Phys., 1976, 64, 718.
44. Lewis, W. B.; Zeltmann, A. H. "Optoacoustic Spectroscopy and the Energy of Photodissociation of Uranium Hexafluoride," J. Photochem., submitted for publication.
45. Freed, S.; Sancier, K. M. "Photochemical Activity of Salts of Uranium in Solutions at the Temperature of Liquid Nitrogen," J. Chem. Phys., 1954, 22, 928.
46. Donohue, T. Naval Research Laboratory, personal communication.
47. Moody, D. C.; Odom, J. D. "The Chemistry of Trivalent Uranium: The Synthesis and Reaction Chemistry of the Tetrahydrofuran Adduct of Uranium Trichloride, $UCl_3(THF)_x$," J. Inorg. Nucl. Chem., 1979, 41, 533; Moody, D. C.; Penneman, R. A.; Salazar, K. V. "The Chemistry of Trivalent Uranium. 2. Synthesis of UCl_3 (18-crown-6) and $U(BH_4)_3$ (18-crown-6)," Inorg. Chem., 1979, 18, 208.
48. Condorelli, G.; Costanzo, L. L.; Pistara, S.; Tondello, E. "Photochemistry of Chloro Complexes of Uranium(IV) and Dioxouranium(VI)," Inorg. Chim. Acta. 1974, 10, 115.
49. Adams, A.; Smith, T. D. "The Formation and Photochemical Oxidation of Uranium(IV) Citrate Complexes," J. Chem. Soc., 1960, 4846.
50. Dainton, F. S.; James, D. G. L. "Photochemical Electron Transfer and Some Related Phenomena in Aqueous Solution of Reducing Ions Containing Polymerizable Monomers," Trans. Farad. Soc., 1957, 53, 333.
51. Andersen, R. A. "Tris((hexamethyldisilyl)amido)uranium(III): Preparation and Coordination Chemistry," Inorg. Chem., 1979, 18, 1507.
52. Sostero, S.; Traverso, O.; Bartocci, C.; DiBernardo, P.; Magon, L.; Carassiti, V. "Photochemistry of Actinide Complexes. III. The Photoproduction Mechanism of Uranium(V) Oxochloro Complexes," Inorg. Chim. Acta., 1976, 19, 229.
53. Miller, S. S.; DeFord, D. D.; Marks, T. J.; Weitz, E. "Infrared Photochemistry of a Volatile Uranium Compound with 10 μm Absorption," J. Amer. Chem. Soc., 1979, 101, 1036.
54. Kalina, D. G., Marks, T. J.; Wachter, W. A. "Photochemical Synthesis of Low-Valent Organothorium Complexes. Evidence for Photoinduced β-Hydride Elimination," J. Amer. Chem. Soc., 1977, 99, 3877.
55. Marks, T. J. Northwestern University, personal communication.
56. Engleman, R. "A New Source of Atomic Uranium for Absorption Spectroscopy and other Applications," Spec. Lett., 1974, 7, 547.

57. Paine, R. T.; Schonberg, P. R.; Light, R. W.; Danen, W. C.; Freund, S. M. "Photochemistry of U(BH4)4 and U(BD4)4," J. Inorg. Nucl. Chem., 1979, 00, 0000.
58. Cox, D. M.; Hall, R. B.; Horsley, J. A.; Kramer, G. M.; Rabinowitz, P.; Kaldor, A. "Isotope Selectivity of Infrared Laser Driven Unimolecular Dissociation of a Volatile Uranyl Compound," Science, 1979, 205, 390.
59. Kaldor, A.; Hall, R. B.; Cox, D. M.; Horsley, J. A.; Rabinowitz, P.; Kramer, G. M. "Infrared Laser Chemistry of Large Molecules," J. Amer. Chem. Soc., 1979, 101, 4465.
60. Kaldor, K.; Cox, D. M. "Energy Localization in CO2 Laser Driven Multiple Photon Dissociation," J. Chem. Phys., submitted for publication.
61. Letokhov, V. S.; Moore, C. B., "Chemical and Biochemical Applications of Lasers," Moore, C. B., Ed. Academic Press, New York, 1977, p. 121.

RECEIVED December 26, 1979.

Multistep Laser Photoionization of the Lanthanides and Actinides

E. F. WORDEN

Lawrence Livermore Laboratory, University of California, Livermore, CA 94550

J. G. CONWAY

Materials and Molecular Research Division, Lawrence Berkeley Laboratory, University of California, Berkeley, CA 94720

Multistep laser photoionization has been applied to determine a number of important physical properties of heavy atoms with complex spectra including ionization potentials,[1,2,3,4,5] energy levels,[4,6] lifetimes of levels,[1,2,4,6,7] branching ratios,[7,8] oscillator strengths,[2,5,7,8] isotope shifts,[9,10,11] hyperfine structure[9,12,13] and autoionization.[2,3,4,5,14] Ionization potentials are useful in the description of systematic trends of the elements and in understanding chemical bonding in gaseous molecules. They are used in the calculation of ion densities in high temperature metal vapors and gaseous mixtures of known temperature. Oscillator strengths are employed for obtaining concentrations of elements in high temperature media and plasmas that emit the spectral lines of the elements. The latter two uses are frequently made by astronomers.

Energy levels are used in determining the electronic structure of atoms. For most of the lanthanides and actinides, this has been accomplished by conventional spectroscopy. However, additional levels in the neutral atoms of these elements are easily found by laser techniques, especially at high excitation energy or near the ionization limit where conventional sources usually fail because of the very low absorption intensities for these levels and their low population in emission sources.

Lifetimes, oscillator strengths, branching ratios, isotope shifts, hyperfine structure and autoionization structure are all critical parameters in atomic vapor laser isotope separation,[2,6,10,15,16,17] while the first three are important in potential laser excited atomic vapor processes. Isotope shifts and hyperfine structure are useful in determining energy levels, in assigning these energy levels to the different electronic configurations of an element and in determining nuclear properties of isotopes of the elements, but we will not discuss these applications here. We should mention that

0–8412–0568–X/80/47–131–381$11.25/0
© 1980 American Chemical Society

lifetimes,[18,19,20] branching ratios, oscillator strengths, energy levels,[18,21] isotope shifts[22] and hyperfine structure[21,22,23,24] have been or can be determined by laser techniques that do not use photoionization for detection. Laser induced fluorescence, absorption and the optogalvonic effect are some of the methods used,[25] but they are not a subject of this review.

We will discuss the application of multistep laser excitation and ionization to determine the physical properties mentioned above in the lanthanides and actinides with emphasis on the determination of accurate ionization potentials. The discussion will point out how the laser techniques can circumvent many of the experimental obstacles that make these measurements difficult or impossible by conventional spectroscopy. The experimental apparatus and techniques described can be employed to measure all the properties and they are typical of the apparatus and techniques employed generally in multistep laser excitation and ionization. We do not claim completeness for literature cited, especially for laser techniques not involving photoionization detection.

Ionization Potentials

Ionization potentials of atoms are usually obtained by the determination of a photoionization threshold or more accurately by the observation of long Rydberg progressions. With the exception of a few of these elements with simple spectra, obtaining such measurements for lanthanides and actinides is difficult if not impossible by conventional spectroscopy. Therefore, very accurate ionization limits were not available for the majority of these elements.[26]

The difficulty in observing Rydberg series arises from the extreme complexity of the electronic structure which results in very dense spectra characterized by weak absorptions into Rydberg levels with large principal quantum numbers. The presence of a number of thermally populated low-lying levels in most of the atoms of these elements together with the great density of potentially perturbing valence levels at high energy so complicates most of the single photon absorption spectra that Rydberg series cannot be identified. Indeed, the only lanthanides where Rydberg series have been observed by conventional spectroscopy[27,28,29,30] have relatively simple spectra and very few low-lying energy levels.[31] The elements lanthanum,[27] europium,[28] thulium,[29] ytterbium,[30] and lutetium[29] all have only one or two well-isolated low levels that are thermally populated at the temperatures needed to produce an atomic vapor and have only a few well-separated ion levels to serve as Rydberg convergence limits. For the remaining elements with complex spectra, we

have applied the more sensitive and flexible methods of multistep laser spectroscopy.[2,3,32,33]

The same arguments apply to the study of ionization thresholds.   While some success has been possible for the elements
with simpler electronic structure (ytterbium, europium, and
thulium),[34,35] for the remainder of the lanthanides it is
nearly an impossible task to unravel the spectra originating
from the many populated metastable levels to accurately determine the ionization potential with confidence.[36]

Recently, stepwise laser photoexcitation and ionization has
been used to identify Rydberg series in atomic uranium.[1]
They allow levels connected by optical transitions to the
ground level or to any of the low-lying thermally populated
metastable levels to be selectively excited.   The excitation
may take place in one, two or three steps to reach the desired
level.   Spectra obtained from these laser prepared excited
levels are not subject to the ambiguities associated with conventional absorption and ionization spectra.   One clearly
avoids the difficulty of sorting out which of the thermally
populated low-lying levels is associated with a specific
feature of the spectrum.   When required, time delaying the ionization step can be used to discriminant and preferentially
detect the long-lived Rydberg levels.   These methods are similar to those used by Dunning and Stebbings.[37]

The Rydberg series and photoionization thresholds obtained
have permitted the accurate determination of ionization limits
for uranium,[1] neptunium,[4] and ten lanthanides.[3]
When these results for the lanthanides are combined with available literature values, accurate experimental ionization
potentials become available for all the lanthanides except
promethium.   These ionization limits, when normalized to
correspond to the energy between the lowest level of the
$f^N s^2$ configuration of the neutral and the lowest level of
the $f^N s$ configuration of the ion, and when plotted against N,
display a connected two-straight-line behavior with a slope
change at the half-filled shell.   Theory predicts such a
behavior for lowest level to lowest level ionization potential
for $f^N s^2 - f^N s$ configurations.[38]

Experimental.   A multistep laser photoionization apparatus
is shown schematically in Fig. 1.   It has been described in
detail previously.[1,3,4,6]   Other investigators have used
basically the same type of instrumentation.   Briefly it is a
crossed beam spectrometer in which the atoms in the atomic beam
are irradiated and eventually ionized by the output of either
two or three pulsed, nitrogen laser-pumped tunable dye lasers.
The resistively heated tungsten tube oven is usually operated
at a temperature sufficient to give an atomic vapor pressure of
roughly 10^{-3} Torr (10^{-1} Pa).   The vapor effuses through a
slit into an interaction chamber where, at an atom density of

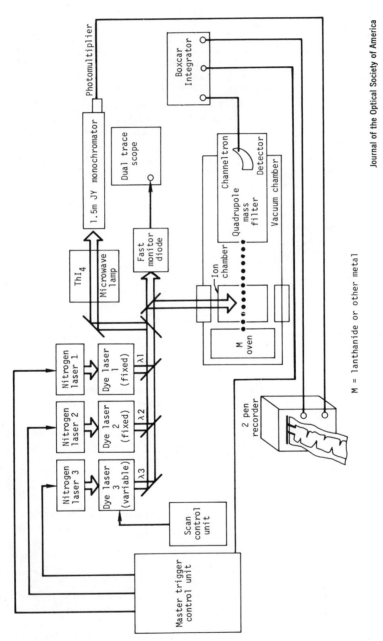

Journal of the Optical Society of America

M = lanthanide or other metal

Figure 1.   Laser spectroscopy apparatus (3)

approximately 10^9 to 10^{11} atoms/cm^3, it is irradiated by
the dye laser pulses. The detector is a channeltron particle
multiplier contained in quadrupole mass analyzer that is tuned
to the mass of the atom under study to discriminate against
detection of oxide or other impurities. The vacuum chamber
background pressure is typically 10^{-7} Torr (10^{-5} Pa).

The interaction chamber-quadrupole setup can be replaced by
a field ionization chamber and channeltron ion detector. The
channeltron detects the ions produced and deflected by the
pulsed electric field. For most of these studies, a pulsed
electric field of 5 kV per cm delayed by 5 μs with respect to
the final laser was used.

The nitrogen pump lasers are triggered by a common master
control unit with delay lines arranged so that each laser fires
at predetermined and well-controlled times with respect to the
others. The dye laser pulses were monitored by a fast vacuum
photodiode and oscilloscope. The dye lasers provided 5-10 ns
pulses having 0.5-2.0 cm^{-1} spectral linewidths with less than
5 ns jitter. A boxcar integrated the signal received from the
particle multiplier.

The first and/or second dye lasers were tuned to the
specific wavelength(s) to populate the desired level(s). The
final laser in the excitation sequence (either the second or
third laser) was then continuously scanned to obtain the
Rydberg or autoionization spectrum. The spectrum and wavelength calibrations were recorded simultaneously on a two pen
recorder. Wavelength calibration was obtained by directing a
portion of the scan laser radiation to a monochromator that was
preset at known U or Th emission lines from an electrodeless
lamp.

The excitation schemes employed to obtain photoionization
and Rydberg spectra are indicated in Fig. 2. A time delay of
10-20 ns was introduced between laser outputs to provide an
unambiguous excitation sequence. The primary excitation (λ_1)
was always a known transition from the ground or low-lying
thermally populated level. (In our notation λ_i is the wavelength of the ith laser in the excitation sequence.) In the
three-step experiments λ_2 was usually a known transition, but
occasionally it was necessary to use a transition obtained by
laser spectroscopy techniques where λ_1 was fixed, λ_2 was
scanned, and λ_3 was set so the energy of $\lambda_1 + \lambda_2 + \lambda_3$ exceeded
the ionization potential of the element. Ion current was
obtained when λ_2 coincided with an allowed transition from the
level populated by λ_1. A more detailed description of this
method has been given previously.[6] Background peaks could
occur in all spectra obtained and the details of how these were
eliminated are given in references [3] and [6].

The photoionization threshold from one or more excited
levels of the atom under study was determined first using the
excitation schemes of Fig. 2. This involved scanning a

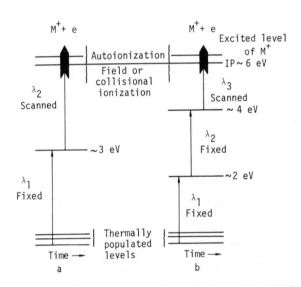

Figure 2.   Excitation schemes used to obtain Rydberg and autoionization spectra
(3)

considerable wavelength range (~100Å) estimated from the
available literature values[26,39,40] of the ionization
potentials that had as their best quoted uncertainties ±200
cm^{-1}. In most cases the thresholds were found within the
ranges estimated from these values. The photoionization
threshold limits obtained were accurate to about 30 cm^{-1}.
From these values, wavelength ranges to search for bound
Rydberg series with field or collisional ionization or to
search for autoionizing series converging to excited states of
the ion were estimated for various parent levels that could be
conveniently populated by one or two-step excitation. The
threshold determinations reduced the search ranges for Rydberg
levels to reasonable values. Scans were made from various
parent levels until series were obtained.

Photoionization Threshold Results. The photoionization
spectra of Nd for two different parent levels are shown in
Fig. 3. The excitation schemes are shown on the figure; the
thresholds are marked by the onset of strong autoionizing transitions. The photoionization threshold for neptunium is shown
in Fig. 4. Representative photoionization threshold results
are given in Table I for two and three-step measurements. The
wavelength(s) in the columns headed "Excitation Wavelengths" in
Table I correspond to transitions from the ground or a thermally populated metastable level to the excited levels in
column 4. The last three columns in the table give the
observed wavelength in Ångstroms of the scanned laser at the
onset of photoionization and the corresponding value of the
ionization threshold from the ground state of the element in
wave numbers and in eV.

Because the ionization potentials obtained from Rydberg
convergence limits are much more accurate, the photoionization
thresholds served mainly to limit the search range to find
Rydberg series. The praseodymium threshold value is an exception because no Rydberg series were obtained for that element.
The photoionization threshold of $5.464^{+0.012}_{-0.002}$ eV is the only
experimental value available for praseodymium.

Janes et al.[2] were the first to apply multistep photoionization to the study of uranium. Their result for the
photoionization threshold of 6.187(2) eV is in good agreement
with the photoionization result of 6.1912(25) eV obtained by
Solarz et al.[1] (Throughout this paper, numbers in
paranthesis following a numerical value indicate the
uncertainty in the last digit of the number.)

Rydberg Series Results. A dysprosium autoionizing Rydberg
spectrum is shown in Fig. 5. This is a double series converging to the $4f^{10}6s$ $^4I_{15/2}$ limit 828.3 cm^{-1} above the
ground level of the ion. Fig. 6 shows a three-step Rydberg
spectrum of neptunium with two series converging to two different limits. Field ionization was used in this case so the

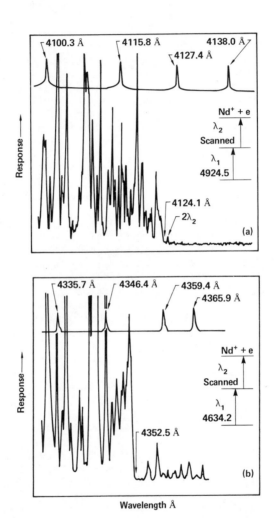

Journal of the Optical Society of America

Figure 3. Photoionization threshold spectra for neodymium. The excitation scheme used in each case is shown on the figure. The scanned laser wavelength calibration is shown at the top of each spectrum. In (a) the 20 300.8 cm⁻¹ level is populated and in (b) the 21 572.6 cm⁻¹ level is populated. The threshold wavelengths indicated yield the same ionization limit value of 5.523 eV. The arrows labeled R. L. indicate the position of the Rydberg convergence limit (3).

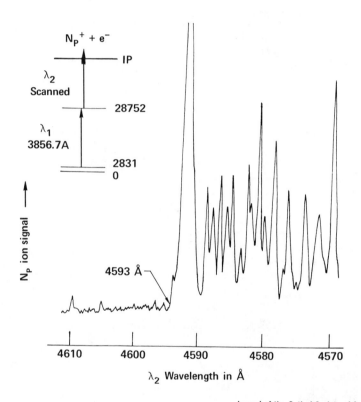

Journal of the Optical Society of America

Figure 4. Neptunium photoionization threshold spectrum. The excitation scheme is shown at the left. The threshold at 4593 Å is marked by the onset of very strong autoionization peaks. It yields an ionization potential of 50 518 cm⁻¹ (6.2624 eV) (4).

Table I. Representative stepwise laser photoionization threshold results

	Excitation Wavelengths[a] λ_1 (Å)	λ_2 (Å)	Excited level used[a] (cm^{-1})	λ_3 Wavelength at threshold (Å)	Ionization threshold[b] (cm^{-1})	(eV)
Ce	6310.10	6996.8	30 131(2)	6883.0	44 656(40)	5.539(5)
	4632.32		21 581.41	4338.2	46 626(40)	5.533(5)
Pr	3945.41		26 715.34	5762.0	44 065(±100)	5.463(±12)
	4939.74		23 085.08	4764.0	44 070(±100)	5.464(±12)
Nd	4634.24		21 572.61	4352.2	44 543(15)	5.523(2)
	4924.53		20 300.84	4124.1	44 542(15)	5.522(2)
Dy	6259.09	6769.79	30 739.79	5831.1	47 884(25)	5.937(3)
	4565.09		21 899.22	3847.5	47 882(25)	5.937(3)
	4612.26		21 675.28	3820.5	47 843(25)	5.931(3)
Ho	6305.36	6576.8	31 056(2)	5720.0	48 534(25)	6.017(3)
	4040.81		24 740.52	4206.0	48 509(25)	6.014(3)
Er	6221.02	6649.06	31 105.66	5516.1	49 230(15)	6.104(2)
	4087.63		24 457.15	4034.2	49 238(15)	6.105(2)
Np	3806.36[c]		26 264.37	4122.6	50 514(3)	6.263(1)
	3849.42		25 970.58	4072.7	50 517(3)	6.263(1)

( ) = uncertainty in last digit of the number.

[a] Excitation wavelengths (λ_1 and λ_2) and excited level values are from Refs. (41), (42) and references therein except for the λ_2 values Ce and Ho which were determined by laser techniques, Ref. (6).

[b] 8065.479 cm^{-1}/eV was used to convert the cm^{-1} values to eV.

[c] Excitation from the 2831.10 cm^{-1} level of Np.

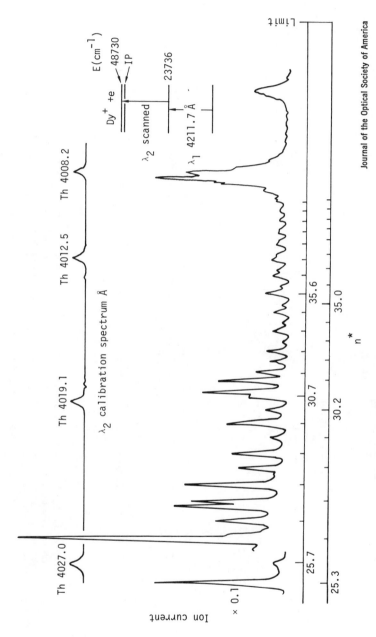

Figure 5. Dysprosium autoionizing Rydberg series converging to the ⁴I₁₅/₂ limit 828.3 cm⁻¹ above the ⁶I₁₇/₂ ground state of the ion. This is a double series converging to the same limit (3).

Journal of the Optical Society of America

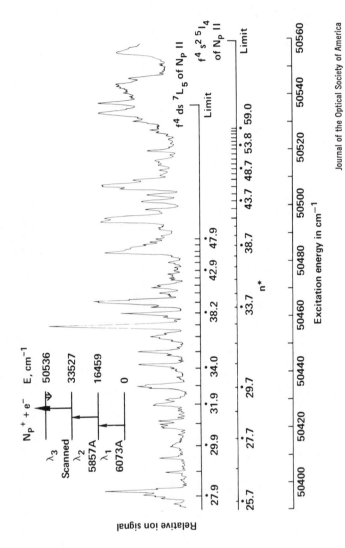

Journal of the Optical Society of America

Figure 6. *Rydberg series in neptunium obtained by field ionization (double arrow) of laser-excited levels. The excitation scheme is shown on the figure. The spectrum contains two series, one converging to the* $5f^46d7s^7L_5$ *ground state and the other converging to the* $^5I_4$ *level at 24.27 cm^{-1} in the ion. For the series converging to the* $^5I_4$ *level, the Rydberg level positions with n* less than 40.7 (short markers) are calculated using a constant fractional defect of 0.3 (4).*

spectrum includes Rydberg levels located below and above the
ionization limit in energy (bound and autoionizing Rydberg
levels, respectively). The bound Rydberg series levels generally converge to the ground state of the ion while autoionizing
Rydberg series converge to excited states of the ion. Thus to
obtain ionization potentials, the convergence limits for autoionization series must be corrected by the energy of the
excited level relative to the ground state of the ion.

The convergence limits were obtained from the series data
using the criteria that the quantum defect (n-n*) is a constant[27] for the correct limit. To do this we determine
the effective quantum number n* of each observed Rydberg level
using the relation

$$n^* = \left[\frac{R}{(\text{assumed limit}) - (\text{level})} \right]^{1/2} \tag{1}$$

for a number of assumed limits and make plots of n-n* vs. n
like in Figs. 7 and 8. The assumed limit that gives the
smoothest and most constant n-n* (zero slope) is taken as the
convergence limit for the series. In equation (1), R is the
Rydberg constant for the element in cm^{-1} and level is the
energy of the observed Rydberg level in cm^{-1}. The intergers
n are not necessarily the principal quantum number of the
Rydberg levels. Equation (1) is derived from the well-known
formula,

$$(\text{Ionization limit}) - (\text{level}) = \frac{R}{(n^*)^2} = \frac{R}{(n-\Delta)^2},$$

where Δ is the quantum defect.

As can be seen from Figs. 7 and 8, the constancy of n-n* is
quite sensitive to the value of the assumed limit. A change of
one wavenumber from the assumed limit of 48731 cm^{-1} (the
limit giving nearly constant n-n*) for dysprosium causes a per-
ceptible slope in the plot.

In the case of cerium, Fig. 8, a greater sensitivity to
change in assumed limit is shown. This results because the
values of n* (and n) are larger for the observed members of the
cerium series and the effect is more pronounced at higher n or
as the limit is approached. Thus the method works best when
applied to the higher members of observed series.

Some lanthanide Rydberg convergence limits derived from
observed series by this method are given in Table II. The
excitation wavelengths and levels from which the series were
observed and the ion levels that they converge to are indicated
in the table. Information on ionic states was obtained from
the literature for Eu,[43] Dy,[44] and Ho,[45] where
autoionizing series were observed. The fact that perturbations
do not affect the measured limits by more than the quoted

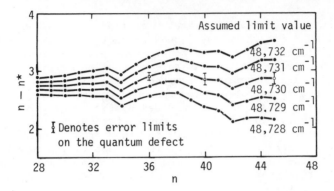

Figure 7.   Variation in quantum defect $(n - n^*)$ vs. n with change in assumed limit for one of the dysprosium double series shown in Figure 5 (n and n* are defined in the text). The assumed limit 48 730 cm^{-1} gives the most constant $(n - n^*)$ value and when corrected by 828.31 cm^{-1} yields the ionization limit for dysprosium (3).

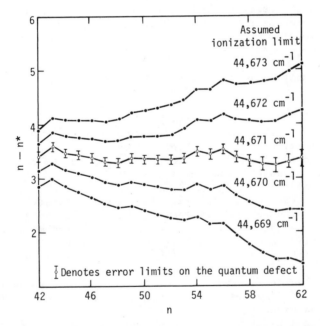

Figure 8.   Variation in quantum defect $(n - n^*)$ vs. n with change in assumed limit for a cerium Rydberg series obtained by field ionization. The assumed limit giving the most constant $(n - n^*)$ value is (44 671 cm^{-1}) and it is taken as the ionization limit since the series converges to the ground state of the ion (3).

Table II. Some Rydberg series limits determined by stepwise laser spectroscopy.

	$\lambda_1{}^a$ (Å)	$\lambda_2{}^a$ (Å)	Excited level used (cm⁻¹)	Convergence energy (cm⁻¹)	Convergence level in ion[b] (cm⁻¹)	First ionization limit[c] (cm⁻¹)	(eV)
Ce	3793.83		26 331.11	44 674(3)	0.00	44 674(3)	5.5389(4)
	3873.03		27 091.56	44 671(3)	0.00	44 671(3)	5.5385(4)
Eu	4661.88		21 444.58	47 405(2)	1669.21	45 736(2)	5.6706(3)
	6291.34	6787.48	30 619.49	47 403(2)	1669.21	45 734(2)	5.6703(3)
Gd	5617.91	6351.72	33 534.71	49 603(5)	0.00	49 603(5)	6.1501(6)
	5701.35	6573.83	32 957.77	49 604(5)	0.00	49 604(5)	6.1502(6)
Dy	4211.72		23 736.60	48 730(5)	828.31	47 902(5)	5.9391(6)
	6259.09	6769.79	30 739.79	48 727(5)	828.31	47 899(5)	5.9388(6)
Ho	4103.38		24 360.55	49 203(8)	637.40	48 566(8)	6.0216(10)
	6305.36	6947.1[d]	30 246(2)	48 567(5)	0.00	48 567(5)	6.0216(6)

( ) in columns = uncertainty in last digit of the value.

[a] Excitation wavelengths (λ_1 and λ_2) and excited level values given to 0.01 are from Ref. (41) and references therein. No value of λ_2 is given for two-step observations when λ_2 is scanned. Values given are λ_2 for three-step results.

[b] Level values are from the following references: Eu, (43); Dy, (44) and Ho, (45).

[c] 8065.479 cm⁻¹/eV used to convert values from cm⁻¹ to eV.

[d] This excited level wavelength and energy determined by laser techniques.

uncertainties is verified by the agreement of the limits determined from different parent levels (different series) of a
given element. In some cases, the series are of different parity. In addition, the ionization limits obtained in holmium
from series converging to different limits agree well within
the quoted uncertainty. The reliability of the method is also
substantiated by the excellent agreement between our value of
45 734(2) cm^{-1} for the ionization potential for europium and
the more accurate value of 45 734.9(2) cm^{-1} determined by
Smith and Tomkins[28] using conventional high-resolution
absorption spectroscopy.

Ionization potentials of 6.1941(5) eV for uranium[1] and
6.2657(6) eV for neptunium[4] have been derived from observed
Rydberg series using laser techniques and the method described
above. These are the most accurate ionization potentials
available for actinide elements. Series converging to the
first excited state and to the ground state of the ion were
observed for both elements. In the case of neptunium, the presence of two series converging to limits 24 cm^{-1} apart (see
Fig. 6) helps to confirm the unpublished value[46] for the
interval between the two lowest levels of neptunium.

Discussion of Lanthanide Ionization Potentials. A summary
of accurate ionization potentials of the lanthanides is given
in the last two columns of Table III. For comparison, values
from electron impact and the semi-emperical spectroscopic
values are given in columns 2 and 3. Although their
uncertainties are much larger, the agreement is quite good.
The exceptions are the electron impact value for erbium and the
spectroscopic values for cerium, praseodymium and neodymium
that are all low.

The photoionization threshold values listed in column 4 are
all lower by some 0.002 to 0.005 eV (15-30 cm^{-1}). Similar
differences were found in uranium and neptunium and remain
unexplained. Electric fields from the ion optics (field ionization) and collisional effects are possible explanations. In
all cases, the Rydberg convergence limits are the most accurate
and they are the preferred values.

No Rydberg series were found for praseodymium, so the
threshold value is the most accurate experimental value. No
measurements were made for promethium. In Table III, interpolated values (in brackets) in the Rydberg convergence column
were obtained for praseodymium and for promethium from an equation derived from a least-squares fit of the experimental
$f^N s^2$ - $f^N s$ ionization potentials, see below. We believe
these are the most accurate values for the ionization
potentials of these two elements. The gadolinium ionization
potential of 6.1494(6) eV determined from Rydberg convergence
by Bekov et al.[5] using the same laser technique developed

Table III.   Summary of lanthanide first ionization potentials.
Values are in eV (1 eV = 8065.479 cm^{-1})

| | | Laser Spectroscopy LLL | | |
Electron Impact[a]	Spectro-scopic[b]	Photoion. Threshold	Rydberg Conver-gence	Rydberg Conver-gence, Others
Ce 5.44(10)	5.466(20)	5.537(5)	5.5387(4)	
Pr 5.37(10)	5.422(20)	5.464(+12)	[5.473(10)]	
Nd 5.49(10)	5.489(20)	5.523(-2)	5.5250(6)	
Pm ---	5.554(20)	---	[5.582(10)]	
Sm 5.58(10)	5.631(20)	5.639(2)	5.6437(6)	
Eu 5.68(10)	5.666(7)	5.666(4)	5.6704(3)	5.67045(3)[c]
Gd 6.24(10)	6.141(20)	---	6.1502(6)	6.1494(6)[d]
Tb 5.84(10)	5.852(20)	---	5.8639(6)	
Dy 5.90(10)	5.927(8)	5.936(3)	5.9390(6)	
Ho 5.99(10)	6.018(20)	6.017(3)	6.0216(6)	
Er 5.93(10)	6.101(20)	6.104(2)	6.1077(10)	
Tm 6.11(10)	6.18(2)	---	---	6.18436(6)[e]
Yb 6.21(10)	6.25(2)	---	---	6.25394(25)[e]

( ) = error in last digit   [ ] = extrapolated value

[a]Ref. (40). This reference also contains a collection of
limits determined by other techniques up until 1975.

[b]Ref. (26). This reference is a collection of the best
available limits derived by spectroscopic techniques up to
the date of publication in 1974.

[c]Ref. (28), conventional absorption spectroscopy.

[d]Ref. (5), by multistep laser excitation and ionization.

[e]Ref. (29) and (30), by conventional absorption
spectroscopy.

by Solarz et al.[1] is in good agreement with our results of
6.1502(6) eV.

Regularities in Ionization Potentials. Except for cerium
and gadolinium, the ionization potential in the lanthanides is
the energy required to remove an s electron from an atom in the
lowest level of $4f^N6s^2$ configuration and produce an ion in
the lowest level of the $4f^N6s$ configuration. Using the known
energy levels in cerium and gadolinium it is possible to
calculate the energy for this ionization process for these
elements. The ionization potential for the process
$f^Ns^2 \rightarrow f^Ns + e^-$ is plotted in Figure 9 as a function
of N. The plot shows clearly a connected two-straight-line
behavior with a change in slope at the half-filled shell (N =
7). The solid line is an unweighted least-squares fit of the
experimental data using a connected two-straight-line. The
resulting numerical fit is shown at the bottom of the figure.
The difference in slope of the two curves is equal to
$G_3(f,s)$, the exchange integral expressing the electrostatic
repulsion between electrons. These regularities have been
theoretically treated for p, d, and f electrons by Rajnak and
Shore.[38] In the case of the actinide series, there are
only two published accurate experimental values (U and Np) and
as Rajnak and Shore point out the uranium value is unreliable
for use in determining regularities because of the influence of
configuration interaction between the low levels. The
neptunium value is also subject to some configuration
interaction.

Experimental values for heavier actinides where configuration interaction is less important (americium and higher) would
be very valuable as they would yield the slope for the
$5f^N7s^2 - 5f^N7s$ ionization potentials for the second half of
the series, N>7. This would allow the determination of extrapolated values for the ionization potentials of actinides
beyond einsteinium where experimental values cannot be obtained
because materials with sufficiently long half-lives are not
available.

Discussion of Actinide Ionization Potentials. The
ionization potentials of actinides determined by laser
techniques are given in Table IV together with values determined by surface ionization, appearance potential and
semi-emperical methods. For uranium, all values are low
compared with the values determined by laser techniques with
the exception of the surface ionization value by Smith and
Hertel.[48] The spectroscopic values by Sugar[51] were
obtained from the $5f^N7s^2 - 5f^N7s8s$ intervals interpolated
from intervals known for the higher actinides. Except for
Sugar's value, all the neptunium ionization potential values
are low relative to the more accurate values determined by
laser methods. The Rydberg series values are the preferred
ionization potentials.

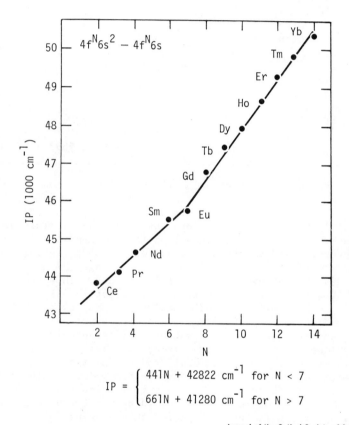

$$IP = \begin{cases} 441N + 42822 \text{ cm}^{-1} & \text{for } N < 7 \\ 661N + 41280 \text{ cm}^{-1} & \text{for } N > 7 \end{cases}$$

Journal of the Optical Society of America

Figure 9.   Normalized ionization potentials of the lanthanides plotted as a function of number of f electrons. Only the cerium and gadolinium points required normalization to the $4f^N6s^2 \rightarrow 4f^N6s + e^-$ process (3).

Table IV. Ionization potentials of uranium and
neptunium determined by various methods.

Method	Uranium (eV)	Ref.	Neptunium (eV)	Ref.
Surface Ionization	6.08(8)	47	6.16(6)	48
Surface Ionization	6.19(6)	48		
Appearance Potential	6.1(1)	49	6.1(1)	52
Appearance Potential	6.11(5)	50		
Spectroscopic	6.05(7)	26,51	6.19(12)	26,51
Semi-emperical	6.00	38	6.193	38
Photoionization; two-step,laser	6.187(2)	2	6.2633(30)	4
Photoionization; three-step,laser	6.1912(25)	1		
Rydberg series; laser	6.1941(5)	1	6.2657(5)	4

Second Ionization Potentials. The determination of
accurate second ionization potentials of the lanthanides and
actinides (ionization potentials of the singly ionized
elements) should be possible by laser techniques. This would
be considerably more difficult than for the neutral elements
for several reasons. Known second ionization potentials for
the lanthanides are in the 10.5 to 14 eV range[26] and the
actinides should have roughly similar values. Thus the laser
energy requirements to photoionize or to populate Rydberg
levels in the singly ionized atoms are roughly twice that
required for the neutral atoms so shorter wavelength lasers or
more steps are necessary. New Raman shifted dye lasers and
eximer lasers (at normal frequencies or Raman shifted) could
prove useful for these studies. Production of a large
population of singly ionized atoms in the ground state or other
known state(s) is necessary. Laser ionization either single-
step, multi-step or multiphoton would be preferred. Eximer
lasers operating in the ultraviolet should be useful for this
application. Mass spectrometer detection of the photoionized
product (M^{2+}) at the correct mass/charge should be straight
forward.

Energy Levels

Energy levels may be determined by numerous variations of
the schemes shown in Fig. 2. For two-laser photonization, λ_1
can be fixed at any known transition and λ_2 scanned to find new
high lying energy levels with energies greater than the known

level energy plus one-half the energy difference between ionization potential and the known level populated by laser 1.
When laser 2 populates a high lying level, another photon is
absorbed by the atom to photoionize it by the mechanism $\lambda_1 +
2\lambda_2$. This is really three-step photoionization. Low lying
levels could be found by fixing λ_2 and scanning λ_1. In this
case, levels could be found for values of λ_1 where the energy
of λ_1 plus λ_2 exceed the ionization potential of the atom. For
three-step photoionization with three lasers, the technique has
been described in the Experimental section under Ionization
Potentials. In all cases, background peaks must be eliminated
by blocking the fixed laser(s) and repeating the scan.
Background peaks of the type $2\lambda_2$, or $2\lambda_1$ for two-laser
photoionization and of the type $2\lambda_2$, $\lambda_2 + \lambda_3$ or $2\lambda_2 + \lambda_3$ for
three-laser photoionization can be eliminated.

The accuracy of level determination depends on the laser
resolution and on how accurately the frequency of the laser is
measured. Not much application of these methods for finding
energy levels has been made to date. Some levels in
uranium[1,6] and in neptunium[4] have been published. A
few levels in the lanthanides were determined for use in three
step searches for Rydberg levels.[3] Miron et al.[18] have
employed multistep laser excitation with fluorescence detection
to determine a number of levels in neutral atomic uranium.
Considerable application of these techniques have been made in
investigations of elements other than the lanthanides and
actinides[25] (see, for example, the work on alkaline earths
by Armstrong, Wynne and Esherick,[53] McIlrath and
Carlsten,[54,55] and Rubbmark, Borgstrom and Bockasten[56]).

Lifetimes, Branching Ratios and Transition Probabilities

Stepwise laser excitation and ionization techinques can be
used to determine lifetimes of excited levels in
atoms.[1,2,4,6,7,8] Oscillator strengths or transition
probabilities may be obtained using a measured lifetime if
suitable intensity data are available to estimate the branching
ratio for the transitions. The branching ratio may also be
determined by laser techniques for certain transitions.[7,8]
Such determinations of the absolute oscillator strengths can be
used to convert relative oscillator strengths obtained by other
methods[57,58,59] to absolute oscillator strengths.

As mentioned in the introduction, transition probabilities
(and lifetimes) are important in astrophysics for determination
of elemental abundances, in plasma physics for diagnostics of
gaseous discharges, in spectrochemistry for abundances and
diagnostics and in laser isotope separation where the choice of
possible transitions and levels depends heavily on transition
probabilities and lifetimes.

Lifetimes. Lifetimes are determined from plots of the
natural logarithm of ion signal vs. delay time between the
laser pulse populating the level and the laser pulse(s) or
pulsed electric field ionizing the level. Fig. 10 is such a
plot for determining the lifetime of the 21783 cm^{-1} level of
dysprosium. Here, two lasers were used with the photoionizing
laser at 3820Å. For a three-step photoionization scheme, the
delay can be between the first and the final two lasers
(low-lying level lifetimes) or between the second and third
laser (high-lying level lifetimes). Details of multi-step
lifetime determination techniques are given in Refs. (1,2,6,
and 8). Table V contains lifetimes determined for four
lanthanides. Similar techniques have been used to measure
lifetimes of a number of levels of U(1,2,6) and Np.(4)
    The chief advantage of lifetime determination by multistep
laser excitation and ionization is that it avoids cascade
problems associated with broad band fluorescence detection and
with conventional excitation techniques.

Branching Ratios. In determining a branching ratio, the
fraction of the atoms returning to the original state after
optical excitation to the upper level of the transition is the
quantity being measured. The laser setup and representative
ion signal are shown schematically in Fig. 11. Precisely timed
pulsed lasers are used. The probe or photoionizing laser(s)
may be fired before or after the pump laser to generate the ion
signal curve shown in the lower part of Fig. 11. The probe
laser(s) must ionize only the lower level of the transition
under study so its wavelength(s) is(are) set at the proper
resonances to strongly photoionize the lower level. The ion
signal is representative of the lower level atom population as
a function of time. When the probe laser is fired before the
pump laser (negative delay) the ion signal is a measure of the
number of atoms initially in the lower level of the transition. The quantity A at zero delay represents the number of
atoms excited to the upper level by the pump pulses. The
quantity B is determined at a delay long compared with the
upper level lifetime so that the excited level population is
essentially zero. The fraction returning to the lower level or
the branching ratio is then B/A. There are some potential
problems associated with this method including excited state
cascade, laser polarization and transit time (motion of the
atoms in the atomic beam relative to the laser volume). These
problems and appropriate solutions have been discussed in some
detail.(7,8) The method works best when the lower level of
the transition is the ground level or metastable thermally
populated level and when the branching ratio is large. In
favorable cases, the branching ratio for a transition can be
determined with an uncertainty less than 10%.(8)
    The advantage of determining branching ratios by this laser

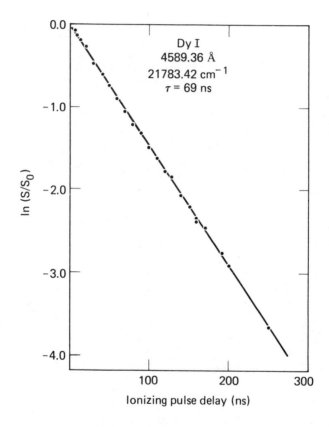

Figure 10.   Plot of the natural logarithm of the ratio of the ion signal (S) to the zero delay ion signal (S₀) vs. ionizing pulse delay time for the 21783 cm⁻¹, J= 7, odd level of dysprosium. Ionizing wavelength, λ₂ = 3820 Å.

Table V. Lifetimes of some lanthanide levels measured
by laser spectroscopy techniques.

	Wavelength (A)	Energy Level (cm^{-1})	Lifetime (ns)		
			This work	Ref. (19)	Ref. (20)
Nd	4634.24	21 572.47	12(1.5)		
	4637.20	21 558.70	81(8)		
	4924.53	20 300.84	10(1.5)		
	4954.78	20 176.90	35(4)		
	5056.89	19 769.49	115(11)		
Sm	4596.74[a]	22 041.02	11(2.0)		
Dy	4194.85	23 832.07	15(1.5)	11(3)	
	4565.09	21 899.22	1160(400)	1205(97)	1200(50)
	4577.78	21 838.53	478(40)	503(40)	489(10)
	4589.36	21 783.42	69(5)	73(2)	75(2)
Er	4087.63	24 457.15	31(6)		
	4190.70	23 855.64	135(14)		

[a]From the 292.58 cm^{-1} level to 22041.02.

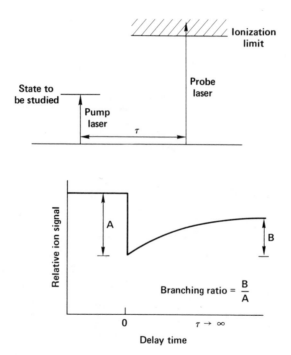

Figure 11.   Schematic of laser sequence and representative ion signal for branching ratio measurement (8)

technique rather than by use of emission spectrum intensities
(discussed below) is that the transition under study is
measured directly and one avoids the need to account for the
intensity of all transitions from the upper level. Many of
these transitions may be in regions where measurements are difficult or they could occur to unknown energy levels of the atom
and be missed entirely.

Transition Probabilities. Absolute transition probabilities of an element may be determined from the lifetime of
the upper level of a transition and the branching ratio,
according to the equation

$$A_{ik} = (B_{ik}/\tau),$$

where A_{ik} is the transition probability, B_{ik} is the branching ratio, and τ is the lifetime of the upper level i. The
branching ratio may be determined by laser methods as described
above. Uranium appears to be the only element where laser
techniques have been used to determine the lifetime and branching ratio for obtaining transition probabilities.[7,8] Three
transitions were investigated and the results are given in
Table VI.

Table VI.   Branching ratios, lifetimes and transition
            probabilities in atomic uranium determined by
            laser techinques.

| Transition | | | | | |
Wave-length (Å)	Levels (cm^{-1})	Lifetime (ns)	Branching ratio	A (in $10^5 s^{-1}$)	Ref.
4362.1	0-22918	69(10)	0.80(10)	117(20)	7
4393.6	0-22754	65(5)	0.68(10)	105(15)	7
6395.4	0-15632	607(20)	0.586(50)	9.6(8)	8

Accurate results of this type are needed for determining
laser utilization in laser isotope separation of uranium. The
application of this technique to obtain results for other elements would be useful in astronomy and plasma physics.

The usual procedure to determine a branching ratio for a
transition is to use line emission intensities in the equation

$$B_{ik} = (I_{ik}/\sum_n I_{in}),$$

where $\sum_n I_{in}$ is the sum of the intensities of all the transitions from the upper level, i, to lower levels, n, and I_{ik} is
the intensity of the transition whose branching ratio is being

determined. This method requires extensive intensity information. It is especially useful when there are only a few
allowed transitions from the upper level of the transition to
the other lower levels. This is the case for the four transitions in dysprosium where the upper level lifetimes have been
well determined.

The lifetimes we measured for these transitions are given
in Table V together with those measured by Hotop and
Marek[19] and by Gustavsson et al.[20] The results are in
good agreement. The extensive intensity and energy level data
available[44] for dysprosium were used to obtain the branching ratios given in Table VII for these four transitions. The
absolute transition probabilities obtained are listed in column
4 of the table. These can be converted to absolute oscillator
strengths (f-values) and used to place the 69 relative
oscillator strengths determined by Perkin et al.[57,58] using
the "hook" method on an absolute scale since the same transitions have been included in both studies. The absolute
transition probabilities were converted to absolute f-values by
the relation

$$f = 1.51 \ \lambda_0^2 \ (g_2/g_1) \ A_{ik},$$

where g_1 and g_2 are the statistical weights (2J + 1) of the
lower and upper levels. Since the constant 1.51 has units of
$cm^{-2}s$, λ_0 is the wavelength of the transition in cm. The
absolute f-values obtained are reported in Table VII. By
multiplying the f_{rel}-values given in Ref. (58) by 3.35 x
10^{-4} (the average value of the ratios f_{abs}/f_{rel} in column
8), one obtains the absolute f-values for 69 lines of
dysprosium with an estimate uncertainty of 20%.

Relative f-values have been reported[57] for transitions
of the other lanthanides with lifetimes given in Table V.
When suitable intensities from emission spectra for branching
ratio determination or when branching ratios by laser
techniques become available, our lifetime measurements may be
used to place the relative f-values on an absolute scale.

Oscillator strengths or absorption cross sections may be
obtained by applying saturation spectroscopy techniques to
multistep photoionization spectroscopy. A few transitions in
uranium have been studied.[6] One of the advantages of saturation spectroscopy is that it can be applied to any one of
the steps in the schemes shown in Fig. 2. The disadvantages
are that the experimental requirements are severe (laser-atomic
beam interaction area,-frequency,-band width and-polarization)
and interpertation of the data can be complex. A detailed discussion will not be given because little application has been
made to the lanthanides and actinides. We will discuss in the
Autoionization section the determination of photoionization
cross sections by a saturation method.

Table VII. Branching ratios, lifetimes, absolute and relative f-values found for dysprosium and the ratio obtained to convert the relative f-values in Ref. (58) to absolute f-values.

Wavelength (Å)	Transition (cm^{-1})	β[a]	τ[b] (ns)	A (in 10^6 s^{-1})	f_{abs}	f_{rel}[c]	f_{abs}/f_{rel} $\times 10^4$
4194	0 - 23832	0.999(1)	13(2)	77(9)	0.21(3)	570(103)	3.58(79)
4565	0 - 21899	0.9(1)	1198(46)	0.75(8)	0.0024(2)	7.5(13)	3.15(65)
4577	0 - 21838	1.000(1)	489(20)	2.1(1)	0.0074(3)	24(4)	3.08(59)
4589	0 - 21783	0.99(1)	73(4)	14(1)	0.039(2)	110(20)	3.57(68)
						Average	3.35(66)

[a]Determined from relative intensities in Ref. (44).

[b]Weighted average of lifetimes in Table V.

[c]Relative f-values from Ref. (58) where estimated uncertainties range from 6 to 18% depending on the number of determinations. We have used 18% as the uncertainty because the number of determinations was not given.

Isotope Shifts and Hyperfine Structure

The measurement of isotope shifts and hyperfine structure (hfs) is possible in multistep laser excitation and ionization if one of the excitation lasers in the excitation schemes shown in Fig. 2 is a narrow band laser and if a collimated atomic beam is used as the source of absorbing atoms. The rest of the aparatus can remain as used for other studies. The narrow band laser(s) may be a pressure tuned pulsed dye laser (~100 MHz, 0.003 cm^{-1}) or a CW dye laser (30 MHz to 30 KHz, 10^{-3} to 10^{-6} cm^{-1}). The atomic beam should be collimated to reduce "Doppler" broaging to the level required to attain the resolution needed for investigating the structure and to fully utilize the narrow band width of the laser. A band width of 10^{-4} cm^{-1} is usually adequate for most investigations of lanthanides and actinides. A portion of the scan laser beam is directed to an etalon and detector (interferometer) to provide relative frequency calibration.

Several lanthanides [10,11,12] and uranium [9,13] have been investigated by this technique. With the exception of the work by Hackel et al.[13], the investigations are for transitions originating from the ground state.

Isotope Shifts. Scans showing the isotope structure in the 5887.9 Å line of neodymium and the 5988 Å line of dysprosium are shown in Fig. 12. The Russian work includes scans for Sm in Refs. [10] and [11] and for Nd, Sm, Eu, Gd, Dy and Er in Ref. (11). These scans may have been obtained using fluorescence (luminescence) detection. It was not defined in Ref. [11]. Much higher resolution structure has been demonstrated for the dysprosium 5988 Å transition by Childs and Goodman,[60] see Fig. 13. They used fluorescence detection but the improvement in resolution was achieved by better atomic beam collimation and narrower scan laser width. Photoionization detection should give similar results, compare Figs. 12b and 14d. A comparison of Figs. 12 and 13 and the texts of Refs. [11] and [60] indicates that the frequency scale in Fig. 12 should have negative values and the labels on the peaks should be transposed.

Published isotope shift data for actinides obtained by multistep laser excitation and ionization is confined to the 5915 Å transition of uranium.[9] Much improved resolution and precision were obtained by Childs, Poulsen and Goodman[22] using their laser-atomic beam fluorescence method.

Hyperfine Structure. An interesting technique for studying the hfs of odd isotopes by use of two step photoionization and mass filtration has been reported by Karlov et al.[12] A high resolution mass filter is set at the mass of the isotope to be studied and the high resolution laser is scanned over the

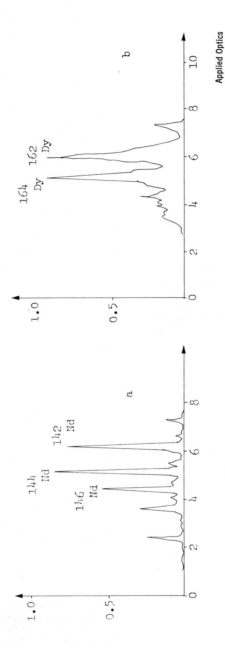

Figure 12. Isotope structure of the 5887.9-Å line of neodymium (a) and the 5988.6-Å line of dys-prosium (b). The abscissa in each figure is relative frequency in gigahertz (11).

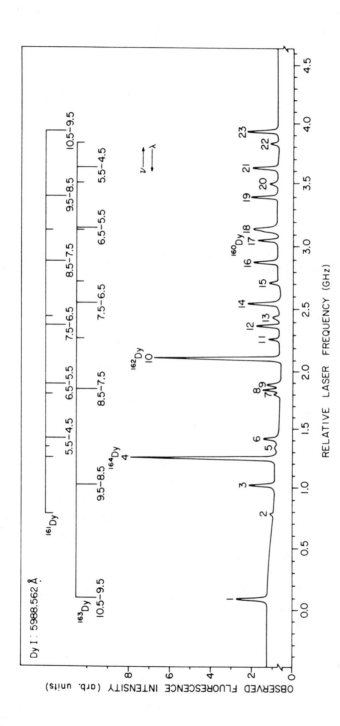

Journal of the Optical Society of America

Figure 13. Structure of the 5988.562-Å line of neutral dysprosium. Three of the lines—4, 10 and 17—are due to even isotopes (no hfs), the remainder are hfs components from the odd I = 5/2 isotopes, dysprosium-161 and dysprosium-163. The key at the top gives the values for F(J = 8) − F′(J = 7) for stronger components and indicates the positions of the weaker components (60).

transition to be investigated. The results of such a study in
Dy are shown in Fig. 14. Scans b and c show the expected six
strong components for the ^{161}Dy and ^{163}Dy isotopes that
both have nuclear spin I = 5/2. By this technique hfs
investigations with natural abundance materials or materials
from nuclear sources can be simplified, eliminating the need
for pure isotopes. Dysprosium and erbium were
investigated.[10,12] For dysprosium and other lanthanide
hfs, the laser-atomic fluorescence and double resonance results
of Childs et al.[23,60,61] are superior. With their improved
resolution, pure isotopes are not needed to separate and assign
the structure for most transitions, see Fig. 13. Similar
resolution should be possible when using photoion detection.

Published studies of hfs in the actinides by laser excitation and ionization have been restricted to a few levels in
uranium.[9,13] Again, the results of Childs et al.[22,24]
are superior for transitions from the ground and 620 cm^{-1}
level. The investigations of Hackel et al.[13], of the hfs
of the uranium transition at 6156.8 Å between the 15632 and
31869 cm^{-1} levels illustrate the application of a high
resolution laser to the second step in the three-step scheme of
Fig. 2. Isotope shifts can be precisely measured by this technique, also.

Autoionization

Multistep laser excitation and ionization is ideally suited
for investigating the autoionization structure from excited
states in atoms. Autoionizing levels lie at energies exceeding
the first ionization potential for the element (that is at
energies more than enough to permit one electron to leave the
atom). Such levels are of electronic configurations with an
"inner" shell electron excited and are referred to as doubly
excited terms. Autoionizing levels usually exhibit very short
lifetimes (about 10^{-12} s) determined mainly by the time for a
non-radiative transition of one of the excited electrons to a
less energetic orbital of the ionized atom and generation of a
free electron.

Recently, Bekov, Letokhov, Matveev and Mishin[5] reported
the observation of an autoionization state in gadolinium with
the relatively long lifetime of 5 x 10^{-10} s. The autoionization spectrum they observed by three-step laser spectroscopy
is shown in Fig. 15. For isolated atoms, the width of the
autoionization resonance is determined by its lifetime, so the
0.07 cm^{-1} half-width yields the estimated 0.5 ns lifetime.
Their explanation for the narrow resonance (long lifetime) is
that the observed state is only 230 cm^{-1} above the ionization
limit so it probably decays to a state consisting of a ground
state gadolinium ion plus an electron. The long lifetime
results because for this state the selection rules forbid such

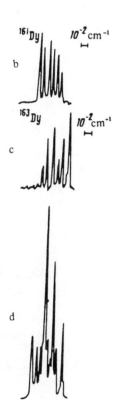

Applied Optics

Figure 14. Structure of the 5988.56-Å line of neutral dysprosium: (b) spectrum with mass spectrometer set to detect dysprosium-161 photoions; (c) spectrum with mass spectrometer set to detect dysprosium-163 photoions; (d) photoionization spectrum without mass selection. The strongest two peaks in (d) correspond to dysprosium-162 and dysprosium-164 transitions (11).

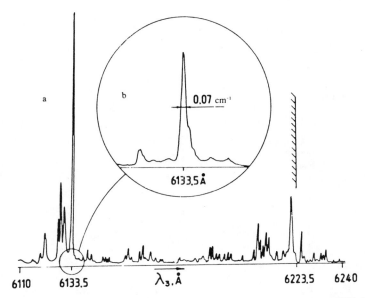

JETP Letters

Figure 15. (a) *Ion signal as a function of wavelength of the third-step laser in gadolinium with* λ_1 *set at 5617.9 Å and* λ_2 *set at 6351.7 Å to populate the 33535 cm^{-1}, 4f^{7}5d 6s7s ^{9}D$^\circ$ level* (λ_2 *scan laser band width, 1 cm^{-1}); (b) scan of a limited region near the strong autoionization resonance at 6133.5 Å with a 0.03-cm^{-1} band width laser (5).*

a decay. This restriction can be lifted by an electric field.
Indeed, a broadening of the autoionization resonance to 0.35
cm^{-1} was observed with a relatively weak field of 100 V/cm.
    Another interesting and important property of this auto-
ionization is the observed intensity or cross section of about
8×10^{-16} cm^2 which is much larger than typical values of
10^{-17} or 10^{-18} cm^2. The cross section was measured by a
saturation technique.[5,14] The ion yield at the resonance
maximum was determined as a function of the energy density E_3
of the ionizing laser pulse to generate the curve shown in Fig.
16 . Saturation is indicated by the flat portion of the curve.
The point A on the curve is taken as the saturation point
characterized by the condition[5]

$$(\sigma_{aut})\ (E_3) \approx 2h\nu_3,$$

where σ_{aut} (in cm^2) is the autoionization cross section to
be determined, E_3 is the laser pulse power in erg/cm^2 and
ν_3 is the frequency of the transition in Hz.
    These results indicate that long lived autoionization
states with excitation cross-sections comparable to those for
excitation of bound high-lying states exist in heavy atoms with
complex spectra. Transitions to these autoionization states
can radically increase the efficiency of photoionization of
atoms, a factor very important in atomic vapor laser isotope
separation.
    The use of autoionizing Rydberg levels converging to
excited states of the ion to determine ionization potentials
has been discussed above. If autoionization resonances as
narrow as those found in gadolinium exist in the actinides, it
should be possible to determine the isotope shifts and hfs of
such features. (Isotope shifts for actinides range up to 0.4
cm^{-1} per mass unit and odd atomic number actinides exhibit
hfs with total widths of 4 to 6 cm^{-1} and hfs component
spacing of 0.2 cm^{-1} or more for some transitions).

Laser Isotope Separation

    Laser isotope separation is one area where multistep
excitation and ionization has great commercial potential. The
research and development efforts in atomic vapor laser enrichment of ^{235}U are a major factor contributing to the current
research activities in laser excitation and ionization processes. The first paper on selective multistep photoionization
of atoms was published in 1971.[62] Since then numerous
review articles[15,16,17,63,64,65] have been written on laser
isotope separation and, in each review, there is a section on
atomic vapor photoionization processes. The subjects of
economics and critical parameters have been well covered in
previous reviews and will not be discussed in detail here. We

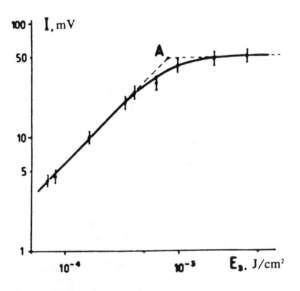

Figure 16. The dependence of the gadolinium photoion yield at the absorption maximum of the autoionization resonance of 6133.5 Å on the pulsed-energy density, E_3, of the ionizing laser. Laser band width 0.03 cm^{-1} (5).

have described in the sections above methods for obtaining most
of the necessary spectroscopic parameters for evaluating
processes.

Ionization potentials, while not critical, give energy
requirements for the lasers to photoionize the atoms. Isotope
shifts are necessary for the selective excitation. By use of
the shift in each step, enhanced selectively can be obtained.
When the isotope is odd as is the case for ^{235}U, knowledge of
the hfs present in each step of a process is required so that
the lasers can be frequency contoured to access all the atoms
vaporized. Lifetimes, branching ratios and cross-sections of
all excitation steps including photoionization are necessary to
evaluate the efficiency of atom and laser utilization.

Another phenomena that is important in laser isotope
separation is excitation energy transfer between isotopes
(resonant energy exchange in like atoms). This has been
studied for Eu isotopes using high resolution laser excitation
and fluorescence detection.[11,66] Atomic beams at various
densities (collision frequencies) were irradiated at the
^{153}Eu frequency and the resulting fluorescence analyzed with
a scanning interferometer. Fluorescence at the ^{151}Eu frequency was studied as a function of density (5 x 10^{12} to 5 x
10^{14} cm^{-3}) while irradiating at the ^{153}Eu frequency
(actually the ratio of the ^{153}Eu to ^{151}Eu fluorescence was
measured). From this, a cross-section of 1.4 x 10^{-13} cm^2
was obtained for excitation transfer of the $4f^76s6p$
$^6P_{7/2}$ 17340 cm^{-1} level (the 5765 Å transition) in
europium. Cross-sections for such energy transfer are
important for determining the atom densities that can be toler-
ated in a laser isotope separation process without loss of the
desired selectively. No studies were made of the effect of
excitation energy and/or electronic configuration on the excitation energy transfer cross-section.

Numerous papers have been presented[67,68,69] and published[2,9,70,71] on the enrichment of ^{235}U by multistep
laser ionization of atomic vapor. To date, only milligram
quantities have been enriched, but facilities to significantly
increase the quantities have been developed.[71]

Laser isotope separation of several lanthanides (Nd, Sm,
Eu, Gd, Dy, and Er) has been demonstrated.[10,11,72] Separation coefficients (undefined) reported in Ref. [11] range from
11 for ^{155}Gd to 726 for ^{164}Dy. These were obtained by
setting the photoselective laser at a frequency absorbed by the
isotope of interest and recording the mass spectrum of the ions
produced. The band width of the selective laser was quoted as
10^{-3} cm^{-1}. Natural abundance metals were used in the evap-
orator. A nitrogen laser or an unfiltered mercury lamp was
used to photoionize the laser excited atoms. A non-selective
background current was present, especially with the mercury
lamp. Better separation coefficients could have undoubtedly

been obtained with second step excitation by a narrow-band
laser to an autoionization level or with three-step excitation
and ionization.

Finally, we mention the application of multistep laser
photoionization to the detection of small numbers of atoms.
Researchers at Oak Ridge National Laboratory and at the
Institute of Spectroscopy, USSR, have published papers on
single atom detection.[73,74,75] In both methods, resonant
multistep excitation and ionization is used. Laser powers are
such that the probability of ionization of a single atom in the
laser volume is one. We will not describe the details of
either technique here but refer the reader to Refs. [73], [74]
and [75] and literature cited there. These techniques can be
used to study spectroscopic properties of very rare and short
lived isotopes of the lanthanides and actinides. In
particular, studies of isotope shifts of nuclear isomers of
heavy elements such as Am should give information on nuclear
volume and deformation of such isomers relative to the ground
states of the various isotopes.

Comments

In this review of multistep laser photoionization of the
lanthanides and actinides, we hope that we have introduced the
reader to a number of laser techniques for determining spectroscopic properties of these elements. We have undoubtedly over-
looked some techniques and some papers on the subjects we did
cover. The importance of laser methods in studying the spectroscopy of the lanthanides and actinides is well established
and future applications should greatly expand our knowledge of
these elements.

Summary

Techniques of stepwise laser excitation and photoionization
have been applied to study spectroscopic properties of neutral
atoms of lanthanides and actinides. The spectroscopic
properties that can be determined include: the ionization
potential, energy levels, isotope shifts, hyperfine structure,
lifetimes of energy levels, branching ratios and oscillator
strengths. We discuss the laser methods used to obtain these
properties (with emphasis on ionization potentials) and give
examples of results obtained for each. The ionization potentials obtained by laser techniques are in eV: Ce, 5.5387(4);
Pr, 5.464(12); Nd, 5.5250(6); Sm, 5.6437(6); Eu,5.6704(3); Gd,
6.1502(6); Tb, 5.8639(6); Dy, 5.9390(6); Ho, 6.0216(6); Er,
6.1077(10); U, 6.1941(5) and Np, 6.2657(6). Regularities in
the $f^N s^2 - f^N s$ ionization limits for the lanthanides have
been found as a result of these accurate values and previously

known accurate values for the others. Multistep photoionization has been employed for laser isotope separation and for
studies of autoionization.

Acknowledgements

We sincerely appreciate the contributions of our co-workers
R. W. Solarz and J. A. Paisner in the determination of the
lanthanide and uranium ionization potentials.

Work performed under the auspices of the U. S. Department
of Energy by the Lawrence Livermore Laboratory under contract
number W-7405-ENG-48.

Literature Cited

1. Solarz, R. W.; May, C. A.; Carlson, L. R.; Worden, E. F.;
    Johnson, S. A.; Paisner, J. A. and Radsiemski, L. J.,
    "Detection of Rydberg states in uranium using time-resolved
    stepwise laser photoionization," Phys. Rev., 1976, A14,
    1129-1136.

2. Janes, G. S.; Itzkan, I.; Pike, C. T.; Levy, R. H. and
    Levin, L.,"Two-Photon Laser Isotope Separation of Atomic
    Uranium: Spectroscopic Studies, Excited-State Lifetimes, and
    Photoionization Cross Sections," IEEE, J. Quantum Electron,
    1976, QE-12, 111-120.

3. Worden, E. F.; Solarz, R. W.; Paisner, J. A. and
    Conway,J. G., "First ionization potentials of lanthanides by
    laser spectroscopy," J. Opt. Soc. Am., 1978, 68, 52-61.

4. Worden, E. F. and Conway, J. G., "Laser spectroscopy of
    neptunium; first ionization potential, lifetimes and new
    high-lying energy levels of Np I," J. Opt. Soc. Am., 1979,
    69, 733-738.

5. Bekov, G. I.; Letokhov, V. S.; Matveev, O. I. and
    Mishin,V. I.,"Discovery of long-lived autoionizion state in
    the spectrum of gadolinium atom," JETP Lett., 1978, 28,
    283-285.

6. Cárlson, L. R.; Paisner, J. A.; Worden, E. F.;
    Johnson, S. A.; May, C. A. and Solarz, R. W., "Radiative
    lifetimes, absorption cross sections, and the observation of
    new high-lying odd levels of ^{238}U using multistep laser
    photoionization," J. Opt. Soc. Am., 1976, 66, 846-853.

7. Carlson, L. R.; Johnson, S. A.; Worden, E. F.; May, C. A.;
    Solarz, R. W. and Paisner, J. A., "Determination of Absolute
    Atomic Transition Probabilities Using Time-Resolved Optical
    Pumping," Opt. Commun., 1977, 21, 116-120.

8. Hackel, L. A. and Rushford, M. C., "Lifetime, branching
    ratio, and absolute transition probability of the 6395.42 Å
    transition of ^{238}U I," J. Opt. Soc. Am., 1978, 68,
    1084-1087.

9. Bohm, H-D.V.; Michaelis, W. and Weitkamp, C., "Hyperfine
    Structure and Isotope Shift Measurements on ^{235}U and
    Laser-Separation of Uranium Isotopes by Two-Step
    Photoionization," Opt. Commun., 1978, 26, 177-182.

10. Karlov, N. V.; Krynetskii, B. B.; Mishin, V. A.;
    Prokhorov, A. M.; Savel'ev, A. D. and Smirnov, V. V.,
    "Separation of samarium isotopes by two-step photoionization
    method," Sov. J. Quantum Electron., 1976, 6, 1363-1364.

11. Karlov, N. V.; Krynetskii, B. B.; Mishin, V. A. and
    Prokhorov, A. M., "Laser isotope separation of rare earth
    elements," Appl. Opt., 1978, 17, 856-862.

12. Karlov, N. V.; Krynetskii, B. B.; Mishin, V. A. and
    Prokhorov, A. M., "Use of the method of two-step
    photoionization and mass filtration for the study of the hfs
    of odd isotopes," JETP Lett. 1977, 25, 294-297.

13. Hackel, L. A.; Bender, C. F.; Johnson, M. A. and
    Rushford, M. C., "Hyperfine structure measurements of
    high-lying levels of uranium," J. Opt. Soc. Am., 1979, 69,
    230-232.

14. Ambartzumian, R. V.; Furzikov, N. P.; Letokhov, V. S. and
    Puretsky, A.A., "Measuring Photoionization Cross-Sections of
    Excited Atomic States," Appl. Phys., 1976, 9, 335-337.

15. Letokhov, V. S. and Moore, C. B., "Laser isotope separation
    (review)," Sov. J. Quant. Electron, 1976, 6, 129-150.

16. Karlov, N. V. and Prokhorov, A. M., "Laser isotope
    separation", Sov. Phys. Usp. 1976, 19, 285-300.

17. Aldridge, J. P.; Birely, J. H.; Cantrell, C. D. and
    Cartwright, D. C., in "Laser Photochemistry, Tunable Lasers
    and Other Topics, Vol 4 of Physics of Quantum Electronics,"
    Jacobs, S. F.; Sargent, M. III; Scully, M. O.; and
    Walker, C. T., Ed.; Addison-Wesley: Reading, MA., 1976.

18. Miron, E.; David, R; Erez, G.; Lavi, S. and Levin, L. A.,
     "Laser spectroscopy of U I using stepwise excitation and
     fluorescence detection," J. Opt. Soc. Am., 1979, 69, 256-264.

19. Hotop, R. and Marek, J., "Lifetime Measurements of Some
     Excited Dy I States," Z. Physik, 1978, A287, 15-17.

20. Gustavsson, M.; Lundberg, H.; Nilsson, . and Svanberg, S.,
     "Lifetime measurements of excited states of rare-earth atoms
     using pulse modulation of a CW dye-laser beam,"
     J. Opt. Soc. Am., 1979, 69, 984-992.

21. Childs, W. J. and Goodman, L. S., "Assignment of
     unclassified lines in Tb I through high-resolution
     laser-fluorescence measurements of hyperfine structure,"
     J. Opt. Soc. Am., 1979, 69, 815-819.

22. Childs, W. J.; Poulsen, O. and Goodman, L. S., "High
     precision measurement of ^{235}U ground-state hyperfine
     structure by laser-rf double resonance," Opt. Letters, 1979,
     4, 35-37.

23. Childs, W. J. and Goodman, L. S., "Hyperfine structure of
     excited, odd-parity levels in ^{139}La by laser-atomic-
     fluorescence," J. Opt. Soc. Am., 1978, 68, 1348-1350.

24. Childs, W. J.; Poulsen, O. and Goodman, L. S.,
     "High-precision measurement of the hyperfine structure of
     the 620 cm^{-1} metastable atomic level of ^{235}U by laser-rf
     double resonance," Opt. Letters, 1979, 4, 63-65.

25. Omenotto, N. Ed., "Analytical Laser Spectroscopy";
     John Wiley and Sons: New York, 1979.

26. Martin, W. C.; Hagan, L.; Reader, J. and Sugar, J., "Ground
     levels and ionization potentials for lanthanide and actinide
     atoms and ions," J. Phys. Chem. Ref. Data, 1974, 3, 771-780.

27. Garton, W. R. S. and Wilson, M., "Autoionization-Broadened
     Rydberg Series in the Spectrum of La I, Astrophys. J., 1966,
     145, 333-336.

28. Smith, G. and Tomkins, F. S., "Autoionization resonances in
     the Eu I absorption spectrum and a new determination of the
     ionization potential," Proc. R. Soc. Lond., 1975, 342,
     149-156.

29. Camus, P., "Etude des spectres d'absorption de l'ytterbium
     du lutecium et du thulium entre 2700 A et 1900 A," Thesis,
     Univ. Paris, Orsay; 1971, 126-334 (unpublished).

30. Camus, P. and Tomkins, F. S., "Spectre d'Absorption de Yb I," J. Phys. (Paris), 1969, 30, 545-550.

31. Blaise, J.; Camus, P. and Wyart, J. F., "Seltenerdelemente, Sc Y, La and Lanthanide", in "Gmelin Handbuch der Anorganishen Chemie, System No. 39,"; Springer: Berlin, 1976.

32. Solarz, R. W.; Paisner, J. A. and Worden, E. F., "Multiphoton Laser Spectroscopy in Heavy Elements", in "Multiphoton Processes,"; Eberly, J. H. and Lambropoulos, P., Eds., John Wiley and Sons: New York, 1978. pp. 267-275.

33. Paisner, J. A.; Solarz, R. W.; Worden, E. F. and Conway, J. G., "Identification of Rydberg States in the Atomic Lanthanides and Actinides," in "Laser Spectroscopy III"; Hall, J. L. and Carlsten, J. L., Eds. Springer-Verlag: Berlin, 1977. pp 160-169.

34. Parr, A. C. and Elder, F. A., "Photoionization of Ytterbium: 1350-2000 A," J. Chem. Phys., 1968, 49, 2665-2667 (1978).

35. Parr, A. C., "Photoionization of Europium and Thulium: Threshold to 1350 A," J. Chem. Phys., 1971, 54, 3161-3167.

36. Parr, A. C. and Inghram, M. G., "Photoionization of samarium in the threshold region," J. Opt. Soc. Am., 1975, 65, 613-614.

37. Dunning, F. B. and Stebbings, R. F., "Role of autoionization in the near-threshold photoionization of argon and krypton metastable atoms," Phys. Rev., 1974, A9, 2378-2382.

38. Rajnak, K. and Shore, B. W., "Regularities in s-electron binding energies in $l^n s^m$ configurations," J. Opt. Soc. Am., 1978, 68, 360-367.

39. Reader, J. and Sugar, J., "Ionization Energies of the Neutral Rare Earths," J. Opt. Soc. Am., 1966, 56, 1189-1194.

40. Ackermann, R. J.; Rauh, E. G. and Thorn, R. J.; "The thermodynamics of ionization of gaseous oxides; the first ionization potentials of the lanthanides metals and monoxides," J. Chem. Phys., 1976, 65, 1027-1031.

41. Meggers, W. F.; Corliss, C. H. and Scribner, B. F., "Tables of Spectral-Line Intensities, Part I Arranged by Elements, Natl. Bur. Stand. U. S. Monogr. 145"; U. S. Government Printing Office: Washington, D. C., 1975.

42.   Fred, M. and Tomkins, F. S., Argonne National Laboratory,
      Argonne, Il.; Blaise, J. E.; Camus, P and Verges, J.
      Laboratorire Aime Cotton, Orsay, France, "The Atomic
      Spectrum of Neptunium," Argonne National Laboratory Report
      ANL-76-68; May, 1976.

43.   Russell, H. N.; Albertson, W. and Davis, D. N., "The spark
      spectrum of europium, Eu II," Phys. Rev., 1941, 60, 641-656.

44.   Conway, J. G. and Worden, E. F., "Preliminary Level Analysis
      of the First and Second Spectra of Dysprosium, Dy I and II,"
      J. Opt. Soc. Am., 1971, 61, 704-726; and University of
      California Radiation Laboratory Report UCRL-19944 (1970)
      available from U. S. Dept. of Commerce.

45.   Livingston, A. E., Jr. and Pinnington, E. H., "Spectra of
      Neutral and Singly Ionized Holmium," J. Opt. Soc. Am., 1971,
      61, 1429-1430.

46.   Blaise, J., Laboratorie Aime Cotton, Orsay, France and Fred,
      M., Argonne National Laboratory, Argonne IL., private
      communication, 1978.

47.   Bakulina, I. N. and Ionov, N.I., "Determination of the
      ionization potential of uranium by a surface ionization
      method," Zh. Eksp. Teor. Fiz., 1959, 35, 1001-1005;
      Sov. Phys. JETP, 1959, 9, 709-712.

48.   Smith, D. H. and Hertel, G. R., "First ionization potentials
      of Th, Np and Pu by surface ionization," J. Chem. Phys.,
      1969, 51, 3105-3107.

49.   Rauh, E. G. and Ackerman, R. J., "The first ionization
      potentials of some refractory oxide vapors," J. Chem. Phys.,
      1974, 60, 1396-1400.

50.   Mann, J. B., "Ionization of U, UO, and UO_2 by electron
      impact," J. Chem. Phys., 1964, 40, 1632-1637.

51.   Sugar, J., "Revised ionization energies of the neutral
      actinides," J. Chem. Phys., 1974, 60, 4103.

52.   Rauh, E. G. and Ackermann, R. J., "The first ionization
      potentials of neptunium and neptunium monoxide,"
      J. Chem. Phys., 1975, 62, 1584.

53.   Armstrong, J. A.; Wynne, J. J. and Esherick, P., "Bound,
      odd-parity, J=1 spectra of the alkaline earths: Ca, Sr and
      Ba," J. Opt. Soc. Am.; 1979, 69, 211-229, and references
      cited.

54. McIlrath, T. J. and Carlsten, J. L., "Production of large numbers of atoms in selected excited state by laser optical pumping: calcium," J. Phys., 1973, B6, 697-708.

55. Carlsten, J. L.; McIlrath, T. J. and Parkinson, W. H., "Absorption spectrum of the laser-populated ^{3}D metastable levels in barium," J. Phys., 1975, B8, 38-51.

56. Rubbmark, J. R.; Borgstrom, S. A. and Bockasten, K., "Absorption spectroscopy of laser-excited barium," J. Phys., 1977, B10, 421-432.

57. Penkin, N. P. and Komaravskii, V. A., "Oscillator strengths of spectral lines and lifetimes of excited levels of atoms of rare earth elements with unfilled 4f shells," J. Quant. Spectrosc. Radiat. Transfer, 1976, 16, 217-252. (in Russian).

58. Penkin, N. P.; Komarovski, V. A. and Smirnov, V. V., "Relative values of oscillator strengths for spectral lines of Dy I," Opt. Spectrosc., 1974, 37, 223-224.

59. Corliss, C. H. and Bozman, R. W., "Experimental Transition Probabilities for Spectral Lines of Seventy Elements. NBS Monograph 53"; U. S. Government Printing Office: Washington, D. C., 1962.

60. Childs, W. J. and Goodman, L. S., "Hyperfine structure and isotope-shift measurements on Dy I 5988.562 using high-resolution laser spectroscopy and an atomic beam," J. Opt. Soc. Am.; 1977, 67, 747-751.

61. Childs, W. J.; Poulsen, O. and Goodman, L. S., "Laser-rf double-resonance spectroscopy in the samarium I spectrum: Hyperfine structure and isotope shifts," Phys. Rev., 1979, 19, 160-167.

62. Ambratzumian, R. V.; Kalinin, V. P. and Letokhov, V.S., "Two-Step Selective Photoionization of Rubidium Atoms by Laser Radiation," JETP Lett., 1971, 13, 217-219.

63. Letokhov. V. S.; Mishin, V.I. and Puretzky, A.A.,"Selective Photoionization of Atoms by Laser Radiation and Its Applications," Prog. Quantum Electron. 1977, 5, 139-203.

64. Letokhov, V.S., "Laser Separation of Isotopes," Ann Rev. Phys. Chem., 1977, 28, 133-159.

65. Moore, C. B., "The Application of Lasers to Isotope Separation," Accounts Chem. Res., 1973, 6, 323-328.

66. Karlov, N.V.; Krynetskii, B.B.; Miohin, V.A., and
     Prokhorov, A.M., "Laser-spectroscopy measurement of the
     cross-section of excitation energy transfer in a gas of like
     atoms," JETP Lett., 1977, 25, 535-537.

67. Tuccio, S. A.; Dubrin, J. W.; Peterson, O.G. and Snavely,
     B. B., paper Q14, VIII International Quantum Electronics
     Conference, San Francisco, June, 1974; IEEE J. Quant.
     Elect., 1974, QE-10, 790.

68. Tuccio, S. A.; Foley, R. J.; Dubrin, J. W. and
     Kirkorian, O., paper 13.B8, CLEA, Washington, D. C. May
     28-30, 1975, IEEE J. Quant. Elect., 1975, QE-11, 101D.

69. Janes, G. S.; Itzkan, I.; Pike, C. T.; Levy, R. H. and
     Levin, L., paper 13.B9, CLEA, Washington, D. C. May, 28-39,
     1975, IEEE J. Quant. Elect., 1975, QE-11, 101D.

70. Snavely, B.B.; Solarz, R. W. and Tuccio, S. A., "Separation
     of Uranium Isotopes by Selective Photoionization," in "Laser
     Spectroscopy, 43, Proceedings of the Second International
     Convcerence, Megere, June 23-27, 1975"; Springer-Verlag:
     Berlin, 1975, pp. 267-274.

71. Davis, J. I. and Davis, R. W., "Some Aspects of the Laser
     Isotope Separation Program at Lawrence Livermore
     Laboratory," in "Developments in Uranium Enrichment, AIChE
     Symposium Series, V.73, No. 169"; Benedict, M., Ed.,
     American Institute of Chemical Engineers: 345 East 47 St.,
     New York, 1977, pp. 69-75.

72. Karlov, N.V.; Kreprelskii, B.B.; Mishin, V.A.;
     Prokhorov, A.M.; Savelev, A.D. and Smirnov. A.D., "Isotope
     Separation of Some Rare-Earth Elements by
     Two-Step-Photoionization," Opt. Commun., 1977, 21, 384-386.

73. Kramer, S.C.; Bemis, C. E., Jr.; Young, J. P. and
     Hurst, G.S., "One-atom detection in individual ionization
     tracks," Opt. Letters, 1978, 3, 16-18.

74. Hurst, G.S.; Payne, M. G.; Kramer, S.D. and Young, J.P.,
     "Resonance Ionization Spectroscopy with Amplification,"
     Chem. Phys. Letters, 1979, 63, 1-4.

75. Bekov, G.I.; Letokhov, V.S.; Matveev, O.I. and Mishin, V.I.,
     "Single-atom detection of ytterbium by selective laser
     excitation and field ionization from Rydberg states,"
     Opt. Letters, 1978, 3, 159-161.

RECEIVED February 29, 1980.

Photoelectron Spectra of Actinide Compounds

B. W. VEAL and D. J. LAM

Argonne National Laboratory, 9700 South Cass Avenue, Argonne, IL 60439

Photoemission spectroscopy applied to chemistry and electronic properties studies is a fairly recent development. The x-ray photoemission spectroscopy (XPS) technique was developed, primarily to be a chemical analysis tool (1). In particular it was observed that the absolute binding energies of the atomic-like electron core levels are dependent on the chemical state of the atom under study. This observation led to the widespread use of XPS for basic and applied chemistry studies. Many studies were also undertaken to better understand the physics of the various excitation processes involved. Consequently, XPS has become a powerful tool for studying electronic structure of the outer electron states in solids.

Proceeding in parallel with XPS was the development of ultra violet photoemission spectroscopy (UPS) (2). This technique exploits low energy photons and must be confined to studies of electron states rather near the Fermi level (E_F). For investigating occupied electron states in the vicinity of E_F, UPS and XPS can serve as excellent complementary spectroscopies.

The actinide element series, like the lanthanide series, is characterized by the filling of an f-electron shell. The chemical and physical properties, however, are quite different between these two series of f-electron elements, especially in the first half of the series. The differences are mainly due to the different radial extension of the 4f- and 5f-electron wavefunctions. For the rare-earth ions, even in metallic systems, the 4f electrons are spatially well localized near the ion sites. Photoemission spectra of the f electrons in lanthanide elements and compounds always show "final state multiplet" structure (3), spectra that result from partially filled shells of localized electrons. In contrast, the 5f electrons are not so well localized. They experience a smaller coulomb correlation interaction than the 4f electrons in the rare earths and stronger hybridization with the 6d- and 7s-derived conduction bands. The 5f's thus

0–8412–0568–X/80/47–131–427$05.00/0
© 1980 American Chemical Society

have a greater tendency toward itinerancy than do the 4f's. The
result is that 5f electrons in actinide elements and compounds may
reveal itinerant, local, or intermediate behavior. Furthermore,
the actinide ions can adopt a variable valence state in chemical
compounds and the 5f electrons may participate in bonding.

In this paper, we shall present a brief overview of the
application of photoelectron spectroscopy to the study of actinide
materials. Some phenomenology will be discussed as will studies
of specific materials. Only illustrative examples will be presented (4).

Theoretical Background

Photoemission is viewed as a process wherein an absorbed
photon excites an electron within the solid to a final energy
state greater than the work function. The electron then migrates
to the surface and escapes. Figure 1 shows a schematic representation of the photoemission process. A photon with energy $\hbar \omega$
may excite an electron from some initial state E_i below the Fermi
level to a final state E. After moving through the solid to the
sample surface, the electron escapes into the vacuum with kinetic
energy E_k after having lost energy equal to the work function in
overcoming the surface potential barrier. Since the electron
kinetic energy E_k is monitored, one has a measure of E and E_i.
Thus, with monochromatic exciting radiation, one is able to
measure transition probabilities between states which are at
known energies relative to E_F. The energy of the exciting
radiation for these experiments varies from several electron
volts to kilo electron volts. The low energy extreme is referred
to as UV photoemission spectroscopy (UPS) and the high energy
extreme as x-ray photoemission spectroscopy (XPS). However, the
spectrum can be continuously scanned using synchrotron radiation.

Core Level and Localized Valence States

One of the most important capabilities of the XPS technique
is the measurement of absolute core level binding energies as a
means of probing the local charge state of the ion under study.
The oxidation state of the ion can sometimes be clearly discerned.
An example is the XPS measurement for the intermediate oxides
of uranium, U_3O_8 and U_2O_5, reported by Verbist et al. (5). Doublet structure in the U 4f lines was attributed to U^{4+} and U^{6+} ions
since the 4f peaks appeared with nearly the same binding energies
as the 4f lines in UO_2 and UO_3.

It should be cautioned that photoemission spectroscopic
measurements look at energy differences between an n-electron
ground state and an n-1 electron excited state. Therefore, photoemission provides a good approximation to ground-state properties

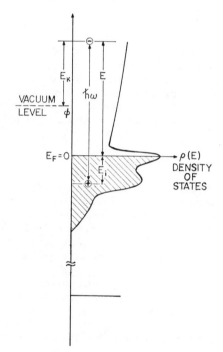

Figure 1. Schematic of photoemission process for a metal. A monochromatic photon $h\omega$ excites an electron from initial state E_i to final (vacuum) state E. The electron escapes with kinetic energy E_k after passing through sample surface having work function ϕ. For fixed $h\omega$, monitoring photoemission intensity I vs. E_k measures transition probabilities between states E_i and E. Valence states as well as core states can be observed.

only in the limit when Koopman's theorem is valid (6), i.e., for
large electronic systems having extended one-electron wavefunctions
like the valence and conduction band states in metallic and semi-
conducting solids. Koopman's theorem is not valid for core electrons since their wavefunctions are well-localized about the atomic site. Observed core level positions are uncertain, relative to
the ground-state core-level energies, by the generally unknown
final state relaxation energy. Similarly if a (localized) electron
is removed from an incompletely filled outer shell of an ion, different final-state configurations of the ion can lead to experimentally observed multiplet structure (7). These different final
states of the ion may be viewed as a form of relaxation which results from intra-atomic correlation effects.

Figure 2 shows XPS data for dioxides of neptunium, plutonium,
and americum compared to the appropriate f^n multiplet calculations
(8). These multiplet spectra do not represent the multiplet structure of either the f^n or the f^{n-1} configurations. They are,
instead, the final state multiplet structure of the f^{n-1} configuration modulated by the transition probability from the f^n ground
state to the f^{n-1} multiplets.

Intensities

For quantitative analysis of surface chemical compositions
using the XPS technique, measurements of subshell photoionization cross-sections (SPC's) are needed. Problems involved in
the determination of relative SPC's from the measurement of relative line intensities in an XPS spectrometer have been discussed
by Cardona and Ley (2). The cross-section measurements require
an appropriate consideration of the sample concentration, energy
dependence of electron analyzer transmission, the angle of the
incoming x-ray beam relative to the outgoing electrons and the
energy dependence of electron escape depths. Using appropriate
compounds, most reported line intensities have been measured
relative to the fluorine 1s cross-section. Reference 2 contains
a comprehensive tabulation of peak intensities and includes representative levels for most of the elements. Evans et al. (9)
have reported the most recent intensity measurements for uranium
compounds. These are integrated intensity measurements and
include corrections for the experimental considerations cited
above.

The multiplet calculations discussed above (see Fig. 2) determine relative intensities within excited multiplets. However,
the f electron intensities are not related to s, p, or d intensities.

Calculations of the expected XPS spectra for the actinide
dioxides uranium through berkelium were reported by Gubanov et al.
(10). Results for UO_2 are shown in Fig. 3 along with experimental
spectra. These calculations, extending about 30 eV below the
Fermi level, are based on a one-electron molecular-cluster approach.

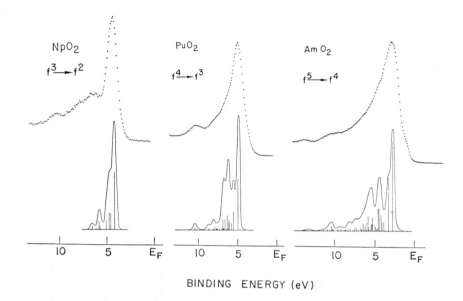

BINDING ENERGY (eV)

Figure 2.   XPS spectra of localized 5f states in three actinide oxides compared with calculated final-state multiplet spectra. The calculated multiplets are broadened to simulate experiment.

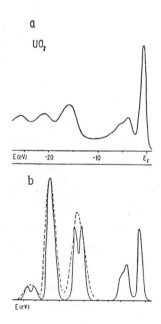

Journal of the Chemical
Society, Faraday Trans.

Figure 3.   Measured (a) and calculated (b) UO₂ XPS spectra. The solid line in (b) has Lorentzian broadening of 0.4 eV, the dashed line, 0.9 eV to simulate experiment.

Intensities were determined using transition probabilities taken
from the calculations of Scofield (11). In general, the (energy)
positions of features in the calculated spectra correspond well
with the experimentally observed features but some calculated
intensities do not show good quantitative agreement with experiment.

5f-electrons and Bonding

A. Oxides.

Due to the photon energy dependence of the photoemission
cross section for electrons with different orbital angular moments, a capability is available for investigating the role of 5f
electrons in bonding. A systematic dependence of the localized
5f electron peak intensity on the degree of oxidation of the
uranium atom was found in a series of uranium binary and ternary
oxides (8). As the oxidation state of uranium in the oxides is
increased, electrons are transferred from the localized 5f states
into the "bonding" molecular orbitals which are predominately
O 2p in character. Results are shown in Fig. 4 for several
binary oxides. The 5f intensities (per electron) are substantially
greater than the O 2p's. With increased oxidation, the uranium
valence state increases and more 5f participation might be expected
in the "bonding orbitals". However, Veal et al. (8) concluded,
from a quantitative study of the valence band XPS intensities,
that 5f electrons do not appear to significantly contribute to
the bonding molecular orbitals. For the hexavalent uranium
compounds, it appears that the 5f levels are pushed above E_F.
However, molecular cluster calculations for actinide oxides (10)
indicate that 5f states do show active participation in the metal-
ligand bond.

Complementary studies of the 5f structure in UO_2 were obtained using UPS. Figure 5 shows photoemission spectra of Evans recorded at 21.2, 40.8, and 1253.6 eV (12). The 5f peak near E_F is
dominant at 1253.6 eV but is barely discernable at 21.2 eV. Additional spectra of UO_2 at 21.2, 40.8, and 48.4 eV were obtained
by Naegele et al. (13). A characteristic of the UO_2 spectra
[noted by B. Brandow in Ref. (13)] is that the high binding energy
side of the "O 2p" band grows in intensity along with the 5f peak
as photon energy is increased. Since the bottom of the 2p band
has the Bloch states with the strongest 2p-5f hybridization, these
results support the view that 5f electrons hybridize with the
O sp's in forming the metal-ligand bond.

B. Intermetallic Compounds.

Binary intermetallic compounds of the light actinides display a wide variety of magnetic and electronic properties that
are not well understood. Physical phenomena associated with

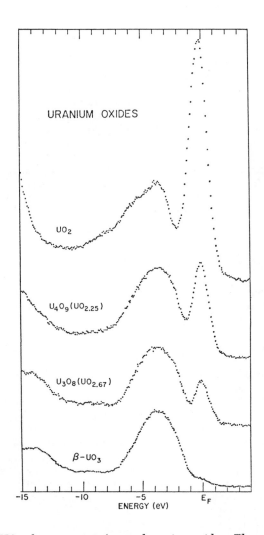

Figure 4.   *XPS valence spectra of several uranium oxides. The uranium 5f peak near E_F is attenuated with increasing uranium oxidation.*

itinerant electron behavior, resonance 5f-electron states, spin
fluctuations and localized electron behavior can all be found.
No single theoretical framework can suitably account for all
phenomena observed.

Several NaCl-type binary compounds, including UN, have been
analyzed (14) using both theoretical band structure (itinerant)
and crystal field (localized) approaches (although the applicability of one of these approaches generally means that the other is
inappropriate). UN may be an intermediate case where neither
approach will yield very satisfactory results. Photoemission
spectroscopy can yield valuable insights into this problem.
Figure 6 shows UPS spectra for UN and ThN (15). The strong
peak near E_F seen in UN is totally missing in ThN. Since ThN
has no 5f occupation but is crystallographically similar to UN,
the 5f nature of the peak near E_F is confirmed. Figure 7 shows
UPS data at 21.2 and 40.8 eV, again work of Norton et al. (15).
The 5f peak near E_F shows the same characteristically strong
photon energy dependence that was observed in UO_2. (The remaining
spectral features are associated with s-p electrons.) The very
narrow 5f peak seen in Fig. 7 implies that the 5f's occupy a
narrow, steeply rising band or that the levels are essentially
localized in which case XPS multiplet theory (see above) should
be applicable. Taking the latter view in Fig. 7, the calculated
multiplet spectrum appropriate for a $5f^3$ ground state is compared
to experiment. There is a remarkably good correspondence between
theory and experiment, particularly at 40.8 eV where the 5f peak
is dominant. For further discussion of these results, see Ref. 4.

Satellite Structure

Satellites appear as peaks on the high binding energy side
of the main peak in an XPS spectrum. The lines are generally
associated with discrete energy losses called "shake-up" or
"shake-off" processes that are attributed to sudden changes in
the local atomic charge that accompany electron ejection. These
processes involve excitations in the n-1 electron system and are
generally described as the low energy excitation of a second
electron "concurrent" with primary electron emission (16). The
"shake-up" core level satellite spectra are generally sensitive
to chemical bonding. Satellite spectra for binary uranium
oxides and fluorides are discussed by Pireaux et al. (17). They
attribute the dominant satellites to an excitation from an
occupied ligand (predominantly O 2p) orbital to an empty or
partially filled metal electronic level. This is a "charge transfer" excitation that accompanies the primary photoejection process.

Satellites may also be observable in photoelectron spectra
if a 2-hole final state, with similar total energy and the same
total angular momentum and parity as the original core hole state,
can occur. Bancroft et al. (18) pointed out that there are several
energetically favorable examples for such (configuration interaction) satellites in light actinides. Their computed satellite

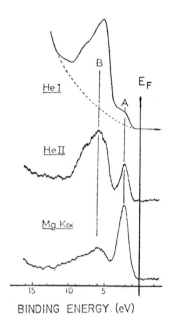

Figure 5. Photoemission spectra for UO$_2$ taken at photon excitation energies of 21.2 eV (He I), 40.8 eV (He II), and 1253.6 eV (MgKα). The uranium 5f electrons (Peak A) have a very different dependence on photon energy than the O 2p's (Peak B).

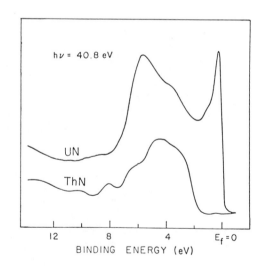

Figure 6. UPS spectra for UN and ThN. The strong peak in UN which appears near E$_F$ corresponds to U 5f electrons.

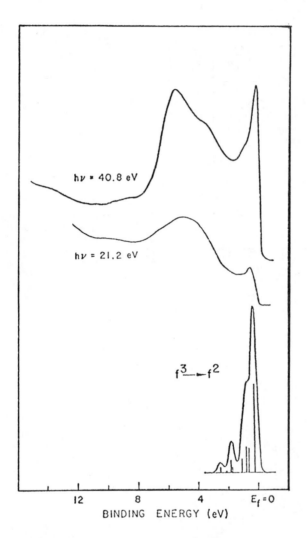

Figure 7. UPS spectra for UN at 21.2 and 40.8 eV. The lower curve is a calculation of the final-state multiplet structure for the 5f² final state.

intensity results, based on the sudden approximation, shows that
the 5s, 5p and 6s levels should produce the most prominent satellites. Depending on which low lying final states are involved in
the excitation, the process may or may not show chemical sensitivity. Kowalczyk (19) argues that when 5f final states are involved,
the effect may be useful for investigating the degree of localization of 5f electrons.

Ligand Field Effects

In a systematic study involving more than twenty uranyl compounds, it was established that the axial ligand field within the
uranyl unit can produce substantial splitting in the XPS spectra
of the actinide $6p_{3/2}$ core level (8). Figure 8 shows examples
of the $6p_{3/2}$ level splittings observed in a sequence of uranyl
samples with different primary U-O separations ($U-O_I$). The
experimental spectra of the uranyl compounds with the smallest
$U-O_I$ separation, including the U $6p_{3/2}$ splitting, is well represented by the characteristic energies obtained from a relativistic molecular cluster calculation (20). An example is shown in
Fig. 9. However, to obtain good agreement between theory and experiment for all the uranyls studied, the effect of the uranium
second-near neighbors had to be included.

Spin-polarized Photoemission

For magnetically ordered materials, photoemitted electrons
have a characteristic spin polarization that reflects the electron spin orientation occurring in the sample before the photoemission process. Recently, techniques have been developed to
measure this photoelectron spin polarization (photo ESP) (21).
When the measured ESP moment is aligned parallel to the total
magnetization, the spin polarization is designated as positive.
Because the ESP technique suffers from low measurable intensities
of polarized photoelectrons, the usual electron energy distribution (EDC) curves are not measured for polarized electrons.
Rather, integrated electron yields for a given photoexcitation
energy are measured and the percentage of polarization of the
integrated yield is determined. The photon energy dependence of
the polarization gives information on the net spin of the
electrons within $\hbar\omega-\phi$ (ϕ is the work function) of the Fermi
level. Measurements are usually reported for photon energies
between 4 and 11 eV.

Photo ESP measurements have been reported for the series of
intermetallic compounds US, USe and UTe (21). The ESP for these
compounds is negative for all $\hbar\omega$. The magnetic moment of
uranium compounds is predominately determined by the occupied
5f electrons. However, the photoyield of the f-electrons at
photon energies less that 11 eV is very small relative to
s, p or d electron yields. Thus, the observed photoelectron spin

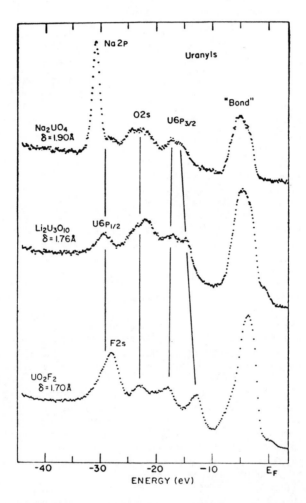

Figure 8.   XPS spectra of three uranyl compounds taken within 40 eV of E_F. The "uranium $6p_{3/2}$ splitting" varies with $U - O_I$ separation δ.

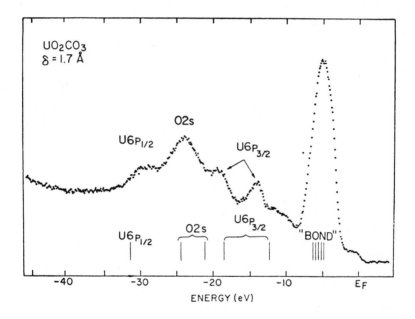

Figure 9. Comparison between experimental XPS spectrum and calculated energy levels for the uranyl compound UO_2CO_3

polarization must result from conduction electrons that are
polarized by the electrons in the partially filled f shell. These
polarization results are consistent with results derived from
magnetization, nuclear magnetic resonance and neutron scattering
measurements (4,22).

Acknowledgment
     This work was supported by the United States Department of
Energy.

Literature Cited
1. Siegbahn, K.; Nordling, C.; Fahlman, A.; Nordberg, R.;
    Hamrin, K.; Hedman, J.; Johansson, G.; Bergmark, T.; Karlsson,
    S.; Lindgren, I.; Lindberg, B. "ESCA Atomic, Molecular, and
    Solid State Structure Studied by Means of Electron Spectros-
    copy" Nova Acta Regiae Societates Scientiarm Upsaliensis,
    Ser. IV, Vol. 20, 1967.
2. Cardona, M.; Ley, L. in "Photoemission in Solids I",
    Cardona, M. and Ley, L., Springer-Verlag 1978, 1-104.
3. XPS studies of lanthanide multiplet structure and fluctuations
    are discussed in Campagna, M.; Wertheim, G. K.; Structure
    and Bonding, 1976, 30, 99.
4. For a more comprehensive discussion of XPS studies of uranium
    compounds, see Veal, B. W.; Lam, D. J.; Vol. "Uran", Gmelin
    Handbuch der Anorganischen Chemie, to be published.
5. Verbist, J.; Rega, J.; Tenret-Noel, C.; Pireaux, J. J.;
    d'Ursel, G.; Caudano, R., and Derouane, E. G. in "Plutonium
    1975 and Other Actinides", ed. H. Blank and R. Lindner,
    North Holland 1976, p. 409.
6. Koopman, T. Physica 1933, 1, 104.
7. Cox, P. A.; Evans, S.; Orchard, A. F. Chem. Phys. Letters,
    1972, 13, 386.
8. Veal, B. W.; Lam, D. J.; Hoekstra, H. R.; Diamond, H.;
    Carnall, W. T. in "Proc. 2nd Int'l. Conf. Elec. Struc. of
    Actinides", ed. J. Mulak, W. Suski and R. Tróc, Wroclaw,
    Poland 1976, p. 145 and references therein.
9. Evans, S.; Pritchard, R.; Thomas, J. J. of Elec. Spectr. and
    Rel. Phenom. 1978, 14, 341.
10. Gubanov, V. A.; Rosén, A.; Ellis, D. E. J. Phys. Chem. Solids,
    1979, 40, 17.
11. Scofield, J. H. J. Electron Spectr., 1976, 8, 129.
12. Evans, S. JCS Faraday II, 1977, 73, 1341.
13. Veal, B. W. in the discussion section of J. de Physique, Colloq.
    C4, Suppl. No. 4, 1979, 40, C4-163.
14. "The Actinides: Electronic Structure and Related Properties",
    ed. A. J. Freeman and J. B. Darby, Academic Press, New York
    (1974), Vols. I and II.
15. Norton, P. R.; Tapping, R. L.; Creber, D. K.; Beyers, W. J. L.;
    private communication.
16. Shirley, D.A. in "Photoemission in Solids I", ed. M. Cardona
    and L. Ley, Springer-Verlag, 1978, pp. 165-195.

17. Pireaux, J. J.; Riga, J.; Thibaut, E.; Tenret-Noel, C.; Caudano, R.; Verbist, J. J. Chem. Phys., 1977, 22, 113.

18. Bancroft, G. M.; Sham, T. K.; Larson, S. Chem. Phys. Lett. 1977, 46, 551.

19. Kowalczyk, S. J. de Physique, Colloq. C4, Suppl. No. 4, 1979, 40, C4-224.

20. Ellis, D. E.; Rósen, A.; Walch, P. F. Int. J. Quantum Chem. Symp., 1975, 9, 351.

21. Eib, W.; Erbudak, M.; Greuter, F.; Reihl, B. J. Phys. C: Solid St. Phys., 1979, 12, 1195; J. de Physique, Colloq. C4 Suppl. No. 4, 1979, 40, C4-72, and references therein.

22. Jena, P.; Emmons, R.; Lam, D. J.; Ray, D. K. Phys. Rev. B, 1978, 18, 3562.

RECEIVED March 10, 1980.

INDEX